Lipid Rafts

METHODS IN MOLECULAR BIOLOGY™

John M. Walker, SERIES EDITOR

419. **Post-Transcriptional Gene Regulation,** edited by *Jeffrey Wilusz, 2008*
418. **Avidin–Biotin Interactions:** *Methods and Applications,* edited by *Robert J. McMahon, 2008*
417. **Tissue Engineering, Second Edition,** edited by *Hannsjörg Hauser and Martin Fussenegger, 2007*
416. **Gene Essentiality:** *Protocols and Bioinformatics,* edited by *Andrei L. Osterman, 2008*
415. **Innate Immunity,** edited by *Jonathan Ewbank and Eric Vivier, 2007*
414. **Apoptosis in Cancer:** *Methods and Protocols,* edited by *Gil Mor and Ayesha Alvero, 2008*
413. **Protein Structure Prediction, Second Edition,** edited by *Mohammed Zaki and Chris Bystroff, 2008*
412. **Neutrophil Methods and Protocols,** edited by *Mark T. Quinn, Frank R. DeLeo, and Gary M. Bokoch, 2007*
411. **Reporter Genes for Mammalian Systems,** edited by *Don Anson, 2007*
410. **Environmental Genomics,** edited by *Cristofre C. Martin, 2007*
409. **Immunoinformatics:** *Predicting Immunogenicity In Silico,* edited by *Darren R. Flower, 2007*
408. **Gene Function Analysis,** edited by *Michael Ochs, 2007*
407. **Stem Cell Assays,** edited by *Vemuri C. Mohan, 2007*
406. **Plant Bioinformatics:** *Methods and Protocols,* edited by *David Edwards, 2007*
405. **Telomerase Inhibition:** *Strategies and Protocols,* edited by *Lucy Andrews and Trygve O. Tollefsbol, 2007*
404. **Topics in Biostatistics,** edited by *Walter T. Ambrosius, 2007*
403. **Patch-Clamp Methods and Protocols,** edited by *Peter Molnar and James J. Hickman, 2007*
402. **PCR Primer Design,** edited by *Anton Yuryev, 2007*
401. **Neuroinformatics,** edited by *Chiquito J. Crasto, 2007*
400. **Methods in Lipid Membranes,** edited by *Alex Dopico, 2007*
399. **Neuroprotection Methods and Protocols,** edited by *Tiziana Borsello, 2007*
398. **Lipid Rafts,** edited by *Thomas J. McIntosh, 2007*
397. **Hedgehog Signaling Protocols,** edited by *Jamila I. Horabin, 2007*
396. **Comparative Genomics,** *Volume 2,* edited by *Nicholas H. Bergman, 2007*
395. **Comparative Genomics,** *Volume 1,* edited by *Nicholas H. Bergman, 2007*
394. **Salmonella:** *Methods and Protocols,* edited by *Heide Schatten and Abe Eisenstark, 2007*
393. **Plant Secondary Metabolites,** edited by *Harinder P. S. Makkar, P. Siddhuraju, and Klaus Becker, 2007*
392. **Molecular Motors:** *Methods and Protocols,* edited by *Ann O. Sperry, 2007*
391. **MRSA Protocols,** edited by *Yinduo Ji, 2007*
390. **Protein Targeting Protocols, Second Edition,** edited by *Mark van der Giezen, 2007*
389. **Pichia Protocols, Second Edition,** edited by *James M. Cregg, 2007*
388. **Baculovirus and Insect Cell Expression Protocols, Second Edition,** edited by *David W. Murhammer, 2007*
387. **Serial Analysis of Gene Expression (SAGE):** *Digital Gene Expression Profiling,* edited by *Kare Lehmann Nielsen, 2007*
386. **Peptide Characterization and Application Protocols,** edited by *Gregg B. Fields, 2007*
385. **Microchip-Based Assay Systems:** *Methods and Applications,* edited by *Pierre N. Floriano, 2007*
384. **Capillary Electrophoresis:** *Methods and Protocols,* edited by *Philippe Schmitt-Kopplin, 2007*
383. **Cancer Genomics and Proteomics:** *Methods and Protocols,* edited by *Paul B. Fisher, 2007*
382. **Microarrays, Second Edition:** *Volume 2, Applications and Data Analysis,* edited by *Jang B. Rampal, 2007*
381. **Microarrays, Second Edition:** *Volume 1, Synthesis Methods,* edited by *Jang B. Rampal, 2007*
380. **Immunological Tolerance:** *Methods and Protocols,* edited by *Paul J. Fairchild, 2007*
379. **Glycovirology Protocols,** edited by *Richard J. Sugrue, 2007*
378. **Monoclonal Antibodies:** *Methods and Protocols,* edited by *Maher Albitar, 2007*
377. **Microarray Data Analysis:** *Methods and Applications,* edited by *Michael J. Korenberg, 2007*
376. **Linkage Disequilibrium and Association Mapping:** *Analysis and Application,* edited by *Andrew R. Collins, 2007*
375. **In Vitro Transcription and Translation Protocols:** *Second Edition,* edited by *Guido Grandi, 2007*
374. **Quantum Dots:** *Applications in Biology,* edited by *Marcel Bruchez and Charles Z. Hotz, 2007*
373. **Pyrosequencing® Protocols,** edited by *Sharon Marsh, 2007*
372. **Mitochondria: Practical Protocols,** edited by *Dario Leister and Johannes Herrmann, 2007*
371. **Biological Aging:** *Methods and Protocols,* edited by *Trygve O. Tollefsbol, 2007*
370. **Adhesion Protein Protocols,** *Second Edition,* edited by *Amanda S. Coutts, 2007*
369. **Electron Microscopy:** *Methods and Protocols, Second Edition,* edited by *John Kuo, 2007*
368. **Cryopreservation and Freeze-Drying Protocols,** *Second Edition,* edited by *John G. Day and Glyn Stacey, 2007*
367. **Mass Spectrometry Data Analysis in Proteomics,** edited by *Rune Matthiesen, 2007*
366. **Cardiac Gene Expression:** *Methods and Protocols,* edited by *Jun Zhang and Gregg Rokosh, 2007*
365. **Protein Phosphatase Protocols:** edited by *Greg Moorhead, 2007*
364. **Macromolecular Crystallography Protocols:** *Volume 2, Structure Determination,* edited by *Sylvie Doublié, 2007*
363. **Macromolecular Crystallography Protocols:** *Volume 1, Preparation and Crystallization of Macromolecules,* edited by *Sylvie Doublié, 2007*
362. **Circadian Rhythms:** *Methods and Protocols,* edited by *Ezio Rosato, 2007*
361. **Target Discovery and Validation Reviews and Protocols:** *Emerging Molecular Targets and Treatment Options, Volume 2,* edited by *Mouldy Sioud, 2007*

METHODS IN MOLECULAR BIOLOGY™

Lipid Rafts

Edited by

Thomas J. McIntosh

*Department of Cell Biology, Duke University Medical Center,
Durham, NC*

HUMANA PRESS ✴ TOTOWA, NEW JERSEY

© 2007 Humana Press Inc.
999 Riverview Drive, Suite 208
Totowa, New Jersey 07512

www.humanapress.com

All rights reserved. No part of this book may be reproduced, stored in a retrieval system, or transmitted in any form or by any means, electronic, mechanical, photocopying, microfilming, recording, or otherwise without written permission from the Publisher. Methods in Molecular Biology™ is a trademark of The Humana Press Inc.

All papers, comments, opinions, conclusions, or recommendations are those of the author(s), and do not necessarily reflect the views of the publisher.

This publication is printed on acid-free paper. ∞
ANSI Z39.48-1984 (American Standards Institute)

Permanence of Paper for Printed Library Materials.
Cover illustration: Fig. 3 from Chapter 6.

Production Editor: Rhukea J. Hussain
Cover design by Karen Schulz

For additional copies, pricing for bulk purchases, and/or information about other Humana titles, contact Humana at the above address or at any of the following numbers: Tel.: 973-256-1699; Fax: 973-256-8341; E-mail: orders@humanapr.com; or visit our Website: www.humanapress.com

Photocopy Authorization Policy:
Authorization to photocopy items for internal or personal use, or the internal or personal use of specific clients, is granted by Humana Press Inc., provided that the base fee of US $30.00 per copy is paid directly to the Copyright Clearance Center at 222 Rosewood Drive, Danvers, MA 01923. For those organizations that have been granted a photocopy license from the CCC, a separate system of payment has been arranged and is acceptable to Humana Press Inc. The fee code for users of the Transactional Reporting Service is: [978-1-58829-729-7 $30.00].

Printed in the United States of America. 10 9 8 7 6 5 4 3 2 1

ISBN 13: 978-1-58829-729-7

e-ISBN 13: 978-1-59745-513-8

Library of Congress Control Number: 2007925519

Preface

In the past several years significant attention has been given to the analysis of the properties and functions of lateral microdomains (rafts) in biological membranes. As described in the overview chapter of this book, as well as in the introductions to many of the other chapters, there are many fundamental unanswered questions concerning the composition, structure, dynamics, and even the very existence of these membrane rafts. Therefore, a variety of sophisticated techniques have been used to study intact cell membranes, as well as model systems composed of specific lipids and proteins thought to be in rafts. The analyses of biological membranes have provided detailed data on intact microdomains, and the complementary studies of model systems have given critical insights on the roles of specific lipid-lipid and lipid-protein interactions in raft formation and function.

The chapters in this book provide detailed information on modern methods that are currently being employed to study lipid rafts. Chapter 1 gives a brief overview of microdomains, emphasizing unanswered questions concerning rafts, Chapters 2 and 3 describe methods to isolate membrane rafts for the purpose of determining their chemical composition, Chapter 4 describes fluorescence techniques to detect rafts in model systems, Chapter 5 uses monolayer techniques to obtain information on intermolecular interactions involved in lipid domain formation, Chapter 6 details methods to form giant unilamellar vesicles and examine them by fluorescence microscopy, Chapter 7 uses fluorescence correlation spectroscopy to determine molecular diffusion in membranes, Chapter 8 describes microscopic techniques to determine the distribution in membranes of dehydroergosterol, an analog to the critical raft component cholesterol, Chapters 9 and 10 describe adaptations of nuclear magnetic resonance (NMR) techniques to detect rafts and measure lateral diffusion, Chapter 11 describes specialized applications of electron paramagnetic resonance (EPR), Chapter 12 uses plasmon-waveguide resonance spectroscopy to analyze lateral segregation in supported model systems, Chapter 13 describes how fluorescence recovery after photobleaching (FRAP) can analyze membrane raft properties, Chapter 14 explains the use of single-molecule tracking to follow the motion of individual lipid and protein molecules in cell membranes, Chapter 15 describes how X-ray diffraction can provide the thickness of raft and non-raft bilayers, Chapter 16 details a novel use of neutron scattering to detect rafts, Chapter 17 illustrates how transmission electron microscopy can observe remodeling events in cell membranes, Chapter 18 describes atomic force microscopy methods for analyzing raft structure in supported bilayers, Chapter 19 provides detailed methods for computer simulations of bilayers containing raft components, and Chapter 20 describes procedures by which microscopic model calculations can provide detailed information on the nature of the interactions between specific lipid molecules.

The many biochemical, biophysical, and computational methods described in this book have already provided detailed information on lipid rafts, and promise to continue to unravel many more mysteries concerning raft formation and properties.

The editing of this book was supported by NIH Grant GM27278.

Thomas J. McIntosh
Durham, North Carolina

Contents

Preface v
Contributors ix

1 Overview of Membrane Rafts
   Thomas J. McIntosh 1
2 Analysis of Raft Affinity of Membrane Proteins by Detergent-Insolubility
   Deborah A. Brown 9
3 Nondetergent Isolation of Rafts
   Mehul B. Shah and Pravin B. Sehgal 21
4 Detecting Ordered Domain Formation (Lipid Rafts)
   in Model Membranes Using Tempo
   Omar Bakht and Erwin London 29
5 Using Monomolecular Films to Characterize Lipid
   Lateral Interactions
   Rhoderick E. Brown and Howard L. Brockman 41
6 Electro-Formation and Fluorescence Microscopy of Giant Vesicles
   With Coexisting Liquid Phases
   Sarah L. Veatch 59
7 Fluorescence Correlation Spectroscopy
   Kirsten Bacia and Petra Schwille 73
8 Multiphoton Laser-Scanning Microscopy and Spatial Analysis
   of Dehydroergosterol Distributions on Plasma Membrane
   of Living Cells
   *Avery L. McIntosh, Barbara P. Atshaves, Huan Huang,
   Adalberto M. Gallegos, Ann B. Kier, Friedhelm Schroeder,
   Hai Xu, Weimin Zhang, Suojin Wang, and Jyh-Charn Liu* 85
9 NMR Detection of Lipid Domains
   Ivan V. Polozov and Klaus Gawrisch 107
10 Lateral Diffusion Coefficients of Raft Lipids From Pulsed Field
   Gradient NMR
   Greger Orädd and Göran Lindblom 127
11 Saturation-Recovery Electron Paramagnetic Resonance
   Discrimination by Oxygen Transport (DOT) Method
   for Characterizing Membrane Domains
   *Witold K. Subczynski, Justyna Widomska, Anna Wisniewska,
   and Akihiro Kusumi* 143

12 Plasmon-Waveguide Resonance Spectroscopy Studies of Lateral
    Segregation in Solid-Supported Proteolipid Bilayers
    Zdzislaw Salamon, Savitha Devanathan, and Gordon Tollin *159*

13 Fluorescence Recovery After Photobleaching Studies
    of Lipid Rafts
    Anne K. Kenworthy *179*

14 Single-Molecule Tracking
    Marija Vrljic, Stefanie Y. Nishimura, and W. E. Moerner *193*

15 X-Ray Diffraction to Determine the Thickness of Raft
    and Nonraft Bilayers
    Thomas J. McIntosh *221*

16 Small-Angle Neutron Scattering to Detect Rafts and Lipid Domains
    **Jeremy Pencer, Thalia T. Mills, Norbert Kucerka, Mu-Ping Nieh,
    and John Katsaras** *231*

17 Exploring Membrane Domains Using Native Membrane Sheets
    and Transmission Electron Microscopy
    **Bridget S. Wilson, Janet R. Pfeiffer, Mary Ann Raymond-Stintz,
    Diane Lidke, Nicholas Andrews, Jun Zhang, Wenxia Yin,
    Stanly Steinberg, and Janet M. Oliver** *245*

18 Atomic Force Microscopy of Lipid Domains
    in Supported Model Membranes
    Alan R. Burns *263*

19 Atomistic and Coarse-Grained Computer Simulations
    of Raft-Like Lipid Mixtures
    Sagar A. Pandit and H. Larry Scott *283*

20 A Microscopic Model Calculation of the Phase Diagram
    of Ternary Mixtures of Cholesterol and Saturated
    and Unsaturated Phospholipids
    R. Elliott, I. Szleifer, and M. Schick *303*

Index *319*

Contributors

ANDREWS, NICHOLAS • *Dept. of Pathology and Cancer Center, University of New Mexico School of Medicine, Albuquerque, NM, USA*
ATSHAVES, BARBARA P. • *Dept. of Physiology and Pharmacology, Texas A&M University, College Station, TX 77843, USA*
BACIA, KIRSTEN • *Institute of Biophysics, Dresden University of Technology, Tatzberg 47-51, D-01307, Dresden, Germany*
BAKHT, OMAR • *Dept. of Biochemistry and Cell Biology, Stony Brook University, Stony Brook, NY 11794-5215, USA*
BROCKMAN, HOWARD L. • *Hormel Institute, University of Minnesota, Austin, MN 55912, USA*
BROWN, DEBORAH A. • *Dept. of Biochemistry & Cell Biology, Stony Brook University, Stony Brook, NY 11794-4215, USA*
BROWN, RHODERICK E. • *Hormel Institute, University of Minnesota, Austin, MN 55912, USA*
BURNS, ALAN R. • *Biomolecular Interfaces and Systems Dept., Sandia National Laboratories, Albuquerque, NM 87185-1413, USA*
DEVANATHAN, SAVITHA • *Dept. of Biochemistry and Molecular Biophysics, University of Arizona, Tucson, AZ 85721, USA*
ELLIOTT, R. • *Materials Research Laboratory, University of California, Santa Barbara, CA 93106-5121, USA*
GALLEGOS, ADALBERTO M. • *Dept. of Pathobiology, Texas A&M University, College Station, TX 77843, USA*
GAWRISCH, KLAUS • *Laboratory of Membrane Biochemistry and Biophysics, NIAAA, National Institutes of Health, Bethesda, MD 20892, USA*
HUANG, HUAN • *Dept. of Physiology and Pharamcology, Texas A&M University, College Station, TX 77843, USA*
KATSARAS, JOHN • *National Research Council, Canadian Neutron Beam Centre, Chalk River Laboratories, Chalk River, Ontario, K0J 1J0, Canada*
KENWORTHY, ANNE K. • *Dept. of Molecular Physiology and Biophysics, Vanderbilt School of Medicine, Nashville, TN 37232, USA*
KIER, ANN B. • *Dept. of Pathobiology, Texas A&M University, College Station, TX 77843, USA*
KUCERKA, NORBERT • *National Research Council, Canadian Neutron Beam Centre, Chalk River Laboratories, Chalk River, Ontario, K0J 1J0, Canada*
KUSUMI, AKIHIRO, *Membrane Mechanisms Project, ICORP, Japan Science and Technology Agency, Kyoto University, Shougoin, Kyoto 606-8507. Japan*

LIDKE, DIANE • *Dept. of Pathology and Cancer Center, University of New Mexico School of Medicine, Albuquerque, NM, USA*
LINDBLOM, GÖRAN • *Dept. of Biophysical Chemistry, Umea University, Umea, Switzerland*
LIU, JYH-CHARN • *Computer Science Department, Texas A&M University, College Station, TX 77843, USA*
LONDON, ERWIN • *Dept. of Biochemistry and Cell Biology, Stony Brook University, Stony Brook, NY 11794-521, USA*
MCINTOSH, AVERY L. • *Dept. of Physiology and Pharmacology, Texas A&M University, College Station, TX 77843, USA*
MCINTOSH, THOMAS J. • *Department of Cell Biology, Duke University Medical Center, Durham, NC 27710, USA*
MILLS, THALIA T. • *Dept. of Physics and Dept. of Molecular Biology, Cornell University, Ithaca, NY, USA*
MOERNER, W. E. • *Dept. of Chemistry, Stanford University Stanford, CA 94305, USA*
NIEH, MU-PING • *National Research Council, Canadian Neutron Beam Centre, Chalk River Laboratories, Chalk River, Ontario, K0J 1J0, Canada*
NISHIMURA, STEFANIE Y. • *Dept. of Chemistry, Stanford University, Stanford, CA 94305, USA*
OLIVER, JANET M. • *Dept. of Pathology and Cancer Center, University of New Mexico School of Medicine, Albuquerque, NM, USA*
ORÄDD, GREGER • *Dept. of Biophysical Chemistry, Umea University, Umea, Switzerland*
PANDIT, SAGAR A. • *Department of Biological, Chemical, and Physical Sciences, Illinois Institute of Technology, 3101 S. Dearborn, Chicago, IL 60616, USA*
PENCER, JEREMY • *National Research Council, Canadian Neutron Beam Centre, Chalk River Laboratories, Chalk River, Ontario, K0J 1J0, Canada*
PFEIFFER, JANET R. • *Dept. of Pathology and Cancer Center, University of New Mexico School of Med., Albuquerque, NM, USA*
POLOZOV, IVAN V. • *Laboratory of Membrane Biochemistry and Biophysics, NIAAA, National Institutes of Health, Bethesda, MD 20892, USA*
RAYMOND-STINTZ, MARY ANN • *Dept. of Pathology and Cancer Center, University of New Mexico School of Med., Albuquerque, NM, USA*
SALAMON, ZDZISLAW • *Dept. of Biochemistry and Molecular Biophysics, University of Arizona, Tucson, AZ 85721, USA*
SCHICK, M. • *Dept. of Physics, University of Washington, Box 351560, Seattle, WA 98195-1560, USA*
SCHROEDER, FRIEDHELM • *Dept. of Physiology and Pharmacology, Texas A&M University, College Station, TX 77843, USA*

Contributors

SCHWILLE, PETRA • *Institute of Biophysics, Dresden University of Technology, Tatzberg 47-51, D-01307, Dresden, Germany*

SCOTT, H. LARRY • *Department of Biological, Chemical, and Physical Sciences, Illinois Institute of Technology, 3101 S. Dearborn, Chicago, IL 60616, USA*

SEHGAL, PRAVIN B. • *Dept. of Cell Biology and Anatomy, New York Medical College, Valhalla, NY 10595, USA*

SHAH, MEHUL B. • *Dept. of Cell Biology and Anatomy, New York Medical College, Valhalla, NY 10595, USA*

STEINBERG, STANLY • *Dept. of Mathematics and Statistics, University of New Mexico School of Medicine, Albuquerque, NM, USA*

SUBCZYNSKI, WITOLD K. • *Dept. of Biophysics, Medical College of Wisconsin, Milwaukee, WI 53226, USA*

SZLEIFER, I. • *Dept. of Chemistry, Purdue University, West Lafayette, IN 47907-1393, USA*

TOLLIN, GORDON • *Dept. of Biochemistry and Molecular Biophysics, University of Arizona, Tucson, AZ 85721, USA*

VEATCH, SARAH L. • *University of British Columbia, Centre for Microbial Disease Research, Lower Mall Research Station, 2259 Lower Mall, Rm 261B, Vancouver, B.C. V6T1Z4, Canada*

VRLJIC, MARIJA • *Dept. of Chemistry, Stanford University, Stanford, CA 94305, USA*

WANG, SUOJIN • *Department of Statistics, Texas A&M University, College Station, TX 77843, USA*

WIDOMSKA, JUSTYNA • *Dept. of Biophysics, Medical College of Wisconsin, Milwaukee, WI 53226, USA*

WILSON, BRIDGET S. • *Dept. of Pathology and Cancer Center, University of New Mexico School of Medicine, Albuquerque, NM, USA*

WISNIEWSKA, ANNA • *Dept. of Biophysics, Faculty of Biotechnology, Jagiellonian University, Krakow, Poland*

XU, HAI • *Computer Science Department, Texas A&M University, College Station, TX 77843, USA*

YIN, WENXIA • *Dept. of Mathematics and Statistics, University of New Mexico School of Med., Albuquerque, NM, USA*

ZHANG, JUN • *Dept. of Pathology and Cancer Center, University of New Mexico School of Medicine, Albuquerque, NM, USA*

ZHANG, WEIMIN • *Department of Statistics, Texas A&M University, College Station, TX 77843, USA*

1

Overview of Membrane Rafts

Thomas J. McIntosh

Summary

Transient lateral microdomains (rafts) in cell membranes have been postulated to perform a number of important functions in normal cells, and are also thought to be critically involved in several pathological conditions. However, there are still a number of fundamental unanswered questions concerning the composition, size, dynamics, and stability of membrane rafts. These questions are currently being addressed by a number of sophisticated biophysical, biochemical, and computational methodologies.

Key Words: Cell functions; membrane structure; methods; microdomains; rafts; bilayers.

1. Introduction

Since the pioneering work of Gorter and Grendel (*1*) and Danielli and Davson (*2*) it has been well understood that the lipid bilayer forms the core of biological membranes. For many years, the structure of cell membranes was often characterized by the "fluid mosaic model" (*3*), which depicted membranes in terms of proteins embedded and freely diffusing in a lipid sea made up of a fluid bilayer. In this paradigm, the bilayer was considered as a uniform, semipermeable barrier that served as a passive matrix for membrane proteins. Long-standing questions about the fluid mosaic model have concerned the reasons for the many classes of lipids found in biological membranes and the differences in lipid compositions of the various organelle membranes. For example, plasma membranes typically contain higher concentrations of cholesterol and sphingomyelin (SM) than do internal membranous organelles (*4,5*). Thus, along the secretion pathway, there are very low concentrations of cholesterol and sphingolipids in the endoplasmic reticulum, but the concentrations of these lipids increase from the *cis*-Golgi to the *trans*-Golgi and then to the plasma membrane (*4,6,7*).

A possible important reason for the relatively large concentrations of SM and cholesterol in the *trans*-Golgi and plasma membranes is their role in the formation of lateral transient microdomains termed "rafts" in these membranes *(8–10)*. Detergent-resistant membranes (DRMs) extracted from cells, thought to be related to rafts, are enriched in cholesterol and SM *(5,9,11)*. Several types of proteins are enriched either in DRMs or membrane microdomains, including acylated proteins *(12–14)*, glycosyl-phosphatidylinositol-anchored proteins *(15,16)*, and certain transmembrane receptors *(17–19)* and channels *(20–22)*. However, many transmembrane proteins, including specific tyrosine phosphatases *(18,23)*, are largely excluded from membrane microdomains or DRMs. Because of their ability to sequester specific lipids and proteins and exclude others, rafts have been postulated to perform critical roles in a number of normal cellular processes, such as signal transduction *(17,18,23,24)*, membrane fusion *(25,26)*, organization of the cytoskeleton *(27,28)*, lipid sorting *(29,30)*, and protein trafficking/recycling *(8,31,32)*, as well as pathological events such as the cellular invasion of influenza, Ebola, and human immunodeficiency-1 viruses *(33–36)*, and formation of the plaques associated with Alzheimer's disease *(37,38)*.

2. Rafts in Lipid Bilayer System

In terms of possible raft involvement in lipid and protein sorting, an early fundamental question was whether rafts could form in lipid mixtures in the absence of proteins. For example, as the necessary first event in the sorting process, Simons and van Meer *(29)* envisaged a lateral separation of glycerolipids and sphingolipids in the Golgi apparatus. In the past several years, many studies have shown that rafts do indeed spontaneously form in lipid bilayers with lipid compositions approximating those found in *trans*-Golgi or plasma membranes. In bilayers made up of a combination of SM, cholesterol, and unsaturated phosphatidylcholine (PC), microdomains have been detected by fluorescence experiments *(39)* and visualized by both atomic force microscopy *(40)* and light microscopy *(41–44)*. Thus, lipid–lipid interactions are thought to play key roles in the formation of membrane rafts. However, lipid–protein, protein–protein, and protein–ligand interactions have also been shown to be involved in sequestering proteins or modifying the lateral dimensions of rafts *(17,20,45–51)*. In this regard, experiments with model membranes have provided valuable insights at the molecular level on the nature of the interactions that sort specific lipids or proteins into or out of rafts *(9,39,52–66)*.

3. Methods to Address Unanswered Questions

Thus, in the past several years models of membrane structure have evolved from the fluid mosaic model and now depict cell plasma membranes as containing transient microdomains that act as platforms or centers for a variety of membrane activities. However, the raft field is still quite controversial. Currently, there are fundamental unsolved questions concerning microdomains in membranes, including their size and stability *(67–69)*, their physical properties *(70,71)*, mechanisms of their formation *(46,50,68,69,72)*, and even their very existence *(70,72–75)*. Detergent extraction procedures were instrumental in the discovery and analysis of rafts, and still are an important tool in determining the composition of membrane microdomains in cells *(9,11–14,23)*. However, this method involves breaking up the membrane and has limitations in terms of defining the size, properties, and dynamics of intact microdomains *(50,71,73,76–78)*. Thus, a variety of sophisticated techniques have recently been used to analyze in detail open questions concerning rafts in cell and model membranes. The chapters in this book provide details on many of these biochemical, biophysical, and computational methodologies with their current applications to the study of rafts.

Acknowledgments

This work was supported by NIH Grant GM27278.

References

1. Gorter, E. and Grendel, F. (1925) On Bimolecular Layers of Lipoids on the Chromocytes of the Blood. *J. Exp. Med.* **41,** 439–443.
2. Danielli, J. F. and Davson, J. (1935) A contribution to the theory of permeability of thin films. *J. Cell Comp. Physiol.* **5,** 495–508.
3. Singer, S. J. and Nicolson, G. L. (1972) The fluid mosaic model of the structure of cell membranes. *Science* **175,** 720–731.
4. Keenan, T. W. and Morre, D. J. (1970) Phospholipid class and fatty acid composition of golgi apparatus isolated from rat liver and comparison with other cell fractions. *Biochemistry* **9,** 19–25.
5. Fridriksson, E. K., Shipkova, P. A., Sheets, E. D., Holowka, D., Baird, B., and McLafferty, F. W. (1999) Quantitative analysis of phospholipids in functionally important membrane domains from RBL-2H3 mast cells using tandem high-resolution mass spectrometry. *Biochemistry* **38,** 8056–8063.
6. Yorek, M. A. (1993) Biological distribution, in *Phospholipids Handbook,* (Cevc, G. S., ed.), Marcel Dekker, Inc., New York, pp. 745–775.
7. van Helvoort, A. and van Meer, G. (1995) Intracellular lipid heterogeneity caused by topolgy of synthesis and specifity in transport. Examples: sphingolipids. *FEBS Lett.* **369,** 18–21.
8. Simons, K. and Ikonen, E. (1997) Functional rafts in cell membranes. *Nature* **387,** 569–572.
9. Brown, D. A. and London, E. (2000) Structure and function of sphingolipid- and cholesterol-rich membrane rafts. *J. Biol. Chem.* **275,** 17,221–17,224.
10. Gkantiragas, I., Brugger, B., Stuven, E., et al. (2001) Sphingomyelin-enriched microdomains at the Golgi complex. *Mol. Biol. Cell* **12,** 1819–1833.

11. London, E. and Brown, D. A. (2000) Insolubility of lipids in Triton X-100: physical origin and relationship to spingolipid/cholesterol membrane domains (rafts). *Biochim. Biophys. Acta.* **1508,** 182–195.
12. Arni, S., Keilbaugh, S. A., Ostermeyter, A. G., and Brown, D. A. (1998) Association of GAP-43 with detergent-resistant membranes requires two palmitoylated cysteine residues. *J. Biol. Chem.* **273,** 28,478–28,485.
13. Melkonian, K. A., Ostermeyer, A. G., Chen, J. Z., Roth, M. G., and Brown, D. A. (1999) Role of lipid modifications in targeting proteins to detergent-resistant membrane rafts. Many raft proteins are acylated, while few are prenylated. *J. Biol. Chem.* **274,** 3910–3917.
14. Moffett, S., Brown, D. A., and Linder, M. E. (2000) Lipid-dependent targeting of G proteins into rafts. *J. Biol. Chem.* **275,** 2191–2198.
15. Benting, J., Rietveld, A., Ansorge, I., and Simons, K. (1999) Acyl and alkyl chain length of GPI-anchors is critical for raft association in vitro. *FEBS Lett.* **462,** 47–50.
16. Sharma, P., Varma, R., Sarasij, R. C., et al. (2004) Nanoscale Organization of Multiple GPI-Anchored Proteins in Living Cell Membranes. *Cell* **116,** 577–589.
17. Field, K. A., Holowka, D., and Baird, B. (1997) Compartmentalized activation of the high affinity immunoglubulin E receptor within membrane domains. *J. Biol. Chem.* **272,** 4276–4280.
18. Janes, P. W., Ley, S. C., and Magee, A. I. (1999) Aggregation of lipid rafts accompanies signaling via the T cell antigen receptor. *J. Cell Biol.* **147,** 447–461.
19. Ridyard, M. S. and Robbins, S. M. (2003) Fibroblast growth factor-2-induced signaling through lipid raft-associated fibroblast growth factor receptor substrate 2 (FRS2). *J. Biol. Chem.* **278,** 13,803–13,809.
20. Schubert, A. L., Schubert, W., Spray, D. C., and Lisanti, M. P. (2002) Connexin family members target to lipid raft domains and interact with caveolin-1. *Biochemistry* **41,** 5754–5764.
21. Wong, W. and Schlichter, L. C. (2004) Differential recruitment of Kv1.4 and Kv4.2 to lipid rafts by PSD-95. *J. Biol. Chem.* **279,** 444–452.
22. Ishikawa, Y., Yuan, Z., Inoue, N., et al. (2005) Identification of AQP5 in lipid rafts and its translocation to apical membranes by activation of M3 mAChRs in interlobular ducts of rat parotid gland. *Am. J. Physiol. Cell Physiol.* **289,** C1303–C1311.
23. Young, R. M., Zheng, X., Holowka, D., and Baird, B. (2005) Reconstitution of regulated phosphorylation of FcepsilonRI by a lipid raft-excluded protein-tyrosine phosphatase. *J. Biol. Chem.* **280,** 1230–1235.
24. Simons, K. and Toomre, D. (2000) Lipid rafts and signal transduction. *Nat. Rev. Mol. Cell Biol.* **1,** 31–39.
25. Chamberlain, L. H., Burgoyne, R. D., and Gould, G. W. (2001) SNARE proteins are highly enriched in lipid rafts in PC12 cells: implications for the spatial control of exocytosis. *Proc. Natl. Acad. Sci. USA* **98,** 5619–5624.
26. Lang, T., Bruns, D., Wenzel, D., et al. (2001) SNAREs are concentrated in cholesterol-dependent clusters that define docking and fusion sites for exocytosis. *EMBO J.* **20,** 2202–2213.
27. Caroni, P. (2001) New EMBO member's review: actin cytoskeleton regulation through modulation of PI(4,5)P(2) rafts. *EMBO J.* **20,** 4332–4336.

28. Laux, T., Fukami, K., Thelen, M., Golub, T., Frey, D., and Caroni, P. (2000) GAP43, MARCKS, and CAP23 modulate PI(4,5)P(2) at plasmalemmal rafts, and regulate cell cortex actin dynamics through a common mechanism. *J. Cell Biol.* **149,** 1455–1472.
29. Simons, K. and van Meer, G. (1988) Lipid sorting in epithelial cells. *Biochemistry* **27,** 6197–6202.
30. Simons, K. and Ikonen, E. (2000) How cells handle cholesterol. *Science* **290,** 1721–1726.
31. Ikonen, E. (2001) Roles of lipid rafts in membrane transport. *Curr. Opin. Cell Biol.* **13,** 470–477.
32. Sharma, D. K., Choudhury, A., Singh, R. D., Wheatley, C. L., Marks, D. L., and Pagano, R. E. (2003) Glycosphingolipids internalized via caveolar-related endocytosis rapidly merge with the clathrin pathway in early endosomes and form microdomains for recycling. *J. Biol. Chem.* **278,** 7564–7572.
33. Duncan, M. J., Shin, J. S., and Abraham, S. N. (2002) Microbial entry through caveolae: variations on a theme. *Cell Microbiol.* **4,** 783–791.
34. Lafont, F., Tran Van Nhieu, G., Hanada, K., Sansonetti, P., and van der Goot, F. G. (2002) Initial steps of Shigella infection depend on the cholesterol/sphingolipid raft-mediated CD44-IpaB interaction. *EMBO J.* **21,** 4449–4457.
35. Campbell, S. M., Crowe, S. M., and Mak, J. (2001) Lipid rafts and HIV-1: from viral entry to assembly of progeny virions. *J. Clin. Virol.* **22,** 217–227.
36. Manes, S., del Real, G., and Martinez, A. C. (2003) Pathogens: raft hijackers. *Nat. Rev. Immunol.* **3,** 557–568.
37. Ehehalt, R., Keller, P., Haass, C., Thiele, C., and Simons, K. (2003) Amyloidogenic processing of the Alzheimer beta-amyloid precursor protein depends on lipid rafts. *J. Cell Biol.* **160,** 113–123.
38. Vetrivel, K. S., Cheng, H., Lin, W., et al. (2004) Association of gamma-secretase with lipid rafts in post-Golgi and endosome membranes. *J. Biol. Chem.* **279,** 44,945–44,954.
39. Ahmed, S. N., Brown, D. A., and London, E. (1997) On the origin of sphingolipid/cholesterol-rich detergent-insoluble cell membranes: physiological concentrations of cholesterol and sphingolipid induce formation of a detergent-insoluble, liquid-ordered lipid phase in model membranes. *Biochemistry* **36,** 10,944–10,953.
40. Rinia, H. A. and deKruijff, B. (2001) Imaging domains in model membranes with atomic force microscopy. *FEBS Lett.* **504,** 194–199.
41. Dietrich, C., Bagatolli, L. A., Volovyk, Z. N., et al. (2001) Lipid rafts reconstituted in model membranes. *Biophys. J.* **80,** 1417–1428.
42. Samsonov, A. V., Mihalyov, I., and Cohen, F. S. (2001) Characterization of cholesterol-sphingomyelin domains and their dynamics in bilayer membranes. *Biophys. J.* **81,** 1486–1500.
43. Veatch, S. L. and Keller, S. L. (2003) Separation of liquid phases in giant vesicles of ternary mixtures of phospholipids and cholesterol. *Biophys. J.* **85,** 3074–3083.
44. Veatch, S. L., Polozov, I. V., Gawrisch, K., and Keller, S. L. (2004) Liquid domains in vesicles investigated by NMR and fluorescence microscopy. *Biophys. J.* **86,** 2910–2922.

45. Elliott, M. H., Fliesler, S. J., and Ghalayini, A. J. (2003) Cholesterol-Dependent Association of Caveolin-1 with the Transducin alpha Subunit in Bovine Photoreceptor Rod Outer Segments: Disruption by Cyclodextrin and Guanosine 5′-O-(3-Thiotriphosphate). *Biochemistry* **42,** 7892–7903.
46. Chini, B. and Parenti, M. (2004) G-protein coupled receptors in lipid rafts and caveolae: how, when and why do they go there? *J. Mol. Endocrinol.* **32,** 325–338.
47. Kusumi, A., Koyama-Honda, I., and Suzuki, K. (2004) Molecular dynamics and interactions for creation of stimulation-induced stabilized rafts from small unstable steady-state rafts. *Traffic* **5,** 213–230.
48. van Meer, G. (2004) Invisible rafts at work. *Traffic* **5,** 211–212.
49. Hammond, A. T., Heberle, F. A., Baumgart, T., Holowka, D., Baird, B., and Feigenson, G. W. (2005) Crosslinking a lipid raft component triggers liquid ordered-liquid disordered phase separation in model plasma membranes. *Proc. Natl. Acad. Sci. USA* **102,** 6320–6325.
50. Douglass, A. D. and Vale, R. D. (2005) Single-molecule microscopy reveals plasma membrane microdomains created by protein-protein networks that exclude or trap signaling molecules in T cells. *Cell* **121,** 937–950.
51. Sieber, J. J., Willig, K. I., Heintzmann, R., Hell, S. W., and Lang, T. (2006) The SNARE Motif Is Essential for the Formation of Syntaxin Clusters in the Plasma Membrane. *Biophys. J.* **90,** 2843–2851.
52. Schroeder, R., London, E., and Brown, D. (1994) Interactions between saturated acyl chains confer detergent resistance on lipids and glycosylphosphatidylinositol (GPI)-anchored proteins; GPI-anchored proteins in liposomes and cells show similar behavior. *Proc. Natl. Acad. Sci. USA* **91,** 12,130–12,134.
53. Brown, R. E. (1998) Sphingolipid organization in biomembranes: what physical studies of model membranes reveal. *J. Cell Sci.* **111,** 1–9.
54. Wang, T. -Y., Leventis, R., and Silvius, J. R. (2000) Fluorescence-based evaluation of the partitioning of lipids and lipidated peptides into liquid-ordered lipid microdomains: a model for molecular partitioning into "lipid rafts." *Biophys. J.* **79,** 919–933.
55. Dietrich, C., Volovyk, Z. N., Levi, M., Thompson, N. L., and Jacobson, K. (2001) Partitioning of Thy-1, GM1, and cross-linked phospholipid analogs into lipid rafts reconstituted in supported model membrane monolayers. *Proc. Natl. Acad. Sci. USA* **98,** 10,642–10,647.
56. Khan, T. K., Yang, B., Thompson, N. L., Maekawa, S., Epand, R. M., and Jacobson, K. (2003) Binding of NAP-22, a Calmodulin-Binding Neuronal Protein, to Raft-like Domains in Model Membranes. *Biochemistry* **42,** 4780–4786.
57. London, E. (2002) Insights into lipid raft structure and formation from experiments in model membranes. *Curr. Opin. Struct. Biol.* **12,** 480–486.
58. McIntosh, T. J., Vidal, A., and Simon, S. A. (2003) Sorting of lipids and transmembrane peptides between detergent-soluble bilayers and detergent-resistant rafts. *Biophys. J.* **85,** 1656–1666.
59. Kahya, N., Scherfeld, D., Bacia, K., Poolman, B., and Schwille, P. (2003) Probing lipid mobility of raft-exhibiting model membranes by fluorescence correlation spectroscopy. *J. Biol. Chem.* **278,** 28,109–28,115.

60. Allende, D., Vidal, A., and McIntosh, T. J. (2004) Jumping to rafts: gatekeeper role of bilayer elasticity. *Trends Biochem. Sci.* **29,** 325–330.
61. Mukherjee, S. and Maxfield, F. R. (2004) Membrane domains. *Annu. Rev. Cell Dev. Biol.* **20,** 839–866.
62. Vidal, A. and McIntosh, T. J. (2005) Transbilayer Peptide Sorting between Raft and Nonraft Bilayers: Comparisons of Detergent Extraction and Confocal Microscopy. *Biophys. J.* **89,** 1102–1108.
63. de Almeida, R. F., Loura, L. M., Fedorov, A., and Prieto, M. (2005) Lipid rafts have different sizes depending on membrane composition: a time-resolved fluorescence resonance energy transfer study. *J. Mol. Biol.* **346,** 1109–1120.
64. Oradd, G., Westerman, P. W., and Lindblom, G. (2005) Lateral Diffusion Coefficients of Separate Lipid Species in a Ternary Raft-Forming Bilayer: A Pfg-NMR Multinuclear Study. *Biophys. J.* **89,** 315–320.
65. London, E. (2005) How principles of domain formation in model membranes may explain ambiguities concerning lipid raft formation in cells. *Biochim. Biophys. Acta.* **1746,** 203–220.
66. Silvius, J. R. (2005) Partitioning of membrane molecules between raft and non-raft domains: Insights from model-membrane studies. Biochim. *Biophys. Acta.* **1746,** 193–202.
67. Kenworthy, A. K. and Edidin, M. (1998) Distribution of glycosylphosphatidylinositol-anchored protein at the apical surface of MDCK cells examined at a resolution of <100 Å using imaging fluorescence resonance energy transfer. *J. Cell Biol.* **142,** 69–84.
68. Anderson, R. G. and Jacobson, K. (2002) A role for lipid shells in targeting proteins to caveolae, rafts, and other lipid domains. *Science* **296,** 1821–1825.
69. Edidin, M. (2003) The state of lipid rafts: from model membranes to cells. *Annu. Rev. Biophys. Biomol. Struct.* **32,** 257–283.
70. Lai, E. C. (2003) Lipid rafts make for slippery platforms. *J. Cell Biol.* **162,** 365–370.
71. Pike, L. J. (2003) Lipid rafts: bringing order to chaos. *J. Lipid Res.* **44,** 655–667.
72. Kenworthy, A. K., Petranova, N., and Edidin, M. (2000) High-resolution FRET microscopy of cholera toxin B-subunit and GPI-anchored proteins in cell plasma membranes. *Mol. Biol. Cell* **11,** 1645–1655.
73. Munro, S. (2003) Lipid rafts: elusive or illusive? *Cell* **115,** 377–388.
74. Laude, A. J. and Prior, I. A. (2004) Plasma membrane microdomains: organization, function and trafficking (Review). *Mol. Membr. Biol.* **21,** 193–205.
75. McMullen, T. P. W., Lewis, R. N. A. H., and McElhaney, R. N. (2004) Cholesterol-phospholipid interactions, the liquid-ordered phase and lipid rafts in model and biological membranes. *Curr. Opin. Coll. Interface Sci.* **8,** 459–468.
76. Heerklotz, H. (2002) Triton promotes domain formation in lipid raft mixtures. *Biophys. J.* **83,** 2693–2701.
77. Heerklotz, H., Szadkowska, H., Anderson, T., and Seelig, J. (2003) The sensitivity of lipid domains to small perturbations demonstrated by the effect of Triton. *J. Mol. Biol.* **329,** 793–799.
78. Lichtenberg, D., Goni, F. M., and Heerklotz, H. (2005) Detergent-resistant membranes should not be identified with membrane rafts. *Trends Biochem. Sci.* **30,** 430–436.

2

Analysis of Raft Affinity of Membrane Proteins by Detergent-Insolubility

Deborah A. Brown

Summary

Isolation of detergent-resistant membranes (DRMs; also known as detergent-insoluble glycolipid-enriched membranes [DIGs] or glycolipid-enriched membranes [GEMs]) that are enriched in proteins and lipids with a high affinity for rafts is one of the simplest and most widely used methods for studying rafts. However, it is essential to understand the limitations as well as the advantages of this method. DRMs do not correspond precisely to rafts in living cells. For this reason, finding a protein enriched in DRMs does not prove that it was in rafts in the living cell. Furthermore, the fraction of a protein found in DRMs provides no quantitative information about the fraction of the protein originally in rafts. In fact, DRMs may be isolated from membranes that did not even contain rafts before detergent extraction. DRM-association is useful because it reflects a high-inherent affinity of a protein for the ordered membrane state found in rafts. Treatments that affect the DRM-association of a protein can thus be inferred to affect its raft affinity. Current models suggest that rafts may form in a regulated manner, often associated with clustering of membrane proteins or lipids, during processes such as signal transduction. DRM-association is a read-out of whether a protein is likely to associate with rafts that form under these conditions.

Key Words: Brij series detergents; DRM; liquid-ordered phase; membrane microdomain; membrane raft; Triton X-100.

1. Introduction

Model membranes in the liquid-ordered (L_o) or gel phases are insoluble in gentle nonionic detergents such as Triton X-100 Sigma-Aldrich (St. Louis, MO) and CHAPS Calbiochem, EMD Biosciences, (La Jolla, CA) *(1)*. By contrast, these detergents fully solubilize liquid-disordered (L_d) membranes. Furthermore, extraction of two-phase model membrane vesicles containing coexisting L_o- and L_d-phase domains generally results in selective solubilization of the L_d-phase domains, leaving the L_o-phase domains as detergent-resistant membranes

(DRMs) *(2–4)*. Detergent-insolubility results from tight lipid acyl chain packing in the L_o and gel phases. These close interactions cause the bilayer to exclude detergent *(2,3)*. Early studies showed that DRMs could also be isolated after extraction of cells with these detergents *(5)*. Furthermore, molecules such as sphingolipids and glycosyl phosphatidyl inositol (GPI)-anchored proteins, expected to have a high affinity for ordered membranes, were enriched in these DRMs, whereas L_d-favoring membrane phospholipids and transmembrane proteins were solubilized *(5)*. This suggested that L_o- and L_d-phase domains might coexist in cell membranes, and that the L_o-phase domains (rafts) might be isolated as DRMs. Isolation of DRMs became (and remains) a popular method for studying rafts, and especially for determining their protein composition.

Later studies showed that the relationship between rafts and DRMs is more complicated than originally thought. For instance, GPI-anchored proteins, which are highly enriched in DRMs *(5,6)*, are either not present in discrete rafts in resting cells *(7–9)* or are present in small, unstable nanoclusters *(10)*. (By contrast, antibody-mediated clustering of these proteins on the cell surface—even to a very small degree—enhances their association with DRMs *(11)* and may result in their association with rafts *[12–14]*). As a further complication, under some conditions, detergent extraction can actually induce phase separation and L_o domain formation in model membranes that were previously in a uniform phase *(15)*.

It should also be noted that in order to isolate DRMs, extraction of cells with Triton X-100 or CHAPS must be performed at low temperatures. Phase behavior is extremely temperature-sensitive, and the amount of plasma membrane present in the L_o phase on ice will always be substantially higher than that at 37°C. Detergents such as those of the Brij series (Sigma-Aldrich) *(16–18)* and Lubrol WX (Serva, Heidelberg, Germany) *(19)* have also been used to isolate DRMs, sometimes at elevated temperatures. Use of these detergents is subject to an important caveat *(20)*. They insert into membranes much less efficiently than Triton X-100 and CHAPS, and may fail to fully solubilize even L_d-phase membranes. Sterols, sphingolipids, and saturated-chain phospholipids make the plasma membrane significantly more ordered than intracellular membranes. Thus, inefficient detergents may solubilize intracellular membranes (and their proteins), whereas failing to completely solubilize the plasma membrane—even if it is present in a single L_d-like phase. With these important caveats in mind, isolation of DRMs and analysis of their protein composition is a powerful tool for determining the affinity of proteins for rafts.

2. Materials
2.1. Cell Culture, Transfection, and Lysis
1. Dulbecco's modified Eagle's medium (Gibco/BRL, Bethesda, MD) supplemented with 10% iron-supplemented calf serum (CS, JRH, Lenexa, KS) (*see* **Note 1**).

2. Lipofectamine 2000 (Invitrogen, Carlsbad, CA) transfection reagent. Store at 4°C.
3. Tris-HCl/NaCl/EDTA (TNE) buffer: 25 mM Tris-HCl, 150 mM NaCl, 5 mM ethylenediamine tetraacetic acid, pH 7.4.
4. Opti-MEM®I (Gibco, Invitrogen, Carlsbad).
5. TNE/TX/P buffer: TNE buffer containing 1% Triton X-100 and the following protease inhibitor (P) cocktail: 0.2 mM phenylmethyl sulfonyl fluoride, 1 µg/mL leupeptin, and 1 µg/mL pepstatin. Triton X-100 is a viscous liquid. Make up 100 mL of a 10% (w/v) stock solution. Triton X-100 is subject to oxidation on prolonged storage and should be replaced regularly. Add protease inhibitors from 1000X stocks in ethanol or methanol just before use. Phenylmethyl sulfonyl fluoride stock in ethanol is stable at room temperature; leupeptin and pepstatin stocks are stored at –20°C.
6. Teflon cell scrapers (Fisher, Pittsburgh, PA).
7. Dounce homogenizer (Fisher).

2.2. Sucrose Gradient Preparation

1. TNE containing sucrose (separate solutions at 80, 35, and 5%). Eighty percent sucrose is close to the solubility limit, and TNE/80% sucrose is conveniently made with mild heating and magnetic stirring on a heating stir plate. Store at 4°C. For some applications, including detergent in the gradient solutions may be necessary (*see* **Note 2**).
2. SW41 ultracentrifuge tubes (Beckman Coulter Inc., Fullerton, CA).

2.3. Ultracentrifugation and Gradient Fractionation

1. Ultracentrifuge, for example, model L8 (Beckman).
2. SW41 rotor (Beckman).
3. Gradient harvester (ISCO, Lincoln, NE) (*see* **Note 3**).

2.4. Sodium Dodecyl Sulfate-Polyacrylamide Gel Electrophoresis

1. 1 M Tris-HCl, pH 8.8. Store at room temperature.
2. 1 M Tris-HCl, pH 6.8. Store at room temperature.
3. 20% Sodium dodecyl sulfate (SDS). Store at room temperature. If room temperature drops and SDS precipitates, warm briefly to redissolve.
4. 30% Acrylamide/0.8% *bis*-acrylamide solution.
5. Ammonium persulfate (AMPS) (10%).
6. N,N,N,N′-Tetramethyl-ethylenediamine (TEMED).
7. Running buffer (1X): 25 mM Tris-HCl, 190 mM glycine, 0.1% SDS. Make a 10X stock; store at room temperature.
8. Sample buffer (4X): 40 mM Tris-HCl pH 6.8, 20% β-mercaptoethanol, 8% SDS, 40% glycerol, and 0.05% bromophenol blue. Store at –20°C; or make without β-mercaptoethanol and store at room temperature, adding β-mercaptoethanol just before use.
9. Water-saturated isobutanol. Store at room temperature.
10. Benchmark™ prestained molecular weight ladder (Invitrogen, Carlsbad).

11. Mini-PROTEAN®II Dual slab gel system (BioRad, Richmond, CA).
12. 3-mL Syringes and 22-gauge needles (Fisher).

2.5. Transfer to Nitrocellulose and Western Blotting for Placental Alkaline Phosphatase and Transferrin Receptor

1. Transfer buffer: 25 mM Tris-HCl, 190 mM glycine, and 20% (v/v) methanol. Store at 4°C.
2. Nitrocellulose membrane (Millipore, Bedford, MA); 3MM chromatography paper (Whatman, Maidstone, UK).
3. Phosphate-buffered saline with Tween (PBS-T), 20 mM phosphate buffer, pH 7.4, 150 mM NaCl, and 0.05% Tween 20.
4. Mini-Trans-Blot Module (BioRad, Richmond) including Bio-Ice® cooling unit. After each use, fill the Bio-Ice cooling unit with water and store at –20°C.
5. 500 mA Power supply.
6. Blocking buffer: 5% (w/v) nonfat dry milk in PBS-T.
7. Primary antibodies: rabbit antiplacental alkaline phosphatase (PLAP) (Signet, Dedham, MA) and mouse antitransferrin receptor (TfR) (Zymed, S. San Francisco, CA).
8. Secondary antibodies: goat antirabbit immunoglobulin (Ig)G and goat antimouse IgG, both conjugated to horse radish peroxidase (Jackson Immunoresearch Laboratories, West Grove, PA).
9. Enhanced chemiluminescence (ECL) reagents (Western Lightning™, Perkin Elmer, Boston, MA) and Bio-Max ML film (Kodak, Rochester, NY).

2.6. Stripping the Blot for Reprobing With Anti-TfR

1. Stripping buffer: 62.5 mM Tris-HCl, 2% (w/v) SDS, pH 6.8,. Store at room temperature. Just before use, warm to 70°C and then add 100 mM β-mercaptoethanol.
2. Wash buffer: PBS-T.

3. Methods

One of the main uses of DRM isolation is to determine whether individual proteins are enriched in them, as a qualitative measure of their raft affinity. To do this, cold Triton X-100 extracts of cells are subjected to sucrose density gradient centrifugation, to separate floating DRMs from detergent-solubilized proteins, which remain in the load fractions near the bottom of the gradient. Gradient fractions are then analyzed by SDS-polyacrylamide gel electrophoresis (PAGE) and Western blotting to determine the distribution of the protein of interest between DRM and detergent-soluble fractions. DRMs are quite forgiving in terms of buffer and salt composition used in their preparation (*see* **Note 4**). These can generally be adjusted according to individual needs. However, DRM preparation is very sensitive to temperature, and is affected by even a slight increase in extraction temperature *(21)*. DRM isolation must be performed at 0–4°C, and gradients may be formed in a cold room. Another consideration is the tendency of DRMs to associate with cytoskeleton. A large fraction of DRMs can be lost by pelleting to the bottom of the ultracentrifuge tube, in association with the cytoskeletal pellet. It is not clear to what degree cytoskeletal is physiologically relevant, and to what degree it represents nonspecific trapping. Both may contribute. Although caveats apply, the yield of floating DRMs can be increased by the inclusion of alkaline carbonate (*see* **Note 5**).

Because the absolute amount of DRM-association of any protein may vary depending on exact conditions (for instance, detergent concentration or the ratio of detergent to proteins and lipids), it is essential to examine a control non-DRM-associated protein under the same conditions, ideally in the same sucrose gradient. The protocol presented describes methods for analysis of DRM-association of the GPI-anchored protein PLAP, examining the non-DRM protein TfR as a control. This provides a control for efficient solubilization of a protein with low raft affinity, and for good separation of DRMs and solubilized proteins.

3.1. Transfection

1. Two 100-mm dishes of confluent-cultured adherent mammalian cells provide enough DRM material for convenient isolation. For exogenously expressed proteins, one 35-mm dish of confluent cells provides enough signal for detection on Western blots. In this case, transfect cells in one 35-mm dish as described herein, and pool the lysate with that of two 100-mm dishes of the same cell type, used as carrier. For examination of endogenous proteins, skip to **step 2**.
2. Seed cells in one 35-mm dish the day before transfection.
3. For each transfection, set up two sterile microfuge tubes (Fisher), one for DNA and the other for Lipofectamine 2000.
4. Place 200 µL Opti-MEM I and 2 µg desired plasmid DNA in one tube.
5. Place 195 µL Opti-MEM I and 5 µL Lipofectamine 2000 in the other tube. Vortex and spin briefly (1–2 s) in a microfuge and then incubate at room temperature for 5 min.
6. Add the Lipofectamine mixture to the tube containing DNA. Vortex, spin briefly in a microfuge (1–2 s) and incubate for 20 min at room temperature.
7. Remove growth media from cells and rinse twice with PBS. Add 1.5 mL Opti-MEM I to the dish.
8. Add the Lipofectamine 2000/DNA mixture to the dish and swirl gently. Incubate at 37°C in a 5% CO_2 incubator for 5 h.
9. Replace media with regular growth media and incubate at 37°C in a 5% CO_2 incubator overnight or up to 48 h.

3.2. Preparation of Samples for Isolation of DRMs from Triton X-100 Lysates and Ultracentrifugation

1. Rinse dishes twice with ice-cold TNE and place on ice.
2. Add ice-cold TNE/TX/P (1-mL/100-mm dish and 0.5-mL/35-mm dish). Incubate 20 min in a covered ice bucket, optionally in the cold room.

3. Scrape lysates with Teflon scraper and transfer to a Dounce homogenizer. Homogenize 10 strokes.
4. Transfer homogenates to disposable glass tubes, avoiding foaming as much as possible. Pool lysates from duplicate 100-mm dishes and (if used) 35-mm dish of transfected cells.
5. Add 1.1X volume of TNE/80% sucrose/P. Cover tubes with Parafilm (Pechiney Plastic Packaging, Menasha, WI). In the cold room, mix well by inversion and pipeting with P1000 Pipetman (Rainin Instrument, Oakland, CA) and large tip. Examine mixture to ensure complete mixing. Avoid vortexing as foam will form.
6. In the cold room, transfer lysate/sucrose mixture to the bottom of an SW41 ultracentrifuge tube placed in an ice bucket. Avoid touching the sides of the tube.
7. Using a 5-mL pipet, carefully layer 5–6 mL TNE/38% sucrose/P over the lysate mixture, without disturbing the interface.
8. Layer about 2 mL TNE/5% sucrose/P over the TNE/38% sucrose layer. The top of the TNE/5% sucrose layer should be about 2 mm below the top of the tube. The exact volume of TNE/5% sucrose/P is not important. Ideally, the TNE/38% sucrose layer should be as large as possible, to maximize the distance between the lysate layer at the bottom and the 38% sucrose/5% sucrose interface.
9. If only one sample is used, prepare a balance tube exactly as in step 8, mixing TNE with TNE/80% sucrose for the bottom layer.
10. Weigh the tubes, minimizing the time the tubes are out of the ice bucket, and balance them using TNE/5% sucrose.
11. Place the tubes (uncapped) in opposite buckets of an SW41 swinging bucket rotor (Beckman). Place buckets (all capped) in all rotor positions.

3.3. Ultracentrifugation and Gradient Harvesting

1. Spin for 3–18 h at 41,000 rpm (280,000g). (DRMs will float to equilibrium by 3 h). Braking at the end of the run does not significantly impair gradient quality.
2. Remove gradients from buckets and place in an ice bucket. The DRMs should be visible as white particulate matter near the position of the 5%/38% sucrose interface, although the interface itself will not be visible. DRMs can vary in their degree of clumping. Once centrifugation is complete, strict maintenance of samples on ice is less crucial than before. However, standard procedures for avoiding prolonged exposure of samples to room temperature should be followed.
3. Place gradient in a gradient harvester. The gradient will be harvested from the bottom, by gravity flow. The gradient tube is clamped in place, resting on a support platform. The gradient harvester contains a built-in hollow needle attached to plastic tubing, situated just below the tube support. After securing the tube on the support, the needle is swung up through a small hole in the support platform to pierce the bottom of the tube. The gradient liquid will start flowing through the needle and the tubing. Flow can be blocked until desired by clamping the plastic tubing with a hemostat.
4. Place 15-numbered microfuge tubes in a rack. Mark the position of 1 mL liquid (determined by placing 1 mL water in a sample tube) on each tube.
5. Release the clamp and allow 1 mL liquid to flow from the bottom of the gradient into tube no. 1. Move the tubing to tube no. 2. The flow rate can be adjusted by raising and lowering the end of the tubing as desired. In this way, fractionate the

gradient into 1 mL fractions. (Set up one or two more tubes than you expect to have fractions; discard unused tubes at the end.)
6. Add 350-μL 4X sample buffer to each fraction. Mix well. This will solubilize suspended DRMs to allow uniform dispersion of DRM proteins and lipids. Samples may be stored at –20°C until use. Either before freezing, or just before loading on the gel, heat samples to 95°C in a heating block for 5 min. Prevent tubes from opening during heating using snap-on microcentrifuge tube holders (LabScientific, Livingston, NJ).

3.4. Sodium Dodecyl Sulfate-Polyacrylamide Gel Electrophoresis

1. These instructions assume the use of a BioRad mini-PROTEAN II slab gel apparatus. Prepare a 1.5-mm thick, 10% gel by mixing 1.67 mL of acrylamide/*bis*-solution, 1.88 mL 1 *M* Tris-HCl pH 8.8, 1.4 mL distilled water, 25 μL 20% SDS, 40 μL 10% AMPS, and 8 μL TEMED (*see* **Note 6**). Add AMPS and TEMED immediately before pouring mixture into preassembled mini-PROTEAN II slab gel apparatus. Carefully overlay with water-saturated isobutanol.
2. After the gel polymerizes, pour off the isobutanol and rinse the top of the gel with distilled water. Remove all the water.
3. Prepare the stacking gel by mixing 0.3 mL acrylamide/*bis*-solution, 0.23 mL 1 *M* Tris-HCl, pH 6.8, 9.4 μL 20% SDS, 1.31 mL distilled water, 15 μL 10% AMPS, and 4 μL TEMED. Pour on top of the gel and immediately insert a 15-well comb (BioRad) between the gel plates.
4. After polymerization, carefully remove the comb and place the unit in a gel tank. Fill the unit with 1X running buffer and add running buffer to the tank to cover the bottom 2–3 cm of the gel unit. Use a 3-mL syringe fitted with a bent 22-gauge needle to wash the wells with running buffer. Carefully straighten the wells with the needle if necessary.
5. Load 15–20 μL of each gradient fraction (mixed with 4X sample buffer). Include prestained molecular weight markers (5 μL) in one well. If possible, avoid the wells at each end of the gel.
6. Cover the tank, attach the built-in leads to a power supply, and run at 200 V until the dye front reaches the bottom of the gel (*see* **Note 7**).

3.5. Transfer of Proteins to Nitrocellulose and Western Blotting for PLAP and TfR

1. Disassemble the gel apparatus and discard running buffer. Open the gel plates and discard stacking gel. Soak the gel in transfer buffer 5–30 min.
2. Prepare a tray of cold transfer buffer that is large enough to hold the gel-holder cassette. Place the gel-holder cassette in the tray, gray side down. Place a fiber pad on the cassette. Place a piece of 3MM paper, cut just larger than the gel, on the fiber pad. Place the gel on the 3MM paper. Make sure the gel is immersed in transfer buffer. Carefully place a piece of nitrocellulose, cut just larger than the gel and briefly presoaked in transfer buffer, on top of the gel. Once the nitrocellulose has touched the gel, do not remove it to realign. Press firmly on the nitrocellulose to remove any bubbles. Place a second sheet of 3MM paper (prewetted in transfer buffer) on

the nitrocellulose. Place a wet fiber pad on top of the 3-mm paper. Close the gel-holder cassette (BioRad) without moving the gel–nitrocellulose sandwich.
3. Place in the modular electrode assembly, with the gray side of the cassette next to the gray side of the electrode assembly. Insert the electrode assembly into a running tank. Remove a Bio-Ice cooling unit from the freezer, and place in the tank. Place a magnetic stir bar in the tank, and place the tank on a stir plate. Turn on the stir plate and make sure the stir bar is spinning properly.
4. Place the lid on the tank and attach the leads to a power supply. Turn on the power supply and adjust to 100 V. Transfer for 1 h. The current will rise from about 250 to 350 mA during the transfer.
5. Once the transfer is complete, remove the cassette from the tank. Remove the top sponge, sheets of 3-mm paper, and gel. Mark the top (gel side) of the nitrocellulose with pencil for orientation. The colored molecular weight markers should be clearly visible on the membrane, and will probably also be visible but fainter on the bottom of the nitrocellulose.
6. Incubate the nitrocellulose in 50-mL blocking buffer for 1 h at room temperature on a rocking platform.
7. Remove and discard blocking buffer and rinse the nitrocellulose in PBS-T.
8. For detection of two proteins of very different sizes, the blot can be cut laterally at a position intermediate between the two proteins, using the prestained molecular weight markers (visible on the blot) as a guide. After cutting the blot, mark each portion for orientation. Similarly sized proteins like PLAP (migrates ~70 kDa) and TfR (migrates ~90 kDa) should be detected sequentially by probing for one, stripping the blot, and then, probing for the other.
9. Add a 1:2000 dilution of anti-PLAP antibodies in blocking buffer for 1 h at room temperature on a rocking platform (*see* **Note 8**).
10. Remove the primary antibody and wash the membrane three times for 5 min each with 50 mL TBS-T.
11. Make a 1:4000 dilution of horse radish peroxidase-conjugated goat antirabbit IgG in blocking buffer and add to the membrane for 30 min at room temperature on a rocking platform.
12. Remove the secondary antibody and wash the membrane three times for 10 min each with PBS-T at room temperature with rocking.
13. Immediately before use, mix 2 mL of each of the two ECL reagents and place on the blot. Rotate by hand for 1 min to ensure even coverage.
14. Remove the ECL reagent and place the blot between the leaves of an acetate sheet protector that has been cut to the size of an X-ray film cassette. Attach a phosphorescent marker (Stratagene) (La Jolla, CA) that has been exposed to light immediately before use. Phosphorescent markers may be wrapped in transparent tape for indefinite reuse. Walk quickly to a dark room.
15. Place the acetate containing the membrane in an X-ray film cassette with film for a suitable exposure time, typically a few minutes. Develop the film. Align it with the blot using the phosphorescent marker as a guide, and mark the positions of the standard proteins on the film. An example is shown in **Fig. 1** (bottom panel).

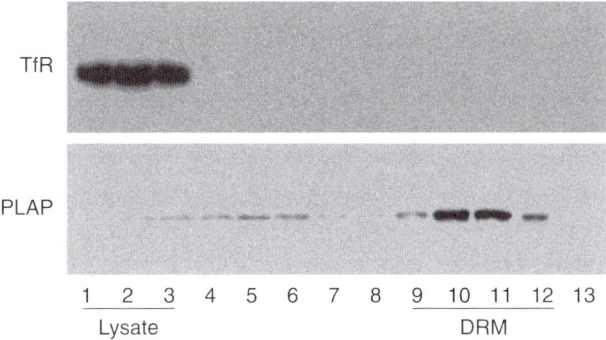

Fig. 1. Detection of PLAP and TfR in sucrose gradient fractions after flotation of DRMs. Mouse B16 melanoma cells transfected with PLAP were extracted with TNE/TX/P, and lysates were subjected to sucrose gradient centrifugation. Thirteen gradient fractions were collected from the bottom (fraction 1). Proteins in each fraction were separated by SDS-PAGE and transferred to nitrocellulose. Blots were probed with anti-TfR (**top**) or anti-PLAP (**bottom**). Fractions 1–3 contain fully solubilized proteins that remained in the load (lysate) position of the gradient. Fractions 9–12 contain DRMs. Although TfR is only detected in the load fractions, most PLAP floats to the DRM position.

3.6. Stripping Blots and Reprobing for TfR

1. To reprobe the blot for detection of TfR or any other protein, strip the first antibodies as described here.
2. Heat stripping buffer (50 mL per blot) to 70°C. Add β-mercaptoethanol, mix, and immediately place the blot in the stripping solution. Put the tray in a water bath heated to 70°C and incubate for 30 min with occasional agitation.
3. Wash the blot in PBS-T (three times 150 mL, each wash for 10 min). In most cases, the blot should not require reblocking.
4. Reprobe the membrane with anti-TfR (1:1000 in blocking buffer) with washes, secondary antibody, and ECL detection as in **Subheading 3.5**. An example is shown in **Fig. 1** (top panel).

4. Notes

1. Methods for cell culture, SDS-PAGE, and Western blotting were loosely adapted from *(22)*.
2. It may be preferable to include detergent in gradient solutions, at the same concentration used to lyse cells. Korzeniowski et al. *(23)* showed that if detergent is omitted from the gradient, fully solubilized proteins may become reconstituted into DRM-like structures during centrifugation.
3. Fractionation of gradients from the bottom of tubes pierced with a gradient harvester is optimal. However, if a gradient harvester is not available, gradients can

be fractionated from the top, by drawing off fractions using a standard Pipetman. If performed carefully, satisfactory results can be obtained using this method.
4. Some workers have reported increased recovery of DRMs after extraction with morpholino ethanesulfonic acid (MES) buffer at mildly acidic pH *(24)*. By contrast, the authors have found little effect of extraction pH on DRM recovery.
5. To optimize yield of floating DRMs, for identification of proteins that associate directly with DRM lipids, cytoskeletal contamination can be minimized by including 0.1 *M* Na carbonate (pH 11.0) in the lysis buffer and in all sucrose gradient solutions. This dissociates many protein–protein interactions, and thus, minimizes interaction of DRMs with detergent-insoluble cytoskeleton. As a result, the amount of DRMs in the sucrose gradient pellet is greatly reduced, whereas the amount in the floating fraction is substantially increased. It is not known to what degree cytoskeletal association with DRMs is physiologically relevant. Furthermore, carbonate is expected to release many proteins that bind DRM lipids indirectly, through binding to another protein that itself binds DRMs lipids directly. Thus, caution should be applied in the use of carbonate treatment. It is most useful for increasing the yield of proteins known to interact directly with DRM lipids in the floating fraction.
6. The relative amounts of acrylamide/*bis*-acrylamide and water may be varied to give final acrylamide concentrations ranging from 5–15%, depending on the size of the protein to be analyzed. Proteins larger than about 10 kDa can be resolved using 15% gels; proteins up to 200–300 kDa or larger may be resolved on 5% gels.
7. For better resolution of large proteins, the gel run can be continued after the dye front leaves the bottom of the gel. For very long runs, change the running buffer if it gets too warm.
8. Signal from inefficient primary antibodies can be increased by incubating the blot with the antibody overnight at 4°C.

Acknowledgment

This work was supported by NIH grant GM47897.

References

1. London, E. and Brown, D. A. (2000) Insolubility of lipids in Triton X-100. Physical origin and relationship to sphingolipid/cholesterol domains (rafts). *Biochim. Biophys. Acta (Biomembranes)* **1508,** 182–195.
2. Xu, X. and London, E. (2000) The effect of sterol structure on membrane lipid domains reveals how cholesterol can induce lipid domain formation. *Biochemistry* **39,** 843–849.
3. Xu, X., Bittman, R., Duportail, G., Heissler, D., Vilcheze, C., and London, E. (2001) Effect of the structure of natural sterols and sphingolipids on the formation of ordered sphingolipid/sterol domains (rafts). Comparison of cholesterol to plant, fungal, and disease-associated sterols and comparison of sphingomyelin, cerebrosides, and ceramide. *J. Biol. Chem.* **276,** 33,540–33,546.

4. Megha and London, E. (2004) Ceramide selectively displaces cholesterol from ordered lipid domains (rafts): implications for lipid raft structure and function. *J. Biol. Chem.* **279,** 9997–10,004.
5. Brown, D. A. and Rose, J. K. (1992) Sorting of GPI-anchored proteins to glycolipid-enriched membrane subdomains during transport to the apical cell surface. *Cell* **68,** 533–544.
6. Hooper, N. M. and Turner, A. J. (1988) Ectoenzymes of the kidney microvillar membrane. Differential solubilization by detergents can predict a glycosylphosphatidylinositol membrane anchor. *Biochem. J.* **250,** 865–869.
7. Kenworthy, A. K., Petranova, N., and Edidin, M. (2000) High-resolution FRET microscopy of cholera toxin B-subunit and GPI-anchored proteins in cell plasma membranes. *Mol. Biol. Cell* **11,** 1645–1655.
8. Kenworthy, A. K., Nichols, B. J., Remmert, C. L., et al. (2004) Dynamics of putative raft-associated proteins at the cell surface. *J. Cell Biol.* **165,** 735–746.
9. Glebov, O. O. and Nichols, B. J. (2004) Lipid raft proteins have a random distribution during localized activation of the T-cell receptor. *Nat. Cell Biol.* **6,** 238–243.
10. Sharma, P., Varma, R., Sarasij, R. C., et al. (2004) Nanoscale organization of multiple GPI-anchored proteins in living cell membranes. *Cell* **116,** 577–589.
11. Harder, T., Scheiffele, P., Verkade, P., and Simons, K. (1998) Lipid-domain structure of the plasma membrane revealed by patching of membrane components. *J. Cell Biol.* **141,** 929–942.
12. Rothberg, K. G., Ying, Y. -S., Kamen, B. A., and Anderson, R. G. W. (1990) Cholesterol controls the clustering of the glycophospholipid-anchored membrane receptor for 5-methyltetrahydrofolate. *J. Cell Biol.* **111,** 2931–2938.
13. Pralle, A., Keller, P., Florin, E. -L., Simons, K., and Hörber, J. K. H. (2000) Sphingolipid–cholesterol rafts diffuse as small entities in the plasma membrane of mammalian cells. *J. Cell Biol.* **148,** 997–1008.
14. Dietrich, C., Yang, B., Fujiwara, T., Kusumi, A., and Jacobson, K. (2002) Relationship of lipid rafts to transient confinement zones detected by single particle tracking. *Biophys. J.* **82,** 274–284.
15. Heerklotz, H. (2002) Triton promotes domain formation in lipid raft mixtures. *Biophys. J.* **83,** 2693–2701.
16. Drevot, P., Langlet, C., Guo, X. -J., et al. (2002) TCR signal initiation machinery is pre-assembled and activated in a subset of membrane rafts. *EMBO J.* **21,** 1899–1908.
17. Vilhardt, F. and Van Deurs, B. (2004) The phagocyte NADPH oxidase depends on cholesterol-enriched membrane microdomains for assembly. *EMBO J.* **23,** 739–748.
18. Magnani, F., Tate, C. G., Wynne, S., Williams, C., and Haase, J. (2004) Partitioning of the serotonin transporter into lipid microdomains modulates transport of serotonin. *J. Biol. Chem.* **279,** 38,770–38,778.
19. Röper, K., Corbeil, D., and Huttner, W. B. (2000) Retention of prominin in microvilli reveals distinct cholesterol-based lipid microdomains in the apical plasma membrane. *Nat. Cell Biol.* **2,** 582–592.

20. Schuck, S., Honsho, M., Ekroos, K., Shevchenko, A., and Simons, K. (2003) Resistance of cell membranes to different detergents. *Proc. Natl. Acad. Sci. USA* **100,** 5795–5800.
21. Melkonian, K. A., Chu, T., Tortorella, L. B., and Brown, D. A. (1995) Characterization of proteins in detergent-resistant membrane complexes from Madin-Darby canine kidney epithelial cells. *Biochemistry* **34,** 16,161–16,170.
22. Mattingly, R. R. (2002) Mitogen-Activated Protein Kinase Signaling in Drug-Resistant Neuroblastoma Cells, in *Cancer Cell Signaling: Methods and Protocols*, (Terrian, D. M., ed.), Humana Press Inc., Totowa, NJ, pp. 71–84.
23. Korzeniowski, M., Kwiatkowska, K., and Sobota, A. (2003) Insights into the association of FcγRII and TCR with detergent-resistant membrane domains: Isolation of the domains in detergent-free density gradients facilitates membrane fragment reconstitution. *Biochemistry* **42,** 5358–5367.
24. Sargiacomo, M., Sudol, M., Tang, Z., and Lisanti, M. P. (1993) Signal transducing molecules and glycosyl-phosphatidylinositol-linked proteins form a caveolin-rich insoluble complex in MDCK cells. *J. Cell Biol.* **122,** 789–807.

3

Nondetergent Isolation of Rafts

Mehul B. Shah and Pravin B. Sehgal

Summary

Raft and caveolar microdomains have been proposed to participate in numerous cellular functions including signal transduction, cholesterol trafficking, and vesicular sorting. Traditional methods of isolation of rafts from cultured cells and tissue samples have exploited the biochemical properties of these microdomains, i.e., their relative resistance to solubilization by nonionic detergents (at 4°C) and their light buoyant density attributable to their high content of cholesterol and sphingolipids. Thus, a common way to isolate raft microdomains has been their separation on a density gradient in the presence of 0.5–1% Triton X-100 (Bochringer Mannheim Roche Applied Sciences Indianapolis, IN or Sigma-Aldrich, St. Louis, MO). This and other detergent-based methods have been discussed. However, the use of detergents may not be favorable because of artifacts that may arise with their use. (The possibility of rafts solely as detergent-induced artifacts appears to have been diffused by a number of biochemical and biophysical studies that strongly demonstrate the presence of a liquid-ordered phase within biological membranes.) In this chapter, three methods are reviewed to isolate rafts from cultured cells without the use of detergents. Two of these, the sodium carbonate and OptiPrep™ (Sigma-Aldrich St. Louis, MO) methods, are based on gradient separation and can be used to isolate rafts in general, whereas the third is a magnetic-bead immunoisolation approach and might be used to isolate subpopulations of rafts enriched for different markers such as caveolin-1, flotillin (reggie proteins), or other suitable markers. Together these methods allow for a detergent-free isolation of rafts for biochemical, proteomic, and microscopic studies.

Key Words: Carbonate; caveolae; detergent-free; immunoisolation; OptiPrep™; rafts.

1. Introduction

Biochemical analyses of cellular membranes and membranes of enveloped viruses (derived from the host cell) since the 1970s and early 1980s have suggested the possibility of lateral heterogeneity in membranes, i.e., the "patchy" or nonuniform distribution of lipids, especially glycosphingolipids and viral glycoproteins within the lipid bilayer *(1,2)*. Subsequently, "lipid rafts" were conceptualized as discrete microdomains within the lipid bilayer that possessed special physical and biochemical properties and which organized the membrane into special platforms or portals for mediating various functions (reviewed in **ref. 3**). Despite extensive research over nearly two decades, a consensus definition of what constitutes a raft or a caveolar domain has remained elusive and for the most part the definition has either been an operational one based on biochemical (detergent-resistant, light buoyant density fractions positive for specific markers) or microscopic (50–100 nm cave-like invaginations) parameters; only very recently does a general agreement on the issue appear to have been reached *(4)*. Rafts have very recently been defined as follows: "Membrane rafts are small (10–200 nm), heterogeneous, highly dynamic, sterol- and sphingolipid-enriched domains that compartmentalize cellular processes. Small rafts can sometimes be stabilized to form larger platforms through protein–protein and protein–lipid interactions" *(4)*.

Whether such a definition will stand the test of time remains to be seen, but an elegant aspect of this definition is that a fraction need not be isolated by detergent-methods alone to qualify as a raft. Broadly speaking, methods to isolate rafts can be classified as detergent- and nondetergent based. Whereas detergent-based methods to isolate rafts have been popular, a growing body of literature has expressed concern about artifacts arising out of the use of nonionic detergents, as well as whether detergent-resistant membranes correspond to or approximate the liquid-ordered phase (rafts) within the membranes of live cells (reviewed in **ref. 5**). A number of researchers advocate the use of "less invasive" methods to isolate rafts, either as an adjunct or as a substitute method. This chapter focuses on three different nondetergent methods to isolate rafts, pioneered by different researchers in the caveolar/raft field. The particular use of one method over another may be subjective and may depend on issues like purity, yield, and the final application for which the obtained fractions are to be used.

2. Materials

All solutions and buffers described here were prepared at room temperature. Unless specified otherwise, all solutions are prepared as weight per volume in distilled water. Commonly used centrifuges, rotors, tubes, and other lab equipment have not been listed under materials.

2.1. Cell Culture and Cell Fractionation

1. Dulbecco's modified Eagle medium supplemented with fetal bovine serum or the appropriate growth media for the cell type to be used.
2. 60- or 100-mm culture plates.
3. Phosphate-buffered saline (PBS): Volume = 500 ml; concentrations NaCl 100mM, KCl 3mM, Na_2HPO_4 10mM, KH_2PO_4 2mM, PH7.4
4. Teflon cell scraper.

5. Hypotonic extract lysis buffer (ELB): 10 mM HEPES, 10 mM NaCl, 3 mM MgCl$_2$, 1 mM dithiothreitol, 0.4 mM phenylmethylsulfonyl fluoride, and 0.1 mM sodium vanadate, pH 7.9. Where necessary, appropriate adjustments with 0.25 M (8.55%) sucrose or other sucrose concentrations should be made.
6. Loose fitting 7-mL Dounce homogenizer (Wheaton, Melville, NJ or Fisher Scientific Pittsburgh, PA).

2.2. Isolation of Rafts Using Sodium Carbonate

1. 0.5 M Na$_2$CO$_3$ solution, pH 11.0.
2. 1X and 2X strength 2-(N-morpholine)-ethane sulphonic acid (MES)-buffered saline (MBS) buffers. 1X corresponds to 25 mM MES, and 0.15 M NaCl, pH 6.5.
3. 5% and 35% sucrose MBS buffers with 0.25 M Na$_2$CO$_3$.
4. 90% Sucrose in 2X MBS.
5. Tekmar tissumizer and Tekmar sonicator. (Tekmar Company, Cinncinnati, OH).

2.3. Isolation of Rafts Using an OptiPrep Gradient

1. 30% Percoll Sigma-Aldrich, St. Louis, MO (v/v) in 0.25 M sucrose in ELB.
2. 5 and 50% OptiPrep™ Sigma-Aldrich, St. Louis, MO solutions (v/v) prepared in 0.25 M sucrose ELB.
3. Tekmar sonicator.
4. Gradient maker.

2.4. Immunoisolation of Rafts

1. 60, 40, 30, and 25% sucrose in ELB.
2. Immunoisolation binding buffer: 25 mM Tris-HCl, 150 mM NaCl, 5 mM ethylene diaminetetraacetic acid, and 1 mM dithiothreitol, pH 7.4.
3. Anticaveolin (Cav)-1 pAb (SantaCruz Biotech; [Santa Cruz, CA] cat. no. sc-894) or other desired antibody for immunoisolation and corresponding nonimmune control.
4. Protein-A magnetic beads (protein-G magnetic beads if a mouse immunoglobulin [Ig]G is used for immunoisolation) (New England Biolabs [Ipswich, MA]).
5. Six tube magnetic separation rack (New England Biolabs [Ipswich, MA]).
6. Nonfat dry milk (NFD) (Bio-Rad, Hercules, CA).
7. PBS.

3. Methods

The methods to isolate rafts by nondetergent procedures described here have been optimized for adherent cell lines and have been tested in a Hep3B hepatoma cell line and a primary bovine arterial endothelial cell strain, but should also be useful for other adherent as well as suspension cell lines with relatively few modifications.

3.1. Isolation of Rafts Using Sodium Carbonate

The isolation of rafts using sodium carbonate was described by Song et al. (*6*). A modification of that method is as follows:

1. Culture plates should be rinsed twice with ice-cold PBS and the cells should be harvested into a total of 2 mL of 0.5 M Na_2CO_3 buffer per experimental group using a T-cell scraper Fisher Scientific, Pittsburgh, PA (*see* **Note 1**).
2. The cell suspension should be homogenized with 10 strokes in a loose-fitting Dounce homogenizer.
3. The cell suspension should be subjected to further homogenization by three 10-s bursts in a tissue grinder (Tekmar tissumizer), and then three 20-s bursts of homogenization in a sonicator (Tekmar sonicator).
4. Using a 90% sucrose solution in 2X MBS buffer, the cell lysate should be adjusted to 45% sucrose and 0.25 M Na_2CO_3 (i.e., mix equal volumes of the cell suspension and 90% sucrose buffer). To prepare the gradient, the resulting mix (~4 mL) should be placed at the bottom of a Beckman SW41 polyallomer tube (Beckman Inc., Palo Alto, CA) and layered with 4 mL of 35% sucrose in 0.25 M Na_2CO_3 and 3 mL of 5% sucrose in 0.25 M Na_2CO_3.
5. The gradients should be subjected to an equilibrium flotation in a Beckman SW41 rotor at 35,000 rpm 150,000 × g for 12–16 h at 4°C. A light scattering band at the 5%/35% interface should be visible and this corresponds to the raft fraction.
6. 0.5–1-mL Aliquot fractions may be collected from the top (manually, using a glass pipet) or from the bottom (using a fraction collector), and equal volumes of these fractions may be used for Western blotting to check for the distribution of the raft and other organellar markers (*see* **Note 4**).
7. Once the distribution of the markers has been confirmed, in subsequent experiments collection of only the raft fraction may be appropriate for biochemical or proteomic studies. For this purpose, the material above the visible fraction should be discarded first and then the raft fraction should be collected, diluted using 40 volumes of MES, and resedimented in a Beckman Ti45 rotor (15,000 rpm 17,000 × g, 4°C, 30 min). The resulting pellet is a raft-enriched fraction and should be resuspended in 100–500 µL of MES, PBS, or any other buffer appropriate for downstream applications of this fraction.

3.2. Isolation of Rafts by Using OptiPrep

A second method to purify rafts using a two-step purification process, which relies on density-based separation in an OptiPrep gradient has been described by Smart et al. *(7)*. (OptiPrep, is a relatively osmotically inert solution of iodixanol, and hence, is preferred over sucrose by some researchers for density-based organellar separation.) Subsequently, a more simplified method using OptiPrep has been described by Pike et al. *(8)*. The following is a modification of the method used by Smart et al. *(7)*, for the preparation of raft fractions from adherent cell cultures.

1. Culture plates should be washed twice with ice-cold PBS and the cells should be harvested by scraping in a total of 2 mL of 0.25 M sucrose ELB (*see* **Notes 1** and **2**).
2. The cells should be lysed using a loose-fitting Dounce homogenizer (30 strokes). Cell breakage and intactness of nuclei should be monitored under a phase-contrast microscope (*see* **Note 3**).

3. The cell lysate should be centrifuged twice at 200g for 3 min (IEC Centra GP8R) to remove nuclei and unbroken nuclei.
4. The postnuclear supernatant should be layered on top of 23 mL of 30% Percoll in 0.25 M sucrose ELB, and centrifuged at 84,000g for 30 min in a Beckman Ti60 rotor. The plasma membrane fraction should be a light-scattering band at nearly one-third of the length of the tube from the bottom. This band should be collected manually using a glass pipet, and diluted with 0.25 M sucrose ELB to make a total of 2 mL.
5. The membrane fraction should be sonicated on ice using a Tekmar sonicator (two pulses, 20 s, power setting at "25"). The sample should then be adjusted to 23% OptiPrep using a 50% solution of OptiPrep in 0.25 M sucrose ELB (i.e., add 2 mL of the sample, 1.84 mL of 50% OptiPrep, and 0.16 mL of 0.25 M sucrose ELB) and then placed at the bottom of a SW41 polyallomer tube. Then 5 mL of a linear 20–10% OptiPrep gradient (prepared by diluting 50% OptiPrep in sucrose ELB with plain 0.25 M sucrose ELB) should be layered on top of the sample, followed by 2 mL of 5% OptiPrep, and the gradient should be centrifuged at 52,000g for 90 min at 4°C in a Beckman SW41 rotor. A light scattering band just above the interface in the 5% OptiPrep region will be visible. This band represents the raft fraction and should be collected using a glass pipet (see **Note 4**).

3.3. Isolation of Rafts by Immunoisolation

For immunoisolation of plasma membrane rafts, plasma membrane fractions may be obtained by any suitable nondetergent method. An equilibrium flotation/sedimentation method has been used to obtain plasma membrane fractions enriched for the markers very late antigen-2α and 5′-nucleotidase (see **steps 1–6**) (9). Purified plasma membrane fractions should then be subjected to immunoisolation (see **steps 7–10**) using an anti-Cav-1 antibody or an antibody of choice to isolate subpopulations of rafts.

1. Culture plates should be rinsed twice with chilled PBS and scraped into a total of 2 mL 0.25 M sucrose ELB (see **Notes 1** and **2**).
2. The cell suspension should be homogenized with 30 strokes in a loose-fitting Dounce homogenizer. To avoid nuclear contamination of the raft preparation, the integrity of nuclei needs to be preserved during this cell-breakage and this should be verified under a phase-contrast microscope (see **Note 3**).
3. Unbroken cells and intact nuclei should be removed from the cell lysate by two rounds of low-speed centrifugation at 1000 rpm 600 × g for 3 min (IEC Centra GP8R) and the postnuclear supernatant should be centrifuged once at 15,000g in a table-top centrifuge (Eppendorf) Beckman Industries, Westbury, NY for 15 min. The pellet (designated "P15") should be washed twice with 1 mL 0.25 M sucrose ELB, resuspended in 30% sucrose ELB to a final volume of 2 mL.
4. The P15 sample should be loaded into a discontinuous sucrose gradient in a Beckman SW41 polyallomer tube as follows: sequentially load layers of 2 mL of 60% sucrose ELB, 3 mL of 40% sucrose ELB, 2 mL of 30% (the P15 sample load),

2 mL of 25% sucrose ELB and 1 mL of ELB. Centrifuge the gradient in an SW41 rotor at 35,000 rpm 150,000 × g for 18 h at 4°C.

5. The following bands should be visible and should be collected (~200 µL of volume each) from the top manually using a glass pipet as follows:
   a. Fraction 1—top of ELB layer.
   b. Fraction 2—the 0/25% sucrose interface.
   c. Fraction 3—the 25/30 interface.
   d. Fraction 4—the 30/40 interface.
   e. Fraction 5—visible band just above the 40/60 interface.
   f. Fraction 6—visible band at 40/60 interface.
   g. Fraction 7—visible band just below the 40/60 interface.
   h. Fraction 8—the pellet (unsuspended materials and unbroken nuclei).
6. Fractions 5 and 6 represent the plasma membrane-enriched fractions. These fractions should be collected, pooled, diluted with 40 volumes of ELB, resedimented (Beckman Ti45 rotor, 15,000g, 4°C, 30 min), and the pellet obtained should be resuspended in minimum volume (200–400 µL) of binding buffer for the immunoisolation (see **Notes 4** and **5**).
7. 2–20 µg of anti-Cav-1 pAb, or an antibody of choice should be added to the samples and the samples should be incubated at 4°C for 2–4 h with continuous mixing. Control immunoisolation using a matching amount of corresponding control normal IgG should be carried out in parallel (see **Note 6**).
8. 25 µL of the protein-A magnetic beads (protein-G if mouse IgG is being used) per sample of immunoisolation should be aliquoted in an Eppendorf tube. (e.g., for 12 samples, use 300 µL of beads.) The beads should be washed twice with the binding buffer in the magnetic rack and resuspended in 1 mL of freshly prepared 5% NFD in PBS. The beads should be blocked in NFD for 1 h at 4°C with continuous mixing. The preblocked beads should be washed thrice with 1 mL binding buffer (with vortexing after each resuspension) in the magnetic rack and should finally be resuspended in binding buffer (25 µL buffer per immunoisolation sample) (see **Notes 7–9**).
9. 25 µL of the bead suspension should be added to each immunoisolation reaction tube and the samples should be incubated (with mixing) at 4°C for a further 1 h.
10. The tubes should be placed in the magnetic rack and the supernatant should be saved in separate tubes (to assess the efficiency of immunoisolation). The beads should be further washed with binding buffer 20 times. For Western blotting of the immunoisolates, the beads may be directly resuspended in Laemmli's sample buffer and boiled for 5 min before separating the supernatant in the magnetic racks. For whole mount electron microscopy, the immunoisolate-carrying beads can be directly spotted on formvar-coated copper grids. For other purposes, elution of the immunoisolates from the beads may be necessary (see **Notes 10–12**).

4. Notes

1. If the experimental conditions require treatment of cells with cytokines, growth factors or other reagents before isolation of rafts, cells may be shifted to serum-free medium prior for 2–12 h before cell fractionation.
2. In traditional harvesting, cells are scraped in a larger volume of PBS or appropriate buffer and then pelleted before Dounce homogenization. With the authors this

method has led to loss of cytoplasmic material. In order to minimize such losses a "serial scraping" method for harvesting cells is recommended. Cells in one culture dish should be scraped into 1–2 mL of the desired buffer (e.g., 0.25 M sucrose ELB). The cell suspension should be gently transferred into the next plate of the same experimental group and that plate should be scraped into the same lysate. This can be continued "serially" so that the total cell suspension is in a very small volume that can be Dounce homogenized directly instead of having to be centrifuged first.
3. The Dounce homogenization protocol suggested here has been optimized for Hep3B cells and pulmonary arterial endothelial cells and may need to be finessed for other cell types. To check for cell breakage, a drop of the cell lysate should be placed from the Dounce homogenizer onto a glass slide, a cover slip should be placed on the sample, and efficiency of cell breakage should be assessed under a phase-contrast microscope. The presence of a large number of intact cells indicates inadequate breakage whereas the presence of broken nuclei indicates that more gentle breakage is needed to avoid nuclear contamination. Additionally, pretreatment of certain cell types with cross-linking agents may "harden" cells resulting in lower yield.
4. All cell fractionation protocols carried out for the first time should be validated using markers for different compartments. Commercial kits for many of these markers are widely available.
5. A method for preparing a high-yield plasma membrane fraction has been provided for the purpose of immunoisolation of rafts. Alternative methods for preparing the plasma membrane fraction (e.g., see **ref. 10**) may be used before immunoisolation of rafts if desired.
6. In lieu of nonimmune IgG as a control for the immunoisolation, a blocking peptide may be used when available. For the SantaCruz Biotech N-20 anti-Cav-1 rabbit pAb (cat. no. sc-894), a blocking peptide is available (cat. no. sc-894P). Preincubation of the antibody with the blocking peptide (SantaCruz recommends a 5:1 volumetric ratio of the peptide solution to the antibody solution), and nonspecific peptide as additional controls are ideal.
7. Magnetic beads are recommended rather than agarose beads for purpose of immunoisolation from membrane fractions. The centrifugation step involved in the use of agarose beads for immunoisolation leads to a centrifugation of unbound membrane pieces together with the agarose beads, thus increasing background.
8. Magnetic beads have a tendency to settle at the bottom of the tube. Hence, extensive vortexing is recommended for each wash before the tubes are placed in the magnetic rack.
9. The preblocking of magnetic bead in 5% NFD in the immunoisolation protocol is recommended to minimize the nonspecific binding of antigens to the beads. This is an alternative to preclearing, which if works well, might be used in place of preblocking.
10. As the methods described here are strictly nondetergent, the protocol for immunoisolation of raft fractions does not recommend the use of Triton. However, it has been found that the presence of a small amount of Triton X-100 (0.05%) in the

binding buffer during the last few washes has reduced background substantially. However, whether this qualifies as a "nondetergent" method is debatable. Additionally, the use of detergent leads to vesicle disruption and may not be suitable for specific uses such as electron microscopy.

11. For immunoisolation using a polyclonal antibody, Western blot detection with a monoclonal antibody is desirable to minimize IgG bands from showing up on the film. Alternatively, detection using an HRP-conjugated primary antibody (such as the SantaCruz HRP-conjugated anti-Cav-1 pAb; cat. no. sc-894HRP) works well in many cases.

12. For select purposes, elution of the immunoisolates may become necessary. In this case, instead of protein-A magnetic beads, the use of Dynabeads (which can be coupled to the desired antibody) from Invitrogen Corporation (Carlsbad, CA) is recommended. The manufacturer's recommended protocol has worked very well in this case. The immunoisolate can then be eluted by incubation using an acidic buffer (e.g., a 0.2 M glycine buffer, pH 2.5) without eluting the antibody. The approach is not recommended unless absolutely required as it carries the risk of breaking protein–protein interactions.

References

1. Thompson, T. E. and Tillack, T. W. (1985) Organization of glycosphingolipids in bilayers and plasma membranes of mammalian cells. *Annu. Rev. Biophys. Biophys. Chem.* **14,** 361–386.
2. Klenk, H. and Choppin, P. W. (1970) Plasma membrane lipids and parainfluenza virus assembly. *Virology* **40,** 939–947.
3. Simons, K. and Toomre, D. (2000) Lipid rafts and signal transduction. *Nat. Rev. Mol. Cell Biol.* **1,** 31–39.
4. Pike, L. J. (2006) Rafts defined: A report on the Keystone Symposium on Lipid Rafts and Cell Function. *J. Lipid Res.* **47,** 1597–1598 (Epub Ahead of Print).
5. Hancock, J. F. (2006) Lipid rafts: contentious only from simplistic standpoints. *Nat. Rev. Mol. Cell Biol.* (7456 –7462).
6. Song, K. S., Shengwen, L., Okamoto, T., Quilliam, L. A., Sargiacomo, M., and Lisanti, M. P. (1996) Co-purification and direct interaction of Ras with caveolin, an integral membrane protein of caveolae microdomains. Detergent-free purification of caveolae microdomains. *J. Biol. Chem.* **271,** 9690–9697.
7. Smart, E. J., Ying, Y. S., Mineo, C., and Anderson, R. G. (1995) A detergent-free method for purifying caveolae membrane from tissue culture cells. *Proc. Natl. Acad. Sci. USA* **92,** 10,104–10,108.
8. Macdonald, J. L. and Pike, L. J. (2005) A simplified method for the preparation of detergent-free lipid rafts. *J. Lipid Res.* **46,** 1061–1067.
9. Sehgal, P. B., Guo, G. G., Shah, M., Kumar, V., and Patel, K. (2002) Cytokine signaling: STATS in plasma membrane rafts. *J. Biol. Chem.* **277,** 12,067–12,074.
10. Aronson, N. N., Jr. and Touster, O. (1974) Isolation of rat liver plasma membrane fragments in isotonic sucrose. *Methods Enzymol.* **31,** 90–102.

4

Detecting Ordered Domain Formation (Lipid Rafts) in Model Membranes Using Tempo

Omar Bakht and Erwin London

Summary

Short-range fluorescence quenching has proven to be an effective method to detect the presence of coexisting ordered and disordered state lipid domains in model membranes. In this approach a fluorescent group and fluorescence-quenching molecule are incorporated into the lipid bilayer of interest. In a typical experiment, the fluorophore chosen partitions into ordered domains to a significant degree, whereas the quencher partitions more favorably into disordered domains. Thus, in the presence of lipid mixtures forming coexisting ordered and disordered domains, fluorophore and quencher segregate so that fluorescence intensity is much stronger than in homogeneous lipid bilayers lacking separate domains. The small nitroxide-labeled molecule tempo (2,2,6,6 tetramethylpiperidine-1-oxyl) is a useful quencher for such experiments. Protocols for using tempo to detect ordered domains and ordered domain thermal stability are described. The advantages and disadvantages of use of tempo as opposed to nitroxide-labeled lipids are also described.

Key Words: Diphenylhexatriene; fluorescence quenching; fluorescence spectroscopy; lipid microdomains; lipid rafts; nitroxides; spin-labels.

1. Introduction

There are many techniques for detecting phase separation in lipid vesicles. Typically, these separations involve the formation of coexisting liquid-disordered and -ordered state domains. The ordered domains are either gel- or liquid-ordered state. Liquid-ordered domains (often called lipid rafts) can be notoriously difficult to detect, especially in biological membranes. One possible reason for this is that these domains may often be submicroscopic. Some reports estimate the size of ordered domains in cells, as commonly no larger than a small cluster of molecules, although they may be large under many conditions *(1–5)*. Owing to the issue of domain size, fluorescence techniques to probe liquid-ordered domains are of great use, and fluorescence-quenching techniques have emerged as an appropriate methodology that should be sensitive enough to detect submicroscopic domains *(6,7)*. Certain types of quenching processes are "short ranged," effective at a distance not significantly longer than the length of single molecule *(6–8)*. Quenching by lipid-attached bromines or lipid-attached nitroxide-labels fall in this category *(8)*. Fastenberg et al. *(7)* investigated the effect of domain size on quenching efficiency for such short-range quenchers, and calculated that circular clusters containing 30–40 molecules could be detected. In contrast, more long-range fluorescence resonance energy transfer (FRET) techniques, which can have effective distances as high as 50–100 Å, would have difficulty detecting such domains.

In past studies, the fluorescence-quenching protocols have made use of the membrane associating fluorophore diphenylhexatriene (DPH). In many conditions, DPH partitions equally between ordered domains and disordered domains *(9–11)*, although there are some exceptions *(12,13)*. As a quencher, the authors have generally used 1-palmitoyl-2-(12-doxyl phosphatidylcholine (12SLPC), a nitroxide-labeled phosphatidylcholine (PC) with a label on the 12-carbon of the 2-position acyl chain. Other labeled lipids can be used, including a nitroxide-labeled lipid labeled on the 7-carbon atom of the 2-position acyl chain *(14)*. 12SLPC tends to form disordered domains, much as do lipids with unsaturated acyl chains (and thus having low gel-to-fluid melting temperatures [Tm]) *(14,15)*. In model membrane samples containing both 12SLPC and a lipid that tends to form ordered domains (e.g., dipalmitoyl PC or sphingomyelin), ordered and disordered domains can coexist. Because the concentration of 12SLPC within the ordered domains is low, DPH residing in the ordered domains will be relatively shielded from 12SLPC, and so fluorescence will be much stronger than in a homogeneous bilayer in which all lipids are mixed, and in which the probability of contact between DPH and 12SLPC is much greater.

By heating samples containing coexisting ordered and disordered domains it is possible to evaluate the thermal stability of ordered domains. As temperature increases, any ordered domains present in the bilayer melt, and the partial segregation of DPH and 12SLPC into different environments is lost, with DPH becoming fully exposed to quencher. As a result, normalized fluorescence intensity decreases dramatically. This technique has been used to ascertain the effects of sterol structure on the thermal stability of ordered domains *(16,17)*. Quenching experiments using 12SLPC have the limitation that they cannot be used to readily investigate the effect of the structure of lipids that tend to form the disordered regions of the membrane on ordered domain stability. The authors have recently used tempo (2,2,6,6 tetramethylpiperidine-1-oxyl), a quencher that is not lipid attached, to circumvent this problem. The protocols in this chapter describe how to carry out tempo-quenching experiments for the detection of ordered domain stability in model membrane vesicles.

2. Materials and Equipment
2.1. Lipid Stock Solutions

1. Lipids (Avanti Polar Lipids, Alabaster, AL). Lyophilized, or dissolved in ethanol, or chloroform. Stored in glass vials at –70–80°C.
2. DPH (Molecular Probes [now Invitrogen], Carlsbad, CA or Sigma-Aldrich Chemical, St. Louis, MO). Dissolved and stored in ethanol.
3. Tempo (Sigma-Aldrich). Tempo should be dissolved in ethanol (at 350 mM) and stored in glass vials in a –20°C freezer (*see* **Note 1**).
4. Aluminum foil. To shield samples from light.
5. A microbalance accurate to 0.01 mg. A Cahn C-33 microbalance (ATI Cahn Instruments, Madison, WI) was used for the studies. An ordinary balance can be used, but will require higher amounts of lipid. Incomplete drying of lipids before weighing becomes increasingly problematic when high amounts of lipid are used.
6. Nitrogen source with regulator capable of generating a gentle stream (~2 psi).
7. Vacuum jar, which can be evacuated (for some types of experiments).

2.2. Sample Preparation

1. Organic solvent-compatible pipetors. For example, Drummond pipetors (Drummond Scientific, Broomall, PA) with disposable glass bores. It is important to use clean bores to avoid cross-contamination.
2. Phosphate-buffered saline (PBS) (1 mM KH$_2$PO$_4$, 10 mM Na$_2$HPO$_4$, 137 mM NaCl, and 2.7 mM KCl, pH 7.4).
3. Nitrogen source with regulator capable of generating a gentle stream (~2 psi).
4. Water bath set at 70°C.
5. A dark place to store samples before measurement (*see* **Note 2**).

2.3. Fluorescence Measurements

1. Fluorometer (e.g., Fluorolog-3) (Horiba, Jovin Yvon, Edison, NJ) with a variable temperature sample holder having the capacity to hold several samples at once.
2. Temperature-controlled circulating water bath, preferably with the ability to increase temperature at a smooth rate (*see* **Note 3**).
3. Electronic digital thermometer VWR, West Chester, PA with minimal size immersible probe (in order to make temperature measurements without affecting cuvet temperature). Make sure the tip material is inert and does not contaminate your sample, especially with fluorescent impurities!

3. Methods

The experiments described in this section use tempo as the quencher. Tempo is a small water-soluble chemical that partitions strongly into lipid bilayers (**Fig. 1**). Tempo also partitions more favorably into disordered lipid phases as opposed to more tightly packed ordered phases. These properties allow addition of tempo to preformed model membrane vesicles in which the disordered lipid

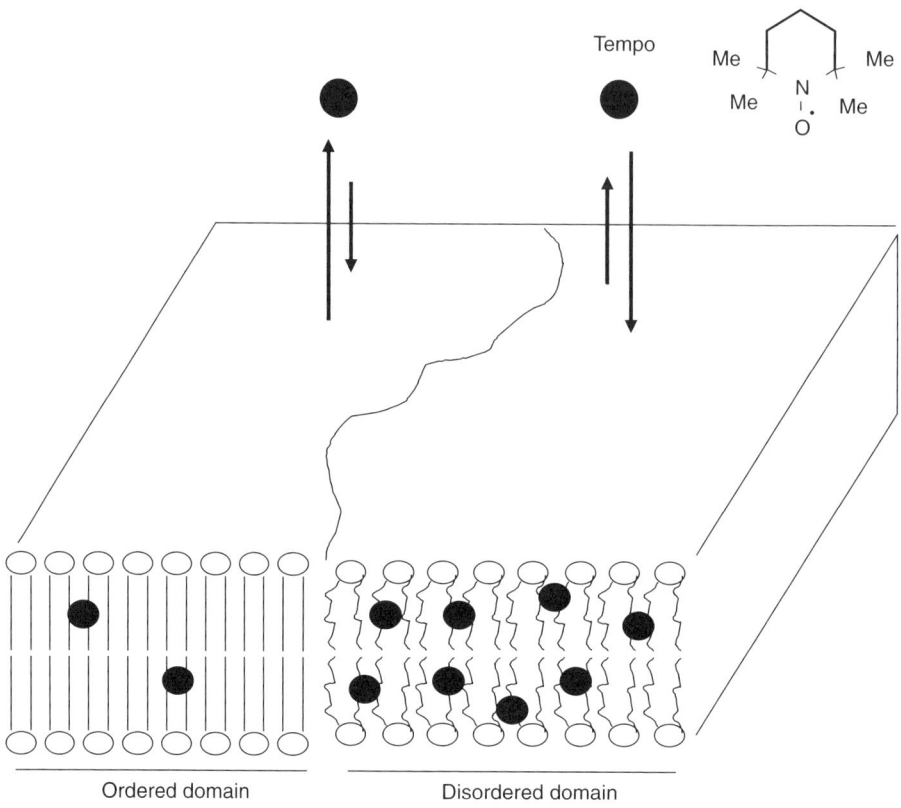

Fig. 1. Schematic illustration of tempo membrane-binding properties. Filled circles represent tempo molecules (chemical structure at top right). Me is the methyl group. Notice that tempo binds to the disordered fluid domains more strongly than ordered domains. As a result, fluorescent molecules that associate with ordered domains are only weakly quenched by tempo, and quenching tends to increase when ordered domains melt at high temperature.

components in the lipid bilayer can be varied. It is much less expensive than nitroxide-labeled lipids, so quenching experiments using tempo can be carried out more economically. The authors have used tempo in studies in which the ability of various cholesterol precursors to stabilize ordered domains was compared *(18)*. Preliminary studies varying the acyl chain structure and polar headgroup of lipids that tend to form disordered domains have also been undertaken. Such studies show tempo is a versatile molecule for quenching experiments.

On the other hand, tempo quenching is dependent on how tightly tempo binds to a lipid mixture. Furthermore, in samples containing a mix of lipids, tempo quenching, unlike that of 12SLPC, is not sensitive to homogeneity of lipid composition in different vesicles. In other words, tempo quenching cannot detect whether or not order- and disorder-favoring lipids form separate vesicles. Appropriate controls to test for proper lipid mixing are described at the end of this section.

In these experiments, model membrane samples typically contain both a lipid for which T_m (the temperature at which the gel state melts and a liquid-disordered state forms) is high (e.g., dipalmitoyl PC or sphingomyelin) and a lipid for which T_m is low. This allows ordered- and disordered domains to coexist at intermediate temperatures. Low T_m lipids typically contain unsaturated acyl chains, whereas high T_m lipids typically contain long saturated acyl chains *(19)*. As in the case of experiments using 12SLPC, DPH is a suitable fluorescent probe. Because the concentration of tempo within the ordered domains is low, DPH residing in the ordered domains will be relatively shielded from tempo. Therefore, DPH fluorescence will be much stronger than in a homogeneous bilayer in which all lipids are mixed and much of the DPH is in contact with tempo. Fluorescence intensity is evaluated using the parameter F/Fo. F/Fo equals the fluorescence intensity in a sample containing quencher (tempo) divided by the fluorescence intensity in the absence of quencher.

By heating samples containing coexisting ordered and disordered domains it is possible to evaluate the thermal stability of ordered domains in a fashion analogous to that used for samples containing 12SLPC. As temperature increases, any ordered domains present in the bilayer melt. As a result, the partial segregation of DPH and tempo into different environments is lost, with DPH becoming fully exposed to tempo. Consequently, F/Fo decreases dramatically. A T_m for the ordered domains can be defined from the temperature at which the slope of F/Fo vs temperature is maximal. The protocol for such experiments is described next.

3.1. Lipid Stock Solutions

The accuracy of the lipid stock concentration is of utmost importance for these experiments. To conserve material one measures the concentrations by drying a known volume of the lipid-containing solution and precisely weighing the dried lipid on a microbalance.

1. Lipids are purchased as lyophilized solids or dissolved in chloroform or ethanol. Solids are dissolved in ethanol. Lipid solutions are stored in glass vials at –70–80°C.
2. Aluminum foil is cut into squares, approx 15 mm to a side, numbered and then shaped into a liquid tight container.
3. The aluminum cups are then weighed on a microbalance out to the closest microgram (*see* **Note 4**).

4. After warming lipids to room temperature, and mixing well to redissolve any precipitated lipid, 50 µL of lipid solution is pipeted into the cup. For best accuracy, three to five such samples are prepared.
5. The solvent is dried under a gentle stream of nitrogen (<2 psi) for 10 min. It is absolutely essential that drying is complete.
6. The cups are placed in a high vacuum for at least 45 min to remove remaining solvent.
7. The cups are then reweighed and the difference in weight is used to calculate the concentration of the solution (*see* **Notes 5** and **6**).
8. Tempo and DPH concentrations can also be derived gravimetrically. However, they can be confirmed by absorbance using an extinction coefficient of 14/cm/M for tempo at 426 nm in water and an extinction coefficient of 88,000/cm/M for DPH at 350 nm in methanol. Absorbance measurements are particularly useful for evaluating concentration in saturated tempo solutions.

3.2. Sample Preparation

After the desired lipids are mixed and dissolved in a minimum volume of ethanol, buffer is rapidly added. Buffer addition causes the lipids to quickly form vesicles. This process of vesicle formation is called ethanol dilution or injection *(20)*. Dilution of lipids from ethanol tends to form small unilamellar vesicles, although they may not be as small as those prepared by sonication *(20)*.

1. The desired amount of each lipid solution needed in the sample is calculated. Each of the lipids is then added to a glass test tube. DPH should also be added at this point. DPH concentration should be equivalent to 0.5 mol% or less, of the total amount of lipid (*see* **Note 7**). Samples are prepared in quadruplicate.
2. Ethanol is then evaporated using a gentle stream (~2 psi) of pure nitrogen for 10 min or until the lipids are dry. Nitrogen is preferred over compressed air because it minimizes the possibility of lipid degradation. Gas flow should not be excessive, or it may splatter the lipid solution on the tube walls. This could give rise to incomplete lipid mixing.
3. The dried samples are then placed into a 70°C water bath for 5 min (*see* **Note 8**).
4. The dried lipids are dissolved in 15 µL of ethanol and shaken gently by swirling.
5. When the lipid is fully dissolved, 980 µL of 70°C PBS is added and vortexed to mix. It is essential that the lipid be fully dissolved in ethanol.
6. Samples are placed in 70°C water bath for 10 min and then vortexed.
7. Samples are allowed to reach room temperature before making fluorescence readings (*see* **Note 9**).
8. To prepare the "F samples" an aliquot of the tempo solution sufficient to achieve a 2 mM final concentration is added to two of the quadruplicate samples prepared as described earlier.
9. To prepare the "Fo samples" an equal volume of ethanol is added to the other two samples from the quadruplicate set.
10. Both sets of samples are vortexed.

3.3. Fluorescence Measurements

1. Excitation and emission monochrometers are set to 358 and 430 nm, respectively.
2. The excitation and emission slits are set to appropriate values (*see* **Note 10**).
3. Samples are then placed into semimicro quartz fluorescence cuvets (10-mm excitation path length and 4-mm emission path length).
4. The cuvets are placed in the sample chamber.
5. The sample holder temperature is then lowered to approx 16°C. The sample temperature should be monitored by a digital thermometer, as there may be a several degree temperature difference between the water bath temperature and the actual sample temperature (*see* **Note 11**).
6. When the temperature of the sample reaches 16°C, a fluorescence intensity measurement is made. (The thermometer tip should be removed from the sample in which temperature is being measured before reading fluorescence.) An instrument setting should be used in which the excitation shutter is only open during measurement as DPH is subject to photoisomerization. A small excitation slit size and short fluorescence acquisition time can also help to minimize photoisomerization.
7. The circulating temperature-controlled water bath is set to increase in temperature gradually and regular readings (at least at every 4°C) are taken, each time noting the sample temperature (*see* **Notes 12–14**).

3.4. Backgrounds and Controls

Background samples lacking fluorophore, but containing all other components, should be prepared with and without tempo. The apparent fluorescent intensity in these samples should be subtracted from the intensity of the samples containing the fluorescent molecule in order to determine the true fluorescence intensity arising from the fluorescent molecule. The fluorescent intensity in background samples is generally much weaker than that in the samples containing the fluorescent molecule. For that reason, it is not necessary to measure background fluorescence at every temperature point. A measurement of background fluorescence at the beginning and end of the temperature series is usually sufficient. The background values at intermediate temperatures can be estimated by linear extrapolation. Of course, one may measure exact background values, but that limits the number of experiments that can be performed simultaneously when the sample holder can only accommodate a few samples.

Lipid-mixing controls are also needed. It is of critical importance that the lipids be mixed homogeneously within the vesicles. If different vesicles have very different lipid compositions, such that lipids that tend to form ordered domains are in one vesicle population and lipids that tend to form disordered domains are in a separate vesicle population, it can lead to serious misinterpretation of the data. This issue does not arise when the lipid 12SLPC is used both as the lipid forming the disordered domains and as the quencher, because the increased quenching on melting of ordered domains automatically shows that the 12SLPC is in the same vesicles as the lipids forming ordered domains. Tempo, on the other hand, interacts with the lipid bilayer from the aqueous phase and gives no information about lipid mixing. For tempo experiments, samples in which the temperature dependence of FRET is measured can be used as a control. These experiments do not use tempo.

One protocol requires preparation of vesicles containing a small amount of donor-labeled and acceptor-labeled lipids that are both largely restricted to disordered domains *(21)*. If the ordered domain- and disordered domain-forming lipids are properly mixed, there will be strong FRET at low temperature because the labeled lipids will be concentrated within the disordered domains. On melting of ordered domains FRET will decrease because the average distance between the donor and acceptor will increase. A fluorescently labeled lipid that favors association with disordered domains and undergoes concentration-dependent self-quenching can be used in an analogous manner *(22)*. In a third protocol, a tryptophan-containing transmembrane peptide is used as the donor. This peptide resides in disordered regions of the membrane *(7)*. The FRET acceptor used is a DPH derivative, long alkyl chain-trimethyl-amino-diphyenylhexatriene. Long alkyl chain-trimethyl-amino-diphyenylhexatriene partitions strongly into ordered domains *(16,23)*. When ordered and disordered domains coexist at low temperature, the FRET donor and FRET acceptor are separated, and thus, FRET is weak. FRET increases on ordered domain melting if the ordered and disordered domain-forming lipids are in the same vesicle.

Another control, important when very exact T_m values are desired, is to evaluate the degree of perturbation of bilayer structure by tempo. The data has indicated, perturbation of bilayer structure by 2 mM tempo is not severe *(18)*. Nevertheless, tempo concentration in the membrane is high. From comparison of the quenching levels obtained with tempo to those observed with 12SLPC, one estimates that there is about one tempo molecule per 10 lipids *(24)*. To take tempo-induced effects into account, one can repeat the experiments using a range of tempo concentrations, and then extrapolate the T_m values obtained to zero tempo concentration. This control is not needed when relative T_m values among a set of different lipid mixtures are more important than the absolute T_m values. It should also be noted that ethanol may affect absolute T_m values.

3.5. Data Analysis

To find the effective T_m for ordered to liquid-disordered transition, graph the quenching ratio (F/Fo) vs temperature. F/Fo is calculated by taking the ratio of the average of the fluorescence intensities with quencher to the average of the fluorescence intensities without quencher at a given temperature. A graph similar to

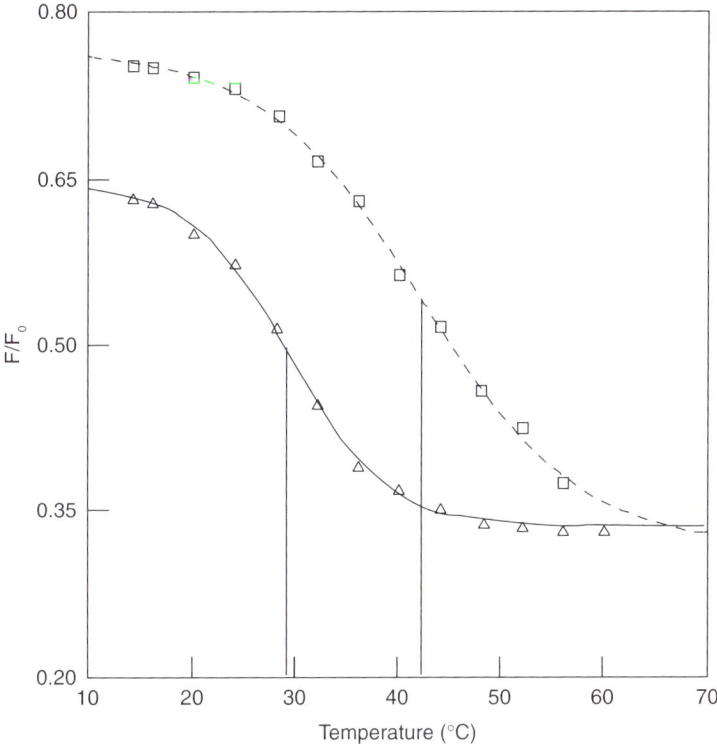

Fig. 2. Melting curves for ordered domains monitored by tempo quenching. Fluorescence measurements were taken from 14 to 60°C. Samples contained lipid vesicles prepared as described in this chapter, and made up of (triangles) dipalmitoyl PC/ dioleoyl PC/cholesterol (3:3:2 mol:mol) or (squares) dipalmitoyl PC/diphytanoyl PC/cholesterol (3:3:2 mol:mol). Each sample also contained 0.5 mol% DPH. Total lipid concentration was 50 μM and the lipids were dispersed in PBS. Notice that ordered domains (formed primarily by dipalmitoyl PC and cholesterol) are more thermally stable in samples containing diphytanoyl PC than in sample containing dioleoyl PC. The estimated temperatures at which melting is half-complete (the T_m of liquid-ordered to liquid-disordered melting transition, vertical lines) are 29.5 and 42°C, for the dioleoyl PC-containing and diphytanoyl PC-containing samples, respectively.

Fig. 2 should be generated. The points are then fit to a sigmoidal function. The program SlideWrite Plus (Advanced Graphics Software Inc., Encinitas, CA) has been used in the lab. The T_m is the point at which the absolute value of the slope of F/Fo vs temperature is at a maximum. **Figure 2** shows an example of an experiment in which the difference in ordered to liquid-disordered T_m for two different lipid mixtures is readily apparent. The actual T_m are 29.5°C for the dioleoyl PC-containing mixture and 42°C for the diphytanoyl PC-containing mixture.

4. Notes

1. Tempo can be dissolved in water. However, preparing a sufficiently concentrated stock solution in water is difficult, so instead tempo can be prepared as a concentrated stock solution in ethanol. The tempo solution should be as concentrated as possible to minimize the amount of ethanol introduced into the final sample.
2. After samples are prepared, they are stored in a dark location (e.g., a cardboard box), until fluorescence is measured in order to limit the amount of photodamage to DPH.
3. Alternatively, samples can be heated stepwise using a temperature-controlled water bath.
4. After weighing, care should be given to not touch the cups to minimize the transfer of hand oils. Precautions include, but are not limited to, wearing gloves and handling aluminum cups with forceps.
5. Care should be taken when transporting the aluminum cups to avoid tipping.
6. Three to five dry weights are taken for each solution and an average is taken.
7. The authors have had some experience using this technique with hydrophilic fluorophores that flip from leaflet to leaflet very slowly. To restrict such a fluorophore to the outer leaflet of a vesicle, add it to preformed vesicles. (This is only practical for fluorophores that can dissolve a bit in water.)
8. This step is included to soften the lipids, and thus, facilitate solubilization in ethanol.
9. At this point, the samples should be kept in a dark place before reading to avoid photodamage.
10. The width settings of the slits will be dependent on your instrument. You may have to adjust the slit widths to get a usable signal.
11. The relative humidity should also be noted. The authors recommend www.weather.com. If the temperature of the sample goes below the dew point, condensation will form on the cuvets and interfere with readings at low temperature. This can be a problem on humid days.
12. It has been found that turning the water cycler (pump) off immediately before a reading slows the warming of the sample enough to get a reading at a stable temperature. Alternatively, temperature of the sample can be read immediately before and after the fluorescence measurement and averaged.
13. The rate in which the water bath heats samples should be programmed to give the desired increase per minute. It was found that the rate of temperature increase in the sample compartment will be affected by the amount of water in the water bath.
14. Instead of a continuous heating protocol, one can increase the temperature in incremental steps and wait for the temperature to stabilize after each step. It has been found that it is quicker to collect data when the water bath temperature is continually increased.

Acknowledgment

This work was supported by NIH grant GM 48596.

References

1. Brown, D. A. and London, E. (1998) Structure and origin of ordered lipid domains in biological membranes. *J. Membr. Biol.* **164**, 103–114.
2. Brown, D. A. and London, E. (2000) Structure and function of sphingolipid- and cholesterol-rich membrane rafts. *J. Biol. Chem.* **275**, 17,221–17,224.
3. Edidin, M. (2001) Shrinking patches and slippery rafts: scales of domains in the plasma membrane. *Trends Cell Biol.* **11**, 492–496.
4. Anderson, R. G. and Jacobson, K. (2002) A role for lipid shells in targeting proteins to caveolae, rafts, and other lipid domains. *Science* **296**, 1821–1825.
5. Grassme, H., Jendrossek, V., Riehle, A., et al. (2003) Host defense against Pseudomonas aeruginosa requires ceramide-rich membrane rafts. *Nat. Med.* **9**, 322–330.
6. London, E. (2002) Insights into lipid raft structure and formation from experiments in model membranes. *Curr. Opin. Struct. Biol.* **12**, 480–486.
7. Fastenberg, M. E., Shogomori, H., Xu, X., Brown, D. A., and London, E. (2003) Exclusion of a transmembrane-type peptide from ordered-lipid domains (rafts) detected by fluorescence quenching: extension of quenching analysis to account for the effects of domain size and domain boundaries. *Biochemistry* **42**, 12,376–12,390.
8. Abrams, F. S. and London, E. (1992) Calibration of the parallax fluorescence quenching method for determination of membrane penetration depth: refinement and comparison of quenching by spin-labeled and brominated lipids. *Biochemistry* **31**, 5312–5322.
9. Florine-Casteel, K. and Feigenson, G. W. (1988) On the use of partition coefficients to characterize the distribution of fluorescent membrane probes between coexisting gel and fluid lipid phase: an analysis of the partition behavior of 1,6-diphenyl-1,3,5-hexatriene. *Biochim. Biophys. Acta* **941**, 102–106.
10. Lentz, B. R., Barenholz, Y., and Thompson, T. E. (1976) Fluorescence depolarization studies of phase transitions and fluidity in phospholipid bilayers. 2 Two-component phosphatidylcholine liposomes. *Biochemistry* **15**, 4529–4537.
11. London, E. and Feigenson, G. W. (1981) Fluorescence Quenching in Model Membranes. An analysis of the local phospholipid environments of diphenylhexatriene and gramicidin A`. *Biochim. Biophys. Acta* **649**, 89–97.
12. Florine, K. I. and Feigenson, G. W. (1987) Influence of the calcium-induced gel phase on the behavior of small molecules in phosphatidylserine and phosphatidylserine-phosphatidylcholine multilamellar vesicles. *Biochemistry* **26**, 1757–1768.
13. Megha, and London, E. (2004) Ceramide selectively displaces cholesterol from ordered lipid domains (rafts): implications for lipid raft structure and function. *J. Biol. Chem.* **279**, 9997–10,004.
14. Ahmed, S. N., Brown, D. A., and London, E. (1997) On the origin of sphingolipid/cholesterol-rich detergent-insoluble cell membranes: physiological concentrations of cholesterol and sphingolipid induce formation of a detergent-insoluble, liquid-ordered lipid phase in model membranes. *Biochemistry* **36**, 10,944–10,953.

15. London, E., Brown, D. A., and Xu, X. (2000) Fluorescence quenching assay of sphingolipid/phospholipid phase separation in model membranes. *Methods Enzymol.* **312,** 272–290.
16. Xu, X., Bittman, R., Duportail, G., Heissler, D., Vilcheze, C., and London, E. (2001) Effect of the structure of natural sterols and sphingolipids on the formation of ordered sphingolipid/sterol domains (rafts). Comparison of cholesterol to plant, fungal, and disease-associated sterols and comparison of sphingomyelin, cerebrosides, and ceramide. *J. Biol. Chem.* **276,** 33,540–33,546.
17. Wang, J., Megha, and London, E. (2004) Relationship between sterol/steroid structure and participation in ordered lipid domains (lipid rafts): implications for lipid raft structure and function. *Biochemistry* **43,** 1010–1018.
18. Megha, Bakht, O., and London, E. (2006) Cholesterol precursors stabilize ordinary and ceramide-rich ordered lipid domains (lipid rafts) to different degrees: Implications for the Bloch hypothesis and sterol biosynthesis disorders. *J. Biol. Chem.* 281, 21,903–21,913.
19. Koynova, R. and Caffrey, M. (1995) Phases and phase transitions of the sphingolipids. *Biochim. Biophys. Acta* **1255,** 213–236.
20. Kremer, J. M., Esker, M. W., Pathmamanoharan, C., and Wiersema, P. H. (1977) Vesicles of variable diameter prepared by a modified injection method. *Biochemistry* **16,** 3932–3935.
21. Wang, T. Y. and Silvius, J. R. (2003) Sphingolipid partitioning into ordered domains in cholesterol-free and cholesterol-containing lipid bilayers. *Biophys. J.* **84,** 367–378.
22. Coste, V., Puff, N., Lockau, D., Quinn, P. J., and Angelova, M. I. (2006) Raft-like domain formation in large unilamellar vesicles probed by the fluorescent phospholipid analogue, C12NBD-PC. *Biochim. Biophys. Acta* **1758,** 460–467.
23. Beck, A., Heissler, D., and Duportail, G. (1993) Influence of the length of the spacer on the partitioning properties of amphiphilic fluorescent membrane probes. *Chem. Phys. Lipids* **66,** 135–142.
24. Kaiser, R. D. and London, E. (1998) Location of diphenylhexatriene (DPH) and its derivatives within membranes: comparison of different fluorescence quenching analyses of membrane depth. *Biochemistry* **37,** 8180–8190.

5

Using Monomolecular Films to Characterize Lipid Lateral Interactions

Rhoderick E. Brown and Howard L. Brockman

Summary

Membrane lipids are structurally diverse in ways that far exceed the role envisioned by Singer and Nicholson of simply providing a fluid bilayer matrix in which proteins reside. Current models of lipid organization in membranes postulate that lipid structural diversity enables nonrandom lipid mixing in each leaflet of the bilayer, resulting in regions with special physical and functional properties, i.e., microdomains. Central to understanding the tendencies of membrane lipids to mix nonrandomly in biomembranes is the identification and evaluation of structural features that control membrane lipid lateral mixing interactions in simple model membranes. The surface balance provides a means to evaluate the lateral interactions among different lipids at a most fundamental level—mixed in binary/ternary combinations that self-assemble at the air–water interface as monomolecular films, i.e., monolayers. Analysis of surface pressure and interfacial potential as a function of average cross-sectional molecular area provide insights into hydrocarbon chain ordering, lateral compressibility/elasticity, and dipole effects under various conditions including those that approximate one leaflet of a bilayer. Although elegantly simple in principle, effective use of the surface balance requires proper attention to various experimental parameters, which are described herein. Adequate attention to these experimental parameters ensures that meaningful insights are obtained into the lipid lateral interactions and enables lipid monolayers to serve as a basic platform for use with other investigative approaches.

Key Words: Ceramide; cholesterol; condensation; dipole moment; film balance; lateral compressibility; phase transition; surface potential; surface pressure; sphingomyelin.

1. Introduction

When membrane lipid amphiphiles are dissolved in a water-insoluble solvent and deposited on a water surface with a microsyringe (*see* **Subheading 3.1.6.**), the solution spreads rapidly to occupy the available area. As the solvent evaporates, the lipid amphiphiles orient to minimize contact of their nonpolar regions with water, while maximizing the water-contact of their polar regions. The resulting one-molecule-thick lipid film, i.e., monolayer, provides a useful model system for studying the lateral packing interactions of lipids in each leaflet of a biomembrane.

Lipid monolayers have a rich history of providing key insights into biomembranes. Perhaps the most famous of such lipid monolayer experiments are those of Gorter and Grendel *(1)*, who used a Langmuir surface balance to deduce the bilayer structure of membranes, by noting that monolayer surface areas produced by the lipid extracts of erythrocytes from several different animals are approximately twice the surface areas of the cells themselves. Since the pioneering work of Gorter and Grendel, it has become clear that many physical phenomena displayed by biomembrane lipids can be observed and modeled in lipid monolayers. These phenomena include phase transitions, and lateral diffusion, as well as mixing interactions that bring about changes in hydrocarbon chain ordering and lateral compressibility, and result in critical points and coexisting lateral phases, i.e., domains *(2–8)*. Over the past decade, interest in such phenomena has enjoyed a renaissance, especially among cell biologists, largely because of the "raft" model proposed for biomembrane structure *(9)*. Studies of lateral interactions among "raft" lipids generally involve phosphatidylcholine (PC) (e.g., 1-palonitoyl-2 oeoyl-phosphatidylcholine [POPC] or dioleoyl-phosphatidylcholine [DOPC]), sphingomyelin (SM), cholesterol, and closely related derivatives in various combinations because of their putative roles in raft microdomain formation *(8,9)*. The focus of this chapter will be on describing the fundamentals of using a surface balance to achieve reliable and reproducible insights into raft lipid lateral interactions.

2. Materials

To obtain meaningful data, serious precautions must be taken to avoid contamination of all materials used in monolayer experiments by surface-active substances. A single human fingerprint contains sufficient surface-active materials to form numerous monolayers in a typical surface balance.

1. Subphase buffer onto which monolayers are spread is kept stored under purified argon or nitrogen after preparation using purified water (*see* **Notes 1** and **2**). If organic-based buffers must be used, for example, 4-(2-hydroxyethyl)-1-piperazine-ethane sulfonic acid (HEPES), they should be of the highest purity available and should be tested for the presence of amphiphilic impurities (*see* **Note 3**).
2. Glassware is acid cleaned, rinsed with purified water, and then treated with hexane/ethanol (95:5) before use.
3. Solvents used to prepare lipid stock solutions and to spread lipids onto the gas–water interface are high-pressure liquid chromatography grade (*see* **Note 3**).
4. Lipid purity is confirmed by thin-layer chromatography using appropriate solvent mixtures (*see* **Note 4**).
5. Lipid stock solutions are kept at −20°C until use in acid-cleaned, borosilicate glass vials (Kimble-Kontes, Vineland, NJ) equipped with Kimble-Kontes Microflex Mininert® push–pull valves (Kimble-Kontes) (*see* **Note 5**).

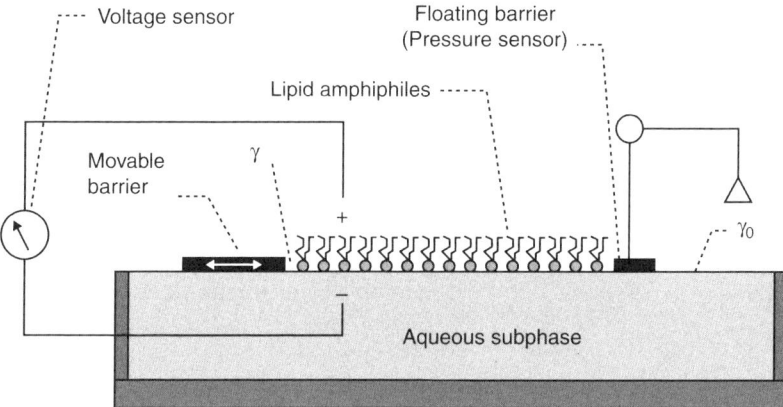

Fig. 1. Langmuir-type surface balance for determining the surface pressure and surface potential as a function of average cross-sectional molecular area. γ is the surface tension of surface occupied by lipid amphiphile and γ_0 is the surface tension of clean aqueous subphase.

3. Methods

The spreading and orienting response of membrane lipid amphiphiles during monolayer compression is a direct consequence of the free energy excess experienced by water molecules at the surface compared with those in the bulk subphase. Water molecules at the surface are more restricted in their hydrogen bonding with other water molecules, i.e., lower entropy, compared with bulk-phase water molecules in the subphase. As a result, water molecules attempt to limit their exposure along the air contact surface by continuously "pulling back" toward the bulk subphase water to maximize hydrogen bonding with neighboring water molecules. The magnitude of the pull by surface water is reflected in its surface tension, which has units of force per unit length. Changes in the surface tension that occur in response to lipid amphiphile addition to the surface provide information about lipid–lipid and lipid–water interactions. When the available area for the lipid monolayer is so large that there is little effect of the lipids on the surface tension of water and the lipid intermolecular lateral interactions are weak, then the monolayer can be regarded as a two-dimensional (2D) gas. When the available surface area of the monolayer is reduced by a movable barrier (*see* **Fig. 1**), the lipid amphiphiles start to exert repulsive effects on each other. This 2D analog of a pressure is called surface pressure (π) and is expressed by the following relationship:

$$\pi = \gamma_0 - \gamma$$

where γ_0 is the surface tension in absence of a monolayer and γ the surface tension with the monolayer present.

Another consequence of the asymmetric orientation of lipid and water molecules at a gas–liquid interface is the generation of a sizeable (hundreds of millivolts) electrical potential perpendicular to the plane of the interface *(3,10)*. Among the roles of this potential is its major contribution in determining the size and shape of lipid rafts *(4,6)*. This dipole potential cannot be measured directly in each half of a bilayer, but like surface pressure, it is readily measured across monolayers using commercially available instrumentation. Its value (ΔV) is expressed as the difference in potential between a monolayer-containing interface and the aqueous phase in the absence of the monolayer,

$$\Delta V = V - V_0$$

For typical raft-forming lipids, the ends of the hydrocarbon moieties equivalent to the center of a bilayer, are at a positive potential relative to the aqueous phase.

In a typical film balance (**Fig. 1**), surface pressure is measured continuously as a function of average molecular area (*see* **Subheading 3.1.**). Control of the available surface area is provided by a movable barrier that can be swept along the subphase surface using computer-controlled motors. No leakage of the monolayer should occur underneath or around the barrier. This is best accomplished when the barrier is made of a hydrophobic material, which is rendered hydrophilic on its lower surface (*see* **Note 6**). The trough holding the subphase is usually made of Teflon® (*see* **Note 6**) and temperature is controlled by circulating water through channels located underneath the trough. Surface pressure is typically monitored by one of the two ways. In the first approach the surface tension is measured directly using a wettable plate, called a Wilhelmy plate, which is vertically suspended and partially immersed into the aqueous subphase. The downward force on the plate, which is attached to a calibrated microbalance, is then converted into surface tension (mN/m or dynes/cm) after taking into account the dimensions of the plate and contact angle (*see* **Note 7**). In the second approach, surface pressure is determined by measuring the translational force acting on a float separating the lipid monolayer-covered area from an adjacent surface free of lipids (**Fig. 1**). Teflon tape connected between the ends of the float and the trough provides a flexible seal for maintaining separation between the clean and monolayer-covered surfaces. A rigid wire, connected on one end to the float and on the other end to a torsion detection system mounted above the subphase, enables accurate detection of the minute translational movement of the float during compression/expansion sweeps of the movable barrier. The detection system used to determine surface pressure distinguishes a Wilhelmy-type film balance from a Langmuir-type film balance but the results obtained are equivalent.

During measurement of the surface pressure, it is possible to simultaneously record the interfacial potential as the movable barrier compresses the lipid monolayer. In reality, it is the potential difference that is measured between the clean water surface and a surface covered by lipid. Typically, an electrode containing either polonium or americium is used in order to ionize the air gap into an electrically conducting medium to achieve potentiometric recording (**Fig. 1**).

3.1. Setup Conditions for Film Balance

1. The *film balance* must be placed in a vibration-free environment to keep the subphase surface stable and wave-free. This can be accomplished by placing the trough on an active or passive vibration isolation table (e.g., Kinetic Systems Vibraplane®, Boston, MA). Commercial manufacturers of film balances include Kibron Inc. (Helsinki, Finland), LSV Ltd. (Helsinki, Finland), and NIMA Technology Ltd. (Coventry, England).
2. The trough should be filled so that the *subphase level* stands slightly above the rim of the trough (*see* **Note 8**). The subphase level is kept constant by enclosing the Langmuir trough with a home-built plexiglass chamber that is continuously purged with purified, humidified argon or nitrogen gas. Avoiding subphase evaporation and setting the subphase level with high reproducibility increases the accuracy and reproducibility of isotherm measurements by ensuring a constant contribution of the meniscus to the total surface area.
3. Accurate *surface pressure calibration* of the film balance is essential for obtaining high-quality data. Calibration methods commonly used for surface balances equipped with Wilhelmy-type (direct measurement of surface tension) and Langmuir-type (surface pressure differences) detection systems have been carefully compared *(11)*. Equilibrium spreading measurements using trioleoylglycerol, oleoylmethanol, trioctanoylglycerol, 1,3-dioleoylglycerol, and oleyl alcohol provide accurate calibration (*see* **Note 9**).
4. Accurate *surface area calibration* as a function of the position of the moving barrier is required to insure data that are independent of the number of molecules spread to form the monolayer. Because of geometric irregularities and meniscus curvature at that end of the trough, the moving barrier proportionality is not normally observed between barrier position and geometrically estimated area. To ensure proportionality, a calibration procedure has been described *(12)* (*see* **Note 9**).
5. *Surface cleaning* before and between runs is accomplished by multiple sweeps of the barrier across the empty surface combined with aspiration of the compressed surface (*see* **Note 10**).
6. *Deposition of lipid* onto the clean subphase surface is accomplished by dissolving it in volatile solvent and carefully applying a precise amount to the surface using a microsyringe. A gastight microsyringe (Hamilton 1700 series) equipped with a Digital Syringe (Hamilton Co., Reno, NV) facilitates manual deposition onto the subphase surface (*see* **Note 11**).
7. *Solvent combinations* particularly effective for spreading phosphoglycerides, sterols, and sphingolipids commonly used for "raft" lipid studies are hexane/isopropanol/water (70:30:2.5). hexane:ethanol (9:1), and toluene/ethanol (5:6).

8. Complete *evaporation of spreading solvent* is needed before initiation of barrier compression. Hexane-based solvents (~50 µL) require 4–5 min to completely evaporate at room temperature.
9. Lipid films are compressed at a rate of ≤4 Å^2/mol/min to minimize the occurrence of metastable phases.
10. *Lipid stock concentrations* must be known with great precision in order to accurately assess the total molecules of deposited lipid. For lipids containing phosphate, the Bartlett method is used *(13)*. When the lipids contain no phosphate, quantitation is achieved by gravimetric determination using a microbalance (e.g., Cahn model 4700 [Cahn Instruments Inc., Cerritos, CA]).

3.2. Acquisition of Monolayer Isotherms Using the Surface Balance

In its most basic configuration, the monolayer film balance measures the surface pressure as a function of available subphase surface area for a known number of molecules of lipid amphiphile. The measurement is performed at constant temperature and the resulting data is commonly referred to as a force-area or surface pressure-area (π-A) isotherm. An isotherm is usually recorded by reducing the available surface area, i.e., laterally compressing the film, with a barrier moving at a constant rate while continuously monitoring the surface pressure. During compression, the isotherm sometimes shows discontinuities that indicate transitions between monolayer phases differing with respect to their lateral packing features. The observed phase behavior of the monolayer is determined mainly by the physicochemical properties of the lipid amphiphile, the subphase temperature, and the subphase composition. The two most commonly observed monolayer states, the liquid-expanded and liquid-condensed monolayer states are analogous to the liquid-crystalline and gel states in bilayers, respectively *(14)*. As with lipid bilayers, the monolayer-phase state for a particular lipid species depends on the length and unsaturation of the hydrocarbon chain and the bulkiness and charge state of the polar headgroup. An increase in the chain length increases the attraction between molecules causing the π-A isotherm to condense. In contrast, ionization of the lipid head groups induces repulsive forces tending to oppose phase transitions.

Representative isotherms collected at 24°C for different "raft" lipids, each in their pure state, are shown in **Fig. 2**. The SM isotherm shows liquid-expanded behavior at molecular areas between 84 and 64 Å^2/mol, and then begins to pass through a transition to a condensed, chain-ordered state. The onset of this horizontal phase transition is very temperature-dependent **(Fig. 3)**. As the temperature is increased the surface pressure value at which the horizontal transition phase occurs will increase and vice versa. The monolayer eventually reaches a collapse point, characterized by either a rapid decrease in the surface pressure

Raft Lipid Monolayers

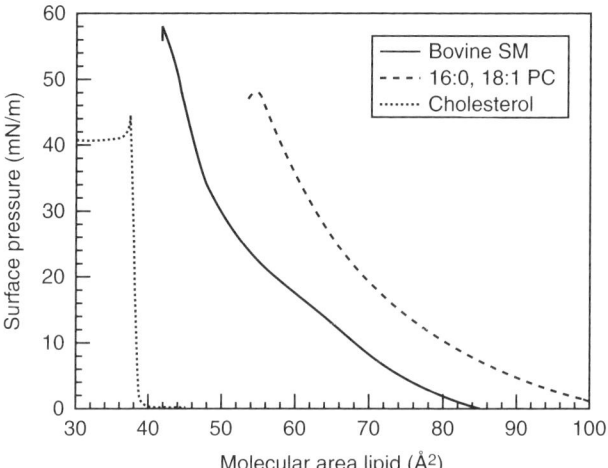

Fig. 2. Representative π-A isotherms for "Raft" Lipids. Dotted line shows the condensed isotherm of cholesterol. Dashed line shows the liquid-expanded (fluid) isotherm of POPC. Solid line shows isotherm of bovine brain SM. The discontinuity occurring near 64 Å^2/mol (14 mN/m) represents the onset of a 2D-phase transition from liquid-expanded-to-condensed behavior with monolayer collapse occurring near 42 Å^2/mol.

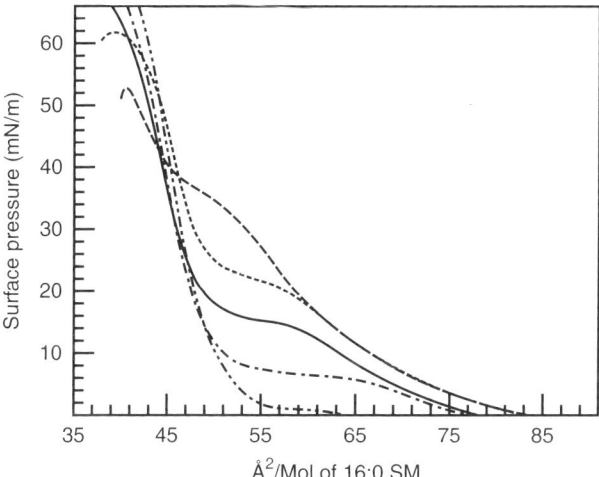

Fig. 3. The effect of temperature on 16:0 SM monolayers. Isotherms (from top to bottom by 2D-phase transition) were measured at 30, 24, 20, 15, and 10°C for SM monolayers containing palmitoyl acyl chains.

or as a horizontal break in the isotherm (*see* **Note 12**). Markedly contrasting responses are displayed by POPC and cholesterol, which show only liquid-expanded and -condensed behavior, respectively, without any evidence of undergoing liquid-expanded-to-condensed phase transitions during compression at room temperature until achieving collapse **(Fig. 2)**.

3.3. Preparation of Monolayers Containing Raft Lipid Mixtures

1. To prepare lipid mixtures for spreading on the film balance, aliquots of the different lipid stock solutions are combined to obtain the desired mole fraction of each lipid component in the mixture (*see* **Note 13**).
2. The total lipid applied to surface of the film balance should allow for spreading to a large average area per molecule after initial positioning of the compression barrier. A general rule of thumb for monolayer spreading is to position the barrier to provide a total surface area that exceeds the "lift off" area by at least 25 Å^2/mol. For a given monolayer, the "lift off" area corresponds to the area per molecule when the surface pressure can be detected, i.e., $\pi \geq 1$ mN/m. It must be determined empirically but is typically 75–125 Å^2/mol.
3. As the barrier sweeps the surface and compresses the lipid film, it is desirable to achieve monolayer collapse before the total area gets too small. Doing so avoids mechanical limitations associated with the surface pressure-detecting system and the approaching barrier mounted to the edge of the trough and increases measurement accuracy. This can be accomplished by applying sufficient lipid to the surface so that monolayer collapse occurs at surface areas ≥25% of the total area available when compression is initiated.

3.4. Analyses of Monolayer Isotherms

π-A Isotherms of raft lipid mixtures differ substantially from those of the pure individual components. To obtain meaningful insights into the lateral packing in the mixed lipid films, various analyses have been developed.

3.4.1. Average Area Molecular vs Composition Analysis

A classic way to detect lateral interactions among lipids is to examine how changing composition affects the average molecular area within the mixed monolayer. The average molecular areas observed in the experimental mixtures are compared with the areas expected when one of the lipids replaces the other at specified mole fraction in the binary mix at a specified surface pressure. The expected areas are calculated by summing the molecular areas of the individual pure components, apportioned by mole fraction in the mixture, using the following equation:

$$A_{av} = X_1(A_1) + (1 - X_1)(A_2)$$

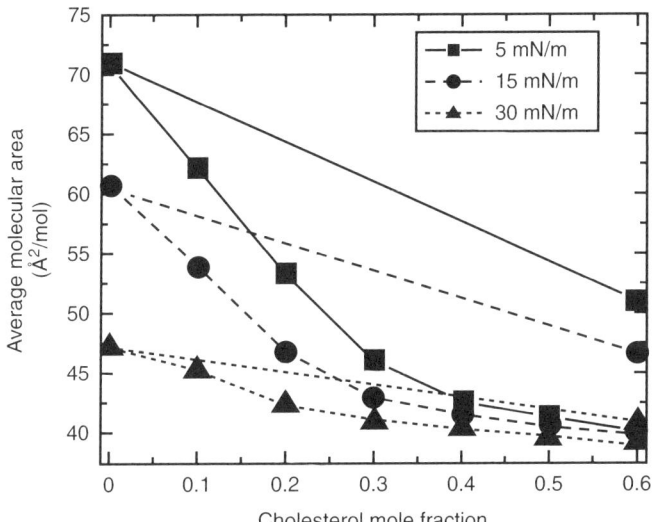

Fig. 4. 16:0 SM-cholesterol average molecular area vs composition analysis. Plots are shown for three different surface pressures in mN/m (5, squares; 15, circles; 30 triangles). The linear plots show the average molecular area obtained by calculation using the molecular areas of pure 16:0 SM and cholesterol, each apportioned by mole fraction. The nonlinear curves represent the experimentally observed areas for the mixtures. Negative deviation from ideal additivity (linear plots) shows the condensing or ordering effect of cholesterol.

where A_{av} is the average molecular cross-sectional area in the mixed monolayer, X_1 is the mole fraction of pure component 1, and A_1 and A_2 are the molecular areas of pure components 1 and 2 at identical surface pressures. Experimental average molecular areas that deviate negatively from the calculated molecular area additivity, as a function of lipid composition reflect cross-sectional area changes beyond what is expected by simply replacing a larger molecule with a smaller molecule (**Fig. 4**). Consequently, negative deviation from area additivity observed in the experimental mixture is often referred to as area condensation and suggests intermolecular accommodation between lipids comprising the mixed monolayer (*see* **Note 14**).

A particularly relevant example of intermolecular accommodation involving lateral interactions of raft lipids is the "condensing effect" by cholesterol after mixing with PC or SM (*see* **refs. 15–19**, and references therein). The π-A isotherm of pure cholesterol is highly condensed (**Fig. 2**), showing almost no change in average molecular area with increasing surface pressure, and a cross-sectional molecular area of approx 36.8 Å^2 at 30 mN/m, a surface pressure that approximates the membrane environment. This response shows that cholesterol is rigid and somewhat bulky compared with the cross-sectional area of an ordered hydrocarbon chain (~19 Å^2). Mixing of the rigid cholesterol with phospholipids significantly reduces the number of possible conformations of the phospholipid acyl chains, thereby increasing chain order and decreasing phospholipid cross-sectional area. For this reason, the negative deviations from area additivity observed in the average molecular areas of cholesterol-PC and cholesterol-SM mixed monolayers can be expressed in terms of the apparent area change of the phospholipid (PL) using the following equation:

$$\text{PL condensation} = A_{PL} - \left[\left(A_{mix} - \left(A_{chol} X_{chol} \right) \right) / X_{PL} \right]$$

where PL condensation units are Å^2/PL molecule, A_{PL} is the area per molecule of the pure phospholipid, A_{mix} is the average area per molecule of the PL/Chol mixed monolayer, A_{chol} is the area per molecule of the pure cholesterol, and X_{chol} is the mole fraction of cholesterol in the mixed monolayer. Application of this analysis to binary combinations of raft lipids (e.g., cholesterol and different molecular species of PC or SM) provides insights into how surface pressure and phospholipid phase state both affect the change in phospholipid cross-sectional area brought about by lateral interaction with cholesterol *(17–19)* (*see* **Note 15**).

3.4.2. Analysis of Lateral Compressibility

The surface compressibility (C_S), i.e., lateral compressibility, of raft lipids, in either pure or mixed monolayers, can be obtained from π-A data using:

$$C_S = -\frac{1}{A} \left(\frac{\partial A}{\partial \pi} \right)_T$$

where A is the area per molecule at the indicated surface pressure (π). Mathematically, C_S values represent the first derivative function, multiplied by the inverse area, and are calculated using routine software packages. When binary combinations of raft lipids are studied, ideal additivity can be modeled by apportioning the C_S value for each lipid (as a pure entity) according to both molecular area fraction and mole fraction *(19–21)* (*see* **Note 16**). Thus, at a given constant surface pressure (π),

$$C_S = (1/A) \left[\left(C_{S^1} A_1 \right) X_1 + \left(C_{S^2} A_2 \right) X_2 \right]$$

where $X_2 = (1 - X_1)$ and C_S is additive with respect to the product ($C_S A_i$) rather than (C_{S^i}) for either ideal or completely nonideal mixing. Deviations of experimental values from calculated additivity provide evidence that the lipid components of the mixed monolayers are partially nonideally mixed *(20)*, in analogous fashion to average area vs cholesterol composition plots.

Raft Lipid Monolayers

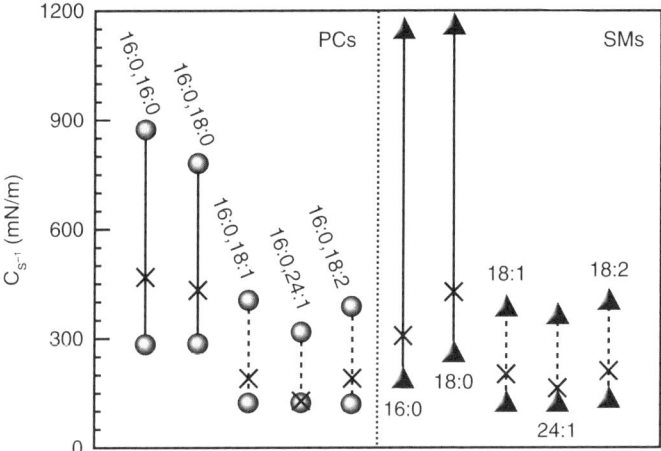

Fig. 5. Changes in surface compressional moduli ($C_{s^{-1}}$) induced on mixing of equimolar cholesterol with PC or SM. Lower symbols represent the $C_{s^{-1}}$ values of the pure lipids in the absence of cholesterol. Upper symbols represent the experimentally observed $C_{s^{-1}}$ values for binary mixtures with equimolar cholesterol. The X along each line represents the ideal $C_{s^{-1}}$ values calculated based on additivity of each pure lipid component in the binary mixtures. All $C_{s^{-1}}$ values were determined at a surface pressure of 30 mN/m. The notation X:Y refers to the acyl chain length in carbon atoms (X) and the number of *cis*-double bonds (Y) present in the different molecular species of PC or SM.

To facilitate comparison with elastic moduli of area compressibility values obtained in bilayer vesicles *(22,23)*, the monolayer C_S values can be expressed in reciprocal form ($C_{s^{-1}}$), originally defined as the surface compressional modulus by Davies and Rideal *(24)*. Thus, ($C_{s^{-1}}$) data provide insights into the lateral packing elasticity, i.e., ease/resistance to lateral compression, within the monolayer. Because $C_{s^{-1}}$ analyses utilize information available in the slopes of the isotherms, they are especially useful for evaluating the effects of cholesterol at high surface pressures (*see* **Note 17**). Moreover, compared with area condensations, $C_{s^{-1}}$ values are more responsive to subtle changes in lipid structure during lateral interaction with cholesterol (**Fig. 5**). For more detailed descriptions of the mechanoelastic properties of model membranes *(22,23,25)* and the application of $C_{s^{-1}}$ analysis to "raft lipids" and their mixtures, readers are referred to **refs. *19* and *26–30***.

3.4.3. Interfacial Potential

For lipids that exhibit fluid π-A isotherms, the dipole potential changes linearly with the 2D concentration, is 1/A of lipid amphiphile, irrespective of the surface pressure. From the slope and intercept of this line, two contributions to the potential can be obtained *(10,12)*. One is the dipole moment, $\mu_\perp$, which arises from the vectorial component of lipid dipoles and associated water normal to the interface. Note that if the lipid carries a formal charge, like diacyl phosphatidylserine or sulfated glycolipids, the charge and its associated counter-ions in solution will also contribute to the value of $\mu_\perp$ (*see* **Note 18**). The second parameter, ΔV_0, is a constant potential difference, relative to water that arises from the epitaxial ordering of all interfacial water by the presence of the lipid-amphiphile in the liquid-expanded and more condensed monolayer phases.

Experimentally, in a lipid mixture showing no monolayer phase transition in its π-A isotherm, the corresponding ΔV – $1/A$ isotherm is linear *(31)*, similar to that of pure lipid amphiphiles (*see* **Note 19**). Thus, an apparent value for each of the two parameters can be obtained for any mixture. These can be compared with ideal values for mixtures of ΔV_0 and $\mu_\perp$ calculated from those of the constituent lipids. Ideal values of the parameters for mixtures have been shown to apportion on the basis of the area fraction and mole fraction of each constituent in the mixture, respectively *(31,32)*. Comparison of apparent parameter values in mixtures with predicted ideal values can reveal interactions not evident from area or compressibility analysis *(31)*.

4. Notes

1. Argon or nitrogen gas is cleaned by passage through a seven-stage series filtration setup consisting of an Alltech-activated charcoal gas purifier (Alltech Associates Inc., Deerfield, IL), a LabClean filter, and a series of Balston disposable filters (Parker Balston Products, Haverhill, MA) consisting of two adsorption (carbon) and three filter units (93 and 99.99% efficiency at 0.1 µm).
2. Water is purified by reverse osmosis, activated charcoal adsorption, mixed-bed de-ionization, and then passage through a Milli-Q UV Plus System (Millipore Corp., Bedford, MA), and filtration through a 0.22-µm Millipak 40 membrane (Millipore Corp., Billerica, MA).
3. Solvent and buffer purity can be assessed by dipole potential measurements using a ^{210}Poionizing electrode (London Co., Cleveland, OH) *(33)*.
4. Primulin is a particularly sensitive and nonspecific detection spray for analyzing lipid purity after thin-layer chromatography. Plates are sprayed with a 0.001% solution of primulin (Sigma-Aldrich, St. Louis, MO), dissolved in acetone:water (4:1), and viewed under long-wave ultraviolet light after evaporation of the acetone from the thin-layer plate.
5. Solvents particularly effective for storing lipids as stock solutions for long periods (>1 yr) at –20°C without any evidence of degradation are hexane:isopropanol:water (70:30:2.5), toluene:ethanol (1:1), and hexane:ethanol (9:1).
6. Occasional etching of Teflon surfaces that are exposed to water with a solution of metallic sodium/naphthalene (Chemgrip Treating Agent, Norton Performance Plastics, Wayne, NJ) increases their wettability and decreases nondesired adsorption with some proteins and lipids *(34,35)*.

7. When using the Wilhelmy plate method for determination of surface tension, the material comprising the plate must be carefully selected to ensure accuracy and avoid artifacts. Momsen et al. *(11)* have shown that nichrome used as a wire configuration provides many advantages over other materials (e.g., platinum, glass, and filter paper) used in the more common plate configuration. An example of the difficulties and artifacts encountered using the Wilhelmy plate method is provided in studies of lignin monolayers *(36)*.
8. Highly reproducible setting of the subphase level is accomplished by slightly overfilling the trough and then reducing the subphase level using a firmly mounted home-built aspirator device consisting of a narrow-gauge tube (e.g., hypodermic needle and tubing attached to a vacuum source). As the subphase level decreases, loss of contact with the sipper needle ensures uniform and reproducible setting of the subphase level *(37)*.
9. Surface pressure and area calibration of the surface balance can be accomplished with great precision using procedures detailed in Momsen et al. *(11)* and Smaby and Brockman *(12)*.
10. Some lipids tend to be more difficult to clean from the surface by simple repetitive sweeps of the barrier. In such instances, repeatedly spreading a fluid-phase lipid (e.g., POPC) and having the barrier/aspirator system sweep it off the surface usually helps remove the residue.
11. To achieve highly reproducible spreading of lipid samples, the film balance can be equipped with a modified high-pressure liquid chromatography auto-injector (Beckman/Altex 500 autosampler, Beckman Coulter Co., Fullerton, CA) for deposition of lipids onto the surface. This also allows for multiple samples to be run in succession when combined with automated cleaning sweeps of the barrier between runs *(37)*.
12. For a comprehensive review of the monolayer properties of various glycosylated and nonglycosylated sphingolipids, readers are referred to Maggio et al. *(38)*.
13. To obtain accurate and reproducible mixing, Hamilton digital syringes (Hamilton Co., Reno, NV) are used.
14. The magnitude of the area condensation should not be used as a sole measure of lateral affinity of cholesterol for different membrane lipids, as is sometimes found in the literature. One must also consider the lipid packing density (area/molecule) before and after mixing with cholesterol to arrive at meaningful conclusions. For further discussion, *see* **ref. 18**.
15. Monolayers at high surface pressures (e.g., $\pi \geq 30$ mN/m) approximate the biomembrane environment *(39)*. However, accurate quantitation of small changes in absolute lipid area at high surface pressures is extremely challenging and beyond the practical capabilities of most surface balances. To circumvent this challenge, investigators sometimes extrapolate to the biomembrane situation using data obtained at low surface pressures because the area condensations are much larger and more easily measured. Such extrapolations are prone to inherent caveats because of the complex and nonlinear relationship that exists between the area condensation and surface pressure for phospholipids with different acyl structures. For further discussion, *see* **ref. 19**.

16. Derivation of the relationships governing the surface compressibility functions in binary mixtures of lipids show that both the mole fraction and area fraction contributions of each pure species must be considered when calculating ideal additivity response *(19,20)*.
17. For π-A isotherms, approx 1200 data points to construct an isotherm are routinely collected. To calculate a $C_{s^{-1}}$ value, a 100-point sliding window that utilizes every fourth π-A data point before advancing the window one point is used. The reliability of the window size for the calculated $C_{s^{-1}}$ values is checked by reducing the window size two- and five-fold. The resulting $C_{s^{-1}}$ values are used to construct $C_{s^{-1}}$ vs average molecular area plots. High $C_{s^{-1}}$ values correspond to low lateral elasticity among packed lipids forming the monolayer. The standard errors of the $C_{s^{-1}}$ values are about 2%. At 30 mN/m pure cholesterol forms highly condensed monolayers characterized by a $C_{s^{-1}}$ value near 1540 mN/m; whereas liquid-expanded POPC has a $C_{s^{-1}}$ value near 122 mN/m. The $C_{s^{-1}}$ values of PC-cholesterol and SM-cholesterol mixed monolayers are strongly affected by PC and SM acyl structure. When mixed with equivalent high-cholesterol mole fractions, PCs and SMs with saturated acyl chains have much higher $C_{s^{-1}}$ values (lower lateral elasticity) than PCs (and SMs) with unsaturated acyl chains consistent with *cis*-double bonds acting as interfacial "springs" that mitigate the capacity of cholesterol to reduce lateral elasticity (**Fig. 5**). For further discussion, *see* **refs.** *19* and *26–29*.
18. The charge contribution to the dipole moment can be almost completely eliminated by raising the p*I* of the subphase buffer (e.g., including ≥100 m*M* NaCl).
19. With highly condensed films consisting of pure raft lipids, for example, cholesterol or ceramide, the analysis of dipole potential within the context of constituent components, $\mu_\perp$ and ΔV_0, becomes complicated because of the small area change associated with large changes in surface potential. In such instances, useful information can still be obtained from simple comparison of dipole potentials without separate component analysis of $\mu_\perp$ and ΔV_0. A recent example involves ceramide and a series of related structural analogs *(40)*. The presence of the 4-5 *trans*-double bond or a triple bond in the sphingosine backbone of ceramide was found to make a large contribution to the dipole potential of the molecule. Interestingly, the biological activities of ceramide and its analogs are largely lost if the double bond is moved even one carbon atom along the sphingosine chain, a change that substantially alters the molecular dipole potential *(40)*.

5. Epilogue

Over the past two decades, surface balances have evolved from free-standing instruments for direct measurement of lipid lateral interactions (π-A) isotherm into basic platforms used in concert with other technological approaches.

1. *Supported monolayers/bilayers:* supported monolayers or even multilayers can be deposited stepwise onto a solid substrate surface by passing it through a lipid monolayer while keeping the surface pressure constant *(41,42)*. Such films have been used in combination with atomic force microscopy to study raft lipids mixtures *(43,44)*.

2. *Epifluorescence microscopy:* mounting of a surface balance onto the stage of an epifluorescence microscope enables the distribution patterns of trace amounts of lipids containing covalently attached reporter fluorophores to be monitored within the lipid monolayer. Because the lipid fluorophores tend to partition nonuniformly among monolayer regions of differing lateral packing density, their lateral distribution patterns can be visualized *(2,4–6,45,46)*. This technology has been especially useful for definitive detection of critical points and domains among various lipid mixtures involving cholesterol and has facilitated construction of phase diagrams for ternary mixtures of "raft" lipids *(47–49)*. For more on the application of epifluorescent microscropy to model membranes, *see* Chapter 12 in this book by Sarah Veatch.
3. *Fluorescence spectroscopy:* surface balances can now be equipped with laser and fiberoptic technology to directly measure lipid fluorophore intensity in lipid monolayers *(50)*. Monolayer studies of 4, 4-difluoro-4-bora-3a, 4a-diaza-s-indacene (BODIPY-PC) indicate that this technology holds promise for providing nanoscale insights into the nonideal mixing behavior of "raft" lipid mixtures, thus considerably increasing resolution compared with epifluorescence microscopy. Moreover, this same technology now provides a means for assessing the real-time adsorption of proteins to lipid monolayers from the subphase by monitoring the increase in the intrinsic emission intensity of tryptophan/tyrosine (or covalently attached extrinsic fluorophores) at the monolayer surface *(51,52)*.

Acknowledgments

We extend our thanks to Bill Momsen, Maureen Momsen, Nancy Mizuno, Jan Smaby, Dmitry Malakov, and Craig Jones for their many contributions to the advancement of monolayer technology used in the study of lipid–lipid and lipid–protein interactions. We are grateful to NIGMS45928, NHLBI49180, and The Hormel Foundation for their past, present, and future support.

References

1. Gorter, E. and Grendel, F. (1925) On bimolecular layers of lipoids on the chromocytes of the blood. *J. Exp. Med.* **41,** 439–443.
2. Rice, P. A. and McConnell, H. M. (1989) Critical shape transitions of monolayer lipid domains. *Proc. Natl. Acad. Sci. USA* **86,** 6445–6448.
3. Möhwald, H. (1990) Phospholipid and phospholipid-protein monolayers at the air/water interface. *Annu. Rev. Phys. Chem.* **41,** 441–476.
4. McConnell, H. M. (1991) Structures and transitions in lipid monolayers at the air-water interface. *Annu. Rev. Phys. Chem.* **42,** 171–195.
5. Keller, S. L., Pitcher, W. H., III., Huestis, W. H., and McConnell, H. M. (1998) Red blood cells lipids form immiscible liquids. *Phys. Rev. Lett.* **81,** 5019–5022.
6. McConnell, H. M. and Vrljic, M. (2003) Liquid-liquid immiscibility in membranes. *Annu. Rev. Biophys. Biomol. Struct.* **32,** 469–492.
7. Silvius, J. R. (2003) Role of cholesterol in lipid raft formation: lessons from lipid model systems. *Biochim. Biophys. Acta* **1610,** 174–183.

8. Simons, K. and Vaz, W. L. C. (2004) Model systems, lipid rafts, and cell membranes. *Annu. Rev. Biophys. Biomol. Struct.* **332,** 269–295.
9. Simons, K. and Ikonen, E. (1997) Functional rafts in cell membranes, *Nature* **387,** 569–572.
10. Brockman, H. L. (1994) Dipole potential of lipid membranes. *Chem. Phys. Lipids* **73,** 57–79.
11. Momsen, W. E., Smaby, J. M., and Brockman, H. L. (1990) The suitability of nichrome for measurement of gas-liquid interfacial tension by the Wilhelmy method. *J. Colloid Interface Sci.* **135,** 547–552.
12. Smaby, J. M. and Brockman, H. L. (1990) Surface dipole moments of lipids at the argon-water interface: Similarities among glycerol-ester-based lipids. *Biophys. J.* **58,** 195–204.
13. Bartlett, G. R. (1959) Phosphorus assay in column chromatography. *J. Biol. Chem.* **234,** 466–468.
14. Kaganer, V. M., Möhwald, H., and Dutta, P. (1999) Structural and phase transitions in Langmuir monolayers. *Rev. Mod. Phys.* **71,** 779–819.
15. Leathes, J. B. (1925) Role of fats in vital phenomena. *Lancet* **208,** 853–856.
16. Phillips, M. C. (1972) The physical state of phospholipids and cholesterol in monolayers, bilayers, and membranes. *Prog. Surf. Membr. Sci.* **5,** 139–221.
17. Smaby, J. M., Brockman, H. L., and Brown, R. E. (1994) Cholesterol's interfacial interactions with sphingomyelins and phosphatidylcholines: Hydrocarbon chain structure determines the magnitude of condensation. *Biochemistry* **31,** 9135–9142.
18. Smaby, J. M., Momsen, M., Kulkarni, V. S., and Brown, R. E. (1996) Cholesterol-induced interfacial area condensations of galactosylceramides and sphingomyelins with identical acyl chains. *Biochemistry* **35,** 5696–5704.
19. Smaby, J. M., Momsen, M. M., Brockman, H. L., and Brown, R. E. (1997) Phosphatidyl-choline acyl unsaturation modulates the decrease in interfacial elasticity induced by cholesterol. *Biophys. J.* **73,** 1492–1505.
20. Ali, S., Smaby, J. M., Brockman, H. L., and Brown, R. E. (1994) Cholesterol's interfacial interactions with galactosylceramides. *Biochemistry* **33,** 2900–2906.
21. Li, X.-M., Momsen, M. M., Smaby, J. M., Brockman, H. L., and Brown, R. E. (2003) Sterol structure and sphingomyelin acyl chain length modulate lateral packing elasticity and detergent solubility in model membranes. *Biophys. J.* **85,** 3788–3801.
22. Evans, E. and Needham, D. (1987) Physical properties of surfactant bilayer membranes: thermal transitions, elasticity, rigidity, cohesion, and colloidal interactions. *J. Phys. Chem.* **91,** 4219–4228.
23. Needham, D. (1995) Cohesion and permeability of lipid bilayer vesicles in *Permeability and Stability of Lipid Bilayers*, (Disalvo, E. A. and Simon, S. A. eds.), CRC Press, Boca Raton, FL, pp. 49–76.
24. Davies, J. T. and Rideal, E. K. (1963) *Interfacial Phenomena, 2nd ed.*, Academic Press, New York, pp. 265.
25. Behroozi, F. (1996) Theory of elasticity in two dimensions and its application to Langmuir-Blodgett films. *Langmuir* **12,** 2289–2291.

26. Smaby, J. M., Kulkarni, V. S., Momsen, M., and Brown, R. E. (1996) The interfacial elastic packing interactions of galactosylceramides, sphingomyelins, and phosphatidylcholines. *Biophys. J.* **70,** 868–877.
27. Li, X. -M., Momsen, M. M., Brockman, H. L., and Brown, R. E. (2002) Lactosylceramide: Effect of acyl chain structure on phase behavior and molecular packing. *Biophys. J.* **83,** 1535–1546.
28. Li, X. -M., Momsen, M. M., Smaby, J. M., Brockman, H. L., and Brown, R. E. (2001) Cholesterol decreases the interfacial elasticity and detergent solubility of sphingomyelins, *Biochemistry* **40,** 5954–5963.
29. Zhai, X., Li, X. -M., Momsen, M. M., Brockman, H. L., and Brown, R. E. (2006) Lactosylceramide: Lateral interactions with cholesterol. *Biophys. J.* **91,** 2490–2500.
30. Allende, D., Vidal, A., and McIntosh, T. J. (2004) Jumping to rafts: Gatekeeper role of bilayer elasticity. *Trends Biochem. Sci.* **29,** 325–330.
31. Ali, S., Brockman, H. L., and Brown, R. E. (1991) Structural determinants of miscibility in surface films of galactosylceramide and phosphatidylcholine: Effect of unsaturation in the galactosylceramide acyl chain. *Biochemistry* **30,** 11,198–11,205.
32. Smaby, J. M. and Brockman, H. L. (1992) Characterization of lipid miscibility in liquid-expanded monolayers at the gas-liquid interface. *Langmuir* **8,** 563–570.
33. Smaby, J. M. and Brockman, H. L. (1991) A simple method for estimating surfactant impurities in solvents and subphases used for monolayer studies. *Chem. Phys. Lipids* **58,** 249–252.
34. Rye, R. R. (1990) Electron irradiation of poly(tetrafluoroethylene): Effect on adhesion and comparison with X-rays. *Langmuir* **6,** 338–344.
35. Kulkarni, V. S. and Brown, R. E. (1994) Interactions of phospholipid bilayer vesicles with monomolecular films at the air-water interface. *Thin Solid Films* **244,** 869–873.
36. Constantino, C. J. L., Dhanabalan, A., and Oliveira, O. N., Jr. (1999) Experimental artifacts in the surface pressure measurement for lignin monolayers in Langmuir troughs. *Rev. Sci. Instr.* **70,** 3674–3680.
37. Brockman, H. L., Smaby, J. M., and Jarvis, D. E. (1984) Automation of surface cleaning and sample addition for surface balances. *J. Phys. E* **17,** 351–353.
38. Maggio, B., Carrer, D. C., Fanani, M. L., Oliveira, R. G., and Rosetti, C. M. (2004) Interfacial behavior of glycosphingolipids and chemically related sphingolipids. *Curr. Opin. Colloid Interface Sci.* **8,** 448–458.
39. Marsh, D. (1996) Lateral pressure in membranes. *Biochim. Biophys. Acta* **1286,** 183–223.
40. Brockman, H. L., Momsen, M. M., Brown, R. E., et al. (2004) The 4,5-double bond of ceramide regulates its dipole potential, elastic properties, and packing behavior. *Biophys. J.* **87,** 1722–1731.
41. Petty, M. C. (1996) *Langmuir-Blodgett Films: An Introduction.* Cambridge University Press.
42. Sackmann, E. (1996) Supported membranes: Scientific and practical approaches. *Science* **271,** 43–48.

43. Yuan, C., Furlong, J., Burgos, P., and Johnston, L. J. (2002) The size of lipid rafts: an atomic force microscopy study of ganglioside GM1 domains in sphingomyelin/DOPC/cholesterol membranes. *Biophys. J.* **82,** 2526–2535.
44. Yuan, C. and Johnston, L. J. (2001) Atomic force microscopy studies of ganglioside GM1 domains in phosphatidylcholine and phosphatidylcholine/cholesterol bilayers. *Biophys. J.* **81,** 1059–1069.
45. Von Tscharner, V. and McConnell, H. M. (1981) An alternative view of phospholipid phase behavior at the air-water interface. Microscope and film balance studies. *Biophys. J.* **36,** 409–419.
46. Weis, R. M. and McConnell, H. M. (1984) Two-dimensional chiral crystals of phospholipid. *Nature (London)* **310,** 47–49.
47. Veatch, S. L. and Keller, S. L. (2002) Organization in lipid membranes containing cholesterol. *Phys. Rev. Lett.* **89,** 268101-1–268101-4.
48. Radhakrishnan, A. and McConnell, H. M. (2002) Critical points in charged membranes containing cholesterol. *Proc. Natl. Acad. Sci. USA* **99,** 13,391–13,396.
49. Stottrup, B. L., Stevens, D. S., and Keller, S. L. (2005) Miscibility of ternary mixtures of phospholipids and cholesterol in monolayers and application to bilayer systems. *Biophys. J.* **88,** 269–276.
50. Dahim, M., Mizuno, N. K., Li, X. -M., Momsen, W. E., Momsen, M. M., and Brockman, H. L. (2002) Physical and photophysical characterization of a BODIPY phosphatidylcholine as a membrane probe. *Biophys J.* **83,** 1511–1524.
51. Momsen, W. E., Mizuno, N. K., Lowe, M. E., and Brockman, H. L. (2005) Real-time measurement of solute partitioning to lipid monolayers. *Anal. Biochem.* **346,** 139–149.
52. Hoang, K. C., Malakhov, D., Momsen, W. E., and Brockman, H. L. (2006) Open, microfluidic flow cell for studies of interfacial processes at gas-liquid interfaces. *Anal. Chem.* **78,** 1657–1664.

6

Electro-Formation and Fluorescence Microscopy of Giant Vesicles With Coexisting Liquid Phases

Sarah L. Veatch

Summary

Giant unilamellar vesicles (GUVs) are routinely used to study coexisting liquid phases in bilayer membranes. Liquid domains are observed in a wide variety of ternary GUV membranes containing phospholipids and cholesterol, and are thought to model raft domains in cell membranes. GUVs are attractive model systems because vesicles are easily prepared using standard and inexpensive laboratory equipment, and phase-separated vesicles can be visualized using optical microscopy. In this chapter, a detailed method is presented to form and view 10–100-μm diameter single-walled vesicles of charged or uncharged lipid mixtures. GUVs can be visualized by fluorescence microscopy and methods are presented to measure miscibility transition temperatures and to distinguish solid (gel) and liquid domains. Numerous experimental artifacts associated with GUV preparation and viewing are discussed, including the effects of nonideal growth conditions and perturbations of fluorescent probes and other impurities.

Key Words: Cholesterol; lipid rafts; liquid immiscibility; liquid-disordered phase; liquid-ordered phase; miscibility transition.

1. Introduction

Over the past several years, many groups have used fluorescence microscopy of giant unilamellar vesicles (GUVs) to study lipid organization in membranes containing coexisting phases *(1–9)*. In 2001, Dietrich et al. *(2)* visualized two macroscopic liquid phases in giant vesicles by fluorescence microscopy, and proposed that such systems provide a useful tool for studying lateral organization in cell membranes. Since then, much effort has gone into characterizing the miscibility transition in vesicles, and a great deal has been learned about phase diagrams *(10–12)* and other membrane physical properties *(7,13,14)*. In addition, investigators are developing methods to prepare vesicles

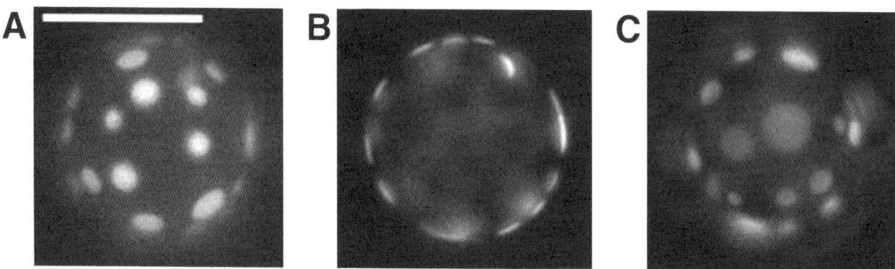

Fig. 1. A single GUV prepared through electroformation and imaged using fluorescence microscopy. Two coexisting phases are visible on the vesicle surface as indicated by the lateral distribution of a fluorescent probe. Using a ×40 objective, the (**A**) top surface, (**B**) equatorial plane, and (**C**) bottom surface of the same vesicle can be imaged independently. Bar = 20 µm.

containing membrane-bound proteins *(15–17)*, actin networks *(18)*, and complex lipid compositions *(19)* in an attempt to construct more biologically relevant model systems. Despite this effort, many questions remain regarding how to draw biological conclusions from results in vesicles *(6,20,21)*. Aside from its potential biological relevance, the miscibility transition in GUVs is a rich field of study from a material physics perspective. Liquid phases in vesicles exhibit many phenomena, like previously studied three-dimensional systems such as spinodal decomposition and critical fluctuations *(11,22)*.

The aim of this chapter is to present a detailed method for preparing electroformed GUVs and imaging them by fluorescence microscopy (**Fig. 1**). This is an attractive experimental strategy for studying lipid organization because it is easy to implement with standard and inexpensive laboratory equipment. Numerous experimental artifacts that are associated with GUV preparation and viewing will also be discussed. In particular, users should be aware that the composition (and thus the phase behavior) of electroformed vesicles could be dependent on the growth conditions. In addition, fluorescent probes and other impurities can modify transition temperatures. Finally, care must be taken to protect vesicles containing unsaturated lipids from oxidation, both in preparation and when viewing by fluorescence methods.

The method described next can be used to prepare 10–100-µm diameter single-walled vesicles of charged or uncharged lipids in water or nonionic solution. This method is based on two original articles *(23,24)*, and has been modified to increase the yield and compositional uniformity of the prepared vesicles *(6,11)*. This assay is easily accomplished using phosphatidylcholine (PC) lipid components in the liquid-crystalline state (e.g., egg PC, POPC, and diphytanol PC). It is also possible to prepare lipid mixtures that contain other lipid headgroups

Fluorescence Microscopy 61

(23), or that undergo phase transitions above room temperature. Giant vesicles can be prepared with significant fraction (>5%) of charged lipids without the application of an AC electric field (e.g., *see* **ref. 4**).

2. Materials

1. Lipids and probes in chloroform stock solutions. Recommended stock concentrations are 10 mg/mL for lipids and 1 mg/mL for probes. Store lipids at ≤ −20°C in tightly sealed glass vial with Teflon-lined lid. Verify lipid concentration before using.
2. Hamilton syringes (Hamilton Co.) (glass) for measuring volumes of organic solvent.
3. Indium tin oxide (ITO)-coated glass microscope slides or cover slips. Slides can be ordered from Delta Technologies, Limited (CG-90IN-S115). ITO-coated slides are also available from multiple other manufactures, or can be produced in-house by sputtering.
4. Multimeter capable of measuring resistance and AC voltage.
5. Simple mechanical vacuum pump with cold trap and chamber.
6. Teflon spacers (~5 × 25 × 0.39 mm^3). If a spacer is used of a different thickness, a different AC voltage may need to be applied.
7. Vacuum grease (e.g., Dow Corning silicone high-vacuum grease).
8. Parafilm. (American National Can)
9. Binder clips.
10. Purified water or nonionic osmolite (e.g., sucrose) for hydrating vesicles. Ionic buffers can be used when preparing vesicles with charged lipids.
11. Signal generator with sign wave output. Minimal power (current) is required.
12. Simple aluminum sheet metal "bus bars" for attaching multiple growth chambers in parallel. These can often be found in an instrument shop scrap bin.
13. Oven or incubator capable of maintaining 60°C.
14. 1-mL syringe with 22+ gauge needle for extracting vesicles from growth chamber.
15. Fluorescence microscope with ×10 and ×40 objective.
16. Digital camera for acquiring images.
17. Peltier-based temperature-controlled microscope stage. These are available commercially, or can be assembled using a peltier device, thermistor, and (proportional integral derivative) PID-based temperature controller *(25)*.

3. Methods

3.1. Preparing GUVs

1. Measure lipids and probe from chloroform stock solutions to desired lipid composition using glass Hamilton syringes. Lipids can be mixed in small glass test tubes that have been rinsed with chloroform. At least 0.25 mg of total lipid should be used for each preparation (*see* **Note 1**). Final concentration in chloroform should be approx 5–10 mg/mL. If excess solvent is present, it should be evaporated at this step. If the volume of solvent used in **step 2** is too great, the slide will cool significantly during lipid deposition. Generally, 0.5 mol% probe is incorporated into vesicles (*see* **Note 2**).

2. Prepare and clean the conducting face of two ITO-coated microscope slides. One 80 × 25 mm² microscope slide can be scored with a diamond-tipped scribe and broken into two 40 × 25 mm² pieces to make the two plates for the capacitor. The slides should be wiped free of dust and glass shards with a lint-free wipe. If desired, slides can be cleaned gently with ethanol, but more aggressive cleaners could damage the conducting layer. Measure the resistance across the surface to determine which side of the slide is conducting. The uncoated side will give a large resistance reading (>1 MΩ) whereas the coated (conducting) side will register a small resistance (<1 kΩ). It is preferable to use a new slide for each preparation. Results may vary if slides are cleaned and reused, and slides should not be reused more than five times, as ITO film degradation occurs.

3. Deposit the lipid/chloroform solution on the conducting face of a clean, ITO-coated slide using a glass pipet rinsed with chloroform. Hold the slide in one hand and push the solution evenly back and forth over the slide using the side of the pipet until it dries. This will spread the lipids into a thin and uniform film. Alternatively, lipids can be deposited by spin-coating, although results are similar with both methods. The deposited film can be rough and crystals can form during the drying process when high-melting temperature lipids are used (especially in the presence of cholesterol) *(26)*. For these samples, heat the slide with the rough dried lipid layer to 60°C for approx 5 min. Add a single drop of chloroform (also heated) and respread the lipid mixture. This results in a more uniform distribution of lipids. The lipid layer is uniform when it is mostly transparent and interference lines can be seen through the lipid layer. The slides can be labeled on the nonconducting face.

4. Place lipids under partial vacuum for at least 30 min to remove any excess solvent. Longer evaporation times are not necessary because only a small quantity of lipid is used. If leaving lipids for longer times (>2 h), backfill with argon or nitrogen to prevent oxidation of unsaturated lipids (*see* **Note 3**).

5. Coat the two long sides on the conducting face of both the lipid-coated and the lipid-free capacitor plates with thin layers of vacuum grease the width of the Teflon spacer (~5 mm) (*see* **Note 4**). The vacuum grease is used to join the slides with the Teflon spacers to reduce evaporation. Assemble capacitor so that the two conducting faces are facing inward and are separated by the two Teflon spacers (**Fig. 2A**). Maintain an overhang of more than 5 mm at the short edge so that slides can be easily attached to voltage. Use the binder clips to clamp the sandwich together.

6. Fill clamped chamber with purified water, sucrose solution, or desired buffer (*see* **Note 5**). Growth chambers can be filled using capillary action and gravity. Fill from one side as in **Fig. 2B**. Do not overfill so that water rinses through the other side. This will wash lipids from the chamber and reduce yield. Once the chamber is mostly full, seal off one open side with vacuum grease. Tap the growth chamber on a hard surface with the recently sealed end pointing down to free any air bubbles. Finish filling the chamber from the remaining open side and seal with vacuum grease without trapping air bubbles. Trapped air bubbles that migrate during vesicle growth will reduce yield. Wrap the recently sealed ends with a thin piece of parafilm to give the sandwich structural support and to further prevent

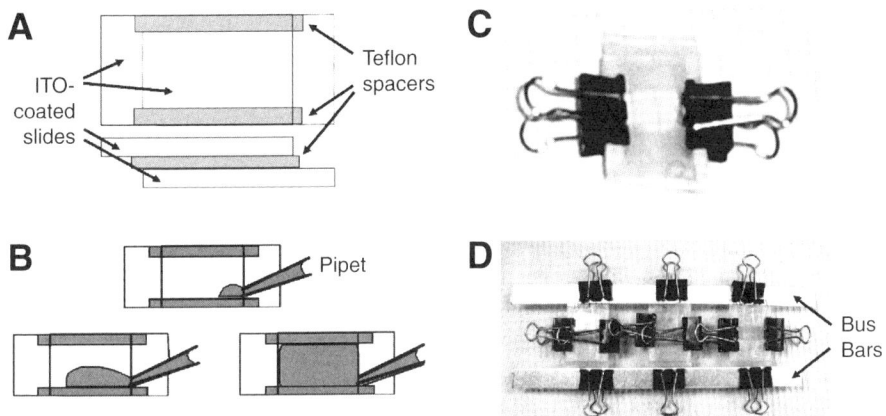

Fig. 2. (**A**) Schematic of capacitor growth chamber made of two ITO-coated glass slides and two Teflon spacers. (**B**) Growth chambers can be filled from one open end using capillary action. (**C**) Image of complete growth chamber. (**D**) Three chambers connected in parallel using aluminum bus bars.

evaporation. Again, ensure that enough space is left uncovered on the two ends to make good electrical connections. Clean excess vacuum grease from chamber ends with ethanol. A completed growth chamber is shown in **Fig. 2C**.

7. Make an electrical connection with either side of the capacitor. If only a single vesicle composition is prepared, an electrical connection can be made by directly attaching alligator-clip leads from the signal generator to the two capacitor places. If multiple compositions are prepared in parallel, line up the capacitors between two long aluminum bus-bars and connect with binder clips as in **Fig. 2D**. Each bus-bar can then be connected to the signal generator leads. Use a multimeter to ensure the connection between bus-bars and capacitor is low ($R < 150\ \Omega$). If the resistance is high, clean the bus-bars and slide ends with ethanol and remove any remaining vacuum grease or parafilm on the capacitor edge. Warped bus-bars can also increase connection resistance and should be replaced.

8. Place growth chambers in a 60–65°C oven or incubator and make final electrical connections to the signal generator (*see* **Note 6**). Set the signal generator to output a sinusoidal wave of 1 V and 10 Hz and verify at the capacitor leads (or bus-bar) with the multimeter. Alternate voltages and frequencies may also be used, but this is a good starting point for optimization. Grow vesicles in oven under this AC electric field for between 1 and 3 h (*see* **Notes 7–9**). Vesicle growth reaches a steady state after approx 1 h for most lipid compositions *(23)*. If desired, the growth progress can be monitored under a microscope (**Fig. 3A**).

9. When vesicle growth is complete, detach growth chamber from voltage and remove binder clips. Vesicles can be slowly removed from growth chamber using a plastic 1-mL syringe with a no. 22 needle (*see* **Note 10**). The 22-gauge needle is big enough to break apart the slides to eliminate suction when removing vesicles,

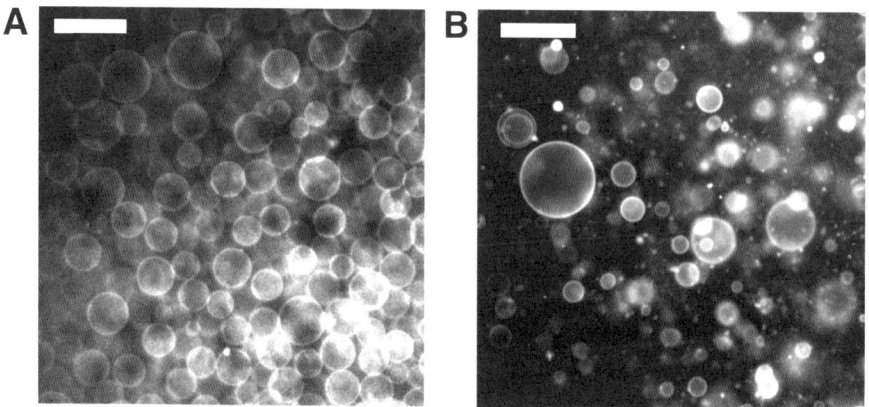

Fig. 3. Fluorescence micrographs of egg PC GUVs (**A**) Growing on an ITO-coated glass slide and (**B**) After extraction and dilution. Vesicles in **A** have a uniform radius because they are imaged at a specific distance from the slide surface. Bars = 50 μm.

whereas leaving the overall structure of the growth chamber intact. Narrower needles may lead to vesicle breakage from larger hydrodynamic forces. Draw the vesicle mixture into the syringe. Put the contents of the syringe into a secondary storage container. Vesicles can be diluted in warm growth solution, or any other isotonic solution heated to the growth temperature. Sudden changes in temperature may break delicate vesicles.

10. Vesicles should be stored in a secondary container at elevated temperature for not more than 4 h (*see* **Note 7**). If longer storage times are necessary, vesicles can be stored in the growth chamber at room temperature for up to 1 wk. Ideally, vesicles should be used immediately after preparation, as lipid degradation or other long-time equilibration effects (such as one component preferentially coating the container walls) can lead to erroneous results.

3.2. Viewing GUVs by Fluorescence Microscopy

1. Vesicles can be viewed by placing 10–60 μL of diluted vesicle solution between two cover slips and sealing with a layer of vacuum grease (*see* **Note 11**). The vacuum grease provides a thin spacer while sealing against evaporation. Thicker spacers will lead to convection currents in the vesicle sample, and gaps in the seal will lead to vesicle motion in a specific direction. If the spacing between cover slips is too small, or vesicles are large, GUVs can rest directly on the slide surface as in **Fig. 4E,F**. If excess pressure is applied, vesicles can burst and deposit on the glass surface.
2. Miscibility transition temperatures can be measured by monitoring the distribution of a fluorescent probe (*see* **Note 12**) while varying the temperature of a microscope stage (*see* **Notes 13–16**). At high temperatures, vesicles are in one uniform phase. When temperature is lowered, certain ternary lipid compositions containing high melting temperature (T_m) lipids, low T_m lipids, and a sterol such

Fluorescence Microscopy

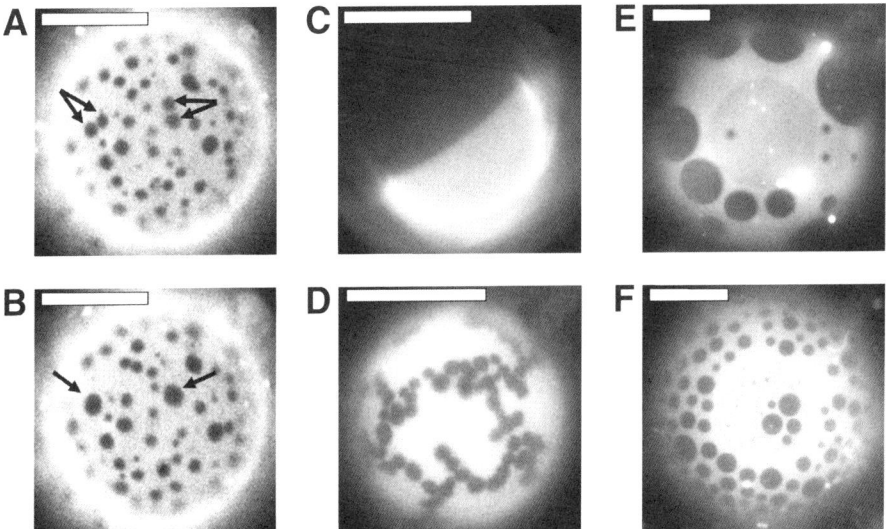

Fig. 4. **(A,B)** Liquid domains collide and coalesce to form larger circular domains. Micrographs are of the same vesicle imaged 1-s apart and colliding domains are indicated by arrows. **(C)** Most vesicles with liquid domains eventually completely phase separate into one bright and one dark region (in minutes to hours). **(D)** Gel domains can be faceted and do not coalesce on collision. **(E,F)** Large vesicles can rest and partially bind to the cover slip surface.

as cholesterol will separate into two liquid phases *(2,6,11)*. Initially, domains are small (~1 µm), and in time, domains collide and coalesce to form larger domains (**Fig. 4A,B**). In most cases, vesicles completely phase separate into one bright and one dark domain as in **Fig. 4C**.
3. Liquid phases are distinguished from solid phases based on domain morphology and lipid mobility. Circular liquid domains coalesce on collision to form a single, larger circular domain within seconds (**Fig. 4A,B**). In contrast, domains that contain gel phase lipids can be faceted and do not coalesce, or coalesce slowly on collision (**Fig. 4D**). Gel domains can be circular; therefore, it is important to view a collision event to distinguish gel and liquid domains.
4. Vesicles are best viewed with either a ×10 or ×40 objective. This makes it possible to view a large area of the vesicle surface (**Fig. 1**) with short integration times. A second ×2 lens can be used before the camera to properly match the image resolution afforded by the microscope objective to the pixel size of the camera CCD chip.

4. Notes
1. The final lipid composition of electroformed vesicles is dependent on the density and uniformity of the lipid film. Sparse and inhomogeneous lipid films lead to

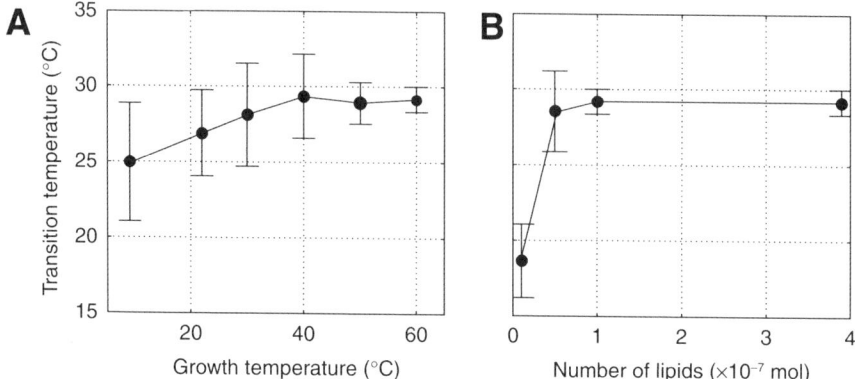

Fig. 5. GUV transition temperatures depend on the method of vesicle preparation for membranes of 1:1 dioleoylphosphatidylcholine/dipalmitoylphosphatidylcholine (DOPC/DPPC) +35% cholesterol. (**A**) Miscibility transition temperatures are lower and errors are larger when vesicles are grown at temperatures significantly below 60°C, particularly below the melting temperature of DPPC (41°C). (**B**) The miscibility transition temperature is lower in vesicles made from scant lipid films. For this lipid composition, 0.25 mg of lipids is equivalent to 3.9×10^{-7} moles. Lipid films were spread over an area of ~5 cm^2.

a broad distribution of compositions between vesicles in the final preparation. Vesicles made from sparse films may also have an average composition, which differs from that of the lipids mixed in solvent (**Fig. 5B**).

2. The presence of fluorescent probes can alter phase behavior of membranes with a miscibility transition, and should be used sparingly. By ^{2}H nuclear magnetic resonance, membranes with 0.5 mol% probe can have higher transition temperatures than membranes without probe, and coexisting phases can have different lipid compositions Veatch, S. L., S. S. Leung, R. E. Hancock, and J. L. Thewalt, *Fluorescent probes alter miscibility phase boundaries in ternary vesicles*, J. Phys Chem B, 2007. 111 (3):P. 502–4. Different probes can have different effects. Currently, the effect of probes is being characterized, but it is likely related to well-known effects of impurities on miscibility transitions in three-dimensional liquid mixtures (e.g., **ref. 27**).

3. Unsaturated lipids contain double bonds in the hydrocarbon chains, which are susceptible to oxidation in the presence of activated oxygen species. Monounsaturated lipids are stable as dried lipid films for hours, but should be kept in an oxygen-free atmosphere as much as possible if being stored for longer time periods. Polyunsaturated lipids must be handled exclusively in a low-oxygen atmosphere. Oxidation of unsaturated bonds can be accelerated in the presence of bright light (photo-oxidation).

4. Vacuum grease is used to seal the growth chamber and protect against water evaporation. It is possible that a small fraction of grease incorporates into GUVs and alters

Fluorescence Microscopy

phase behavior, although vacuum grease was not observed in vesicles analyzed by mass spectrometry (less than ~1%; unpublished observations). Also *see* **Note 11**.

5. Vesicles prepared through electroformation must be grown in nonionic solutions. In the presence of salts, vesicle yields are vanishingly low and salts can deposit on the capacitor plates. Vesicles can sometimes be grown in low-salt conditions and in low-p*I* buffers (such as HEPES as in **ref. 9**). Alternately, vesicles can be grown in a nonionic osmolite (such as sucrose), and then diluted in an isotonic salt or buffer solution (e.g., 300 m*M* sucrose diluted in 150 m*M* NaCl). Vesicles grown in sucrose and diluted in buffer sink in a gravitational field and give vesicles contrast in differential interference contrast (DIC) images *(28,29)* (*see also* **Note 17**).

6. Vesicles should be grown at a temperature well above the transition temperature of the highest T_m lipid in the sample. In some cases, miscibility transition temperature can exceed this T_m, and vesicles should be grown at still higher temperatures. If growth temperatures are comparable with transition temperatures, the vesicles produced have less uniform lipid compositions and the mean transition temperature can be altered as shown in **Fig. 5A**. When vesicles are grown at too high a temperature, water can significantly degas and form bubbles in the chamber. Also lipid degradation is accelerated at elevated temperature. Water (or buffer) can be degassed before use when higher growth temperatures are required (*see* **Note 18**).

7. Vesicles should not be grown or stored at elevated temperature for extended time periods, as degradation can occur. Vesicles ideally should be used 1 h after applying voltage, but can be held at 60°C for about 4 h. Vesicles stored at high temperature for longer time periods can have altered transition temperatures, and in some cases, gel domains form in vesicles that contain only liquid phases after short storage times. Some GUV protocols require overnight incubation at 60°C (e.g., *see* **refs. 4** and **17**). It is found that vesicles stored overnight at high temperature often have altered phase behavior.

8. When preparing vesicles with charged lipids, the preceding procedure can be used without the application of the AC field. As a result, vesicles with about more than 5 mol% charged lipids can be grown on nonconducting glass microscope slides with high yields. An advantage of the protocol outlined here for charged lipids is that yields are high and swelling is accomplished after shorter incubation times (1–2 h) than in previously published protocols *(4,17)*. AC fields can be used to accelerate swelling of lipid films with charged lipids, although in many cases yields are lowered and different vesicle morphologies are observed *(23)*. Charged vesicles swelled in the absence of AC fields can be grown in ionic solutions.

9. GUVs can be made of Phosphatidyl-ethanolamine PE lipids, but care should be taken to grow vesicles at a temperature well below the inverse hexagonal transition temperature. In some cases, incorporating Phosphatidyl-ethanolomine PE lipids reduces overall yield of GUV preparations *(23)*.

10. Vesicles should be detached from growth surface before investigating, as there is evidence that GUVs remain in contact with the lipid film and with neighboring vesicles *(24)*.

11. It is not known if the presence of vacuum grease influences the phase behavior of GUV membranes. In some cases, phase-transition temperatures differ between

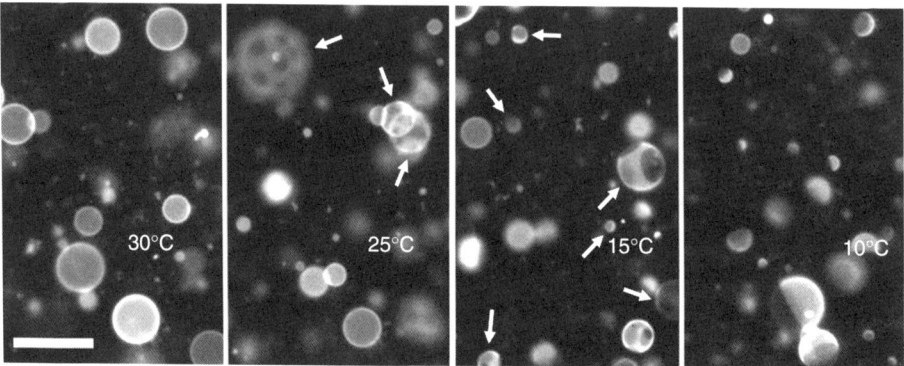

Fig. 6. Miscibility transition temperatures vary between vesicles in a single GUV preparation owing to slight variations in lipid composition. At 30°C, all vesicles of 2:1 diphytanoyl PC/DPPC + 50% cholesterol are in one uniform phase. At 25 and 15°C, a fraction of vesicles contain coexisting liquid phases (arrows). In this sample, all vesicles are phase separated at 10°C. Bar = 50 µm.

vesicles in the center and edges of the cover slip. This could be owing to the presence of impurities in membranes, to increased light levels near the edges from reflection, or some other process. Transition temperatures should be measured only in vesicles away from coverslip edges.

12. Most fluorescent probes preferably partition into the liquid-disordered (L_d) phase when L_d and liquid-ordered (L_o) phases are present. In general, short chain probes (e.g., DiIC12, Molecular Probes), probes with large fluorescent groups in the headgroup region (e.g., Texas-Red, Molecular Probes; DPPE), or lipids labeled in the hydrocarbon chains (e.g., Bodipy-PC), partition strongly into L_d, whereas longer chain probes (e.g., [Molecular Probes] DiIC18, Molecular Probes) and lipids labeled with small fluorescent groups in the headgroup region (e.g., [nitrobenzox-adiazole] NBD-DPPE, Avanti Polar Lipids), partition more weakly between phases. It is reported that certain hydrophobic probes preferably partition into the L_o phase (e.g., perylene), although partitioning is generally weaker and it is acknowledged that partition coefficients can be composition and temperature dependent *(7,17)*. Cholera toxin bound GM1 is often used as a marker of L_o phase lipids, but cholera toxin binding can alter miscibility transition temperatures *(17)*, and GM1 partitioning between phases *(30)*. It is also possible to use probes that partition into both phases, but for which spectral properties depend on the local lipid environment (e.g., Laurdan, Molecular Probes) *(31,32)*.

13. There is an inherent distribution of compositions between vesicles in a single GUV preparation *(6,10)*, thus care must be taken when measuring and reporting transition temperatures. The temperature range is reported over which vesicles phase separate in a given sample preparation (**Fig. 6**).

14. In GUVs, photo-oxidation of lipids in the presence of light can raise the miscibility transition temperature and induce phase separation in vesicles held at constant

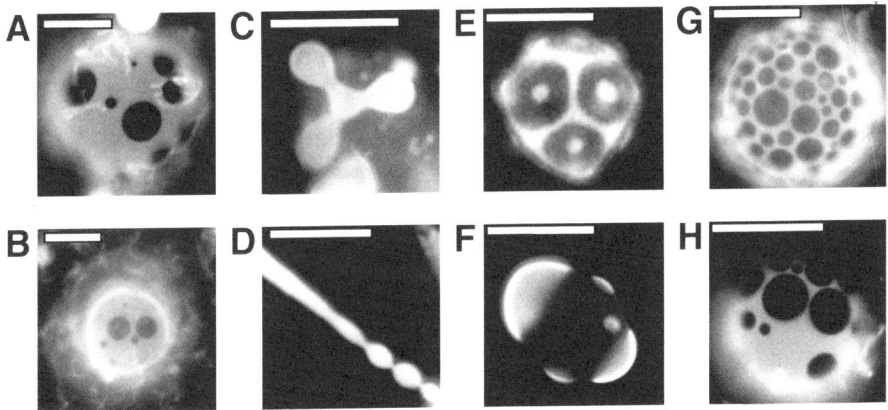

Fig. 7. GUV morphologies that are not typically discussed in the literature. (**A,B**) Strings or tubes are often connected to electroformed vesicles, and can increase in number with storage. (**C,D**) GUVs can form complex shapes even in the absence of phase separation. (**E,F**) Phase-separated GUVs can have interesting morphologies, owing either to differences in bending rigidity between phases and line tension effects. (**G**) Occasionally, domains do not coalesce even in the absence of visible curvature. (**H**) In select cases, domains can be unevenly distributed in vesicles for an unknown reason. Bars = 20 μm.

temperature *(6,10)*. Care must be taken to avoid photo-oxidation related effects, especially in the presence of bright light (as in confocal microscopes). For example, minimize light exposure time and use the lowest possible light levels. Also, ensure that transition temperatures are the same in different regions of the sample and avoid the use of unsaturated lipid probes. Oxygen scavengers can be used to decrease photo-oxidation, but controls should be conducted to ensure that these molecules do not alter vesicle phase behavior. Photo-oxidation makes it difficult to observe critical fluctuations for extended time periods in membranes containing unsaturated lipids *(22)*.

15. It is common to observe "strings" attached to vesicles after electroformation as in **Fig. 7A,B**. Strings may attach the vesicles to the growth surface during swelling, or can form after membranes are extracted. Nonspherical vesicles are also observed in some cases, even in the absence of phase separation as in **Fig. 7C,D**. Phase separation can stabilize complex structures in GUVs as in **Fig. 7E,H** *(7)*.
16. Immediately after preparation, GUV membranes are under tension *(24)*. In time (hours) vesicles relax and mechanical fluctuations are observed. Vesicle surface pressure can be modulated by varying the osmolarity of the surrounding liquid *(24)*. When vesicles are under tension, liquid domains rarely bud from the vesicle surface. Budding is more likely to occur if vesicles prepared in water are stored for extended times below their transition temperature, or if vesicles are immersed in a solution, which reduces surface tension.
17. GUVs grown in sucrose and resuspended in philological salt will deposit onto untreated glass surfaces as in **Fig. 8**. Cover slips can be blocked with bovine serum

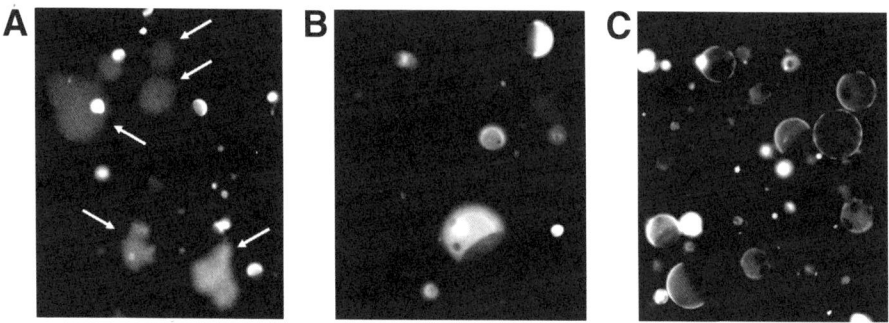

Fig. 8. (**A**) Vesicles grown in sucrose and resuspended in phosphate-buffered saline quickly deposit on untreated glass surfaces, whereas (**B**) Vesicles suspended in sucrose do not. Arrows in **A** indicate deposited vesicles. (**C**) Blocking the surface with bovine serum albumin or skim milk prevents vesicle deposition in the presence of physiological salt.

albumin, skim milk, or hydrophilic polymers to reduce interaction with the surface. Blocking agents can be lipophilic and can alter transition temperatures.

18. No published study has carefully characterized the composition of electroformed GUVs over a range of experimental conditions. Investigators should be conscious that certain lipids may not completely partition into vesicles. For example, it is difficult to grow vesicles containing certain lipids well above their chain-melting temperatures (such as long chain sphingomyelin (SM) lipids or ceramides), and low-growth temperatures can lead to a broader distribution of lipids between vesicles and altered mean transition temperatures (**Fig. 5**). Also, there is a limited solubility of certain lipids (e.g., some sterols) in bilayers that can depend on the method used to prepare vesicles *(33)*. Finally, there are concerns that the presence of the AC electric field itself could lead to degradation of lipid components. Lipid compositions are verified by mass spectrometry and thin-layer chromatography *(11,26)*, but more rigorous studies are necessary to probe for minor components in electroformed vesicles.

Acknowledgments

Many thanks to Sarah Keller and Ben Stottrup for assistance with this chapter.

References

1. Bagatolli, L. A. and Gratton, E. (2000) Two Photon Fluoreescence Microscopy of Coexisting lipid Domains in Giant Unilamellar Vesicles of Binary Phospholipid Mixtures. *Biophys. J.* **78(1),** 290–305.
2. Dietrich, C., Bagatolli, L. A., Volovyk, Z. N., et al. (2001) Lipid Rafts Reconstituted in Model Membranes. *Biophys. J.* **80,** 1417–1428.
3. Feigenson, G. W. and Buboltz, J. T. (2001) Ternary phase diagram of dipalmitoyl-PC/dilauroyl-PC/cholesterol: nanoscopic domain formation driven by cholesterol. *Biophys. J.* **80(6),** 2775–2788.

4. Korlach, J., Schwille, P., Webb, W. W., and Feigenson, G. W. (1999) Characterization of lipid bilayer phases by confocal microscopy and fluorescence correlation spectroscopy. *Proc. Natl. Acad. Sci. USA* **96(15)**, 8461–8466.
5. Parasassi, T., Gratton, E., Yu, W. M., Wilson, P., and Levi, M. (1997) Two-photon fluorescence microscopy of laurdan generalized polarization domains in model and natural membranes. *Biophys. J.* **72(6)**, 2413–2429.
6. Veatch, S. L. and Keller, S. L. (2005) Seeing Spots: Complex phase behaviour in simple membranes. *Biochem. Biophys. Acta* **1746(3)**, 172–185.
7. Baumgart, T., Hess, S. T., and Webb, W. W. (2003) Imaging coexisting fluid domains in biomembrane models coupling curvature and line tension. *Nature* **425(6960)**, 821–824.
8. Kaizuka, Y. and Groves, J. T. (2004) Structure and dynamics of supported intermembrane junctions. *Biophys. J.* **86(2)**, 905–912.
9. Puff, N., Lamaziere, A., Seigneuret, M., Trugnan, G., and Angelova, M. I. (2005) HDLs induce raft domain vanishing in heterogeneous giant vesicles. *Chem. Phys. Lipids* **133(2)**, 195–202.
10. Veatch, S. L. and Keller, S. L. (2002) Lateral Organisation in Lipid Membranes Containing Cholesterol. *Phys. Rev. Lett.* **89(26)**, 268,101.
11. Veatch, S. L. and Keller, S. L. (2003) Seperation of Liquid Phases in Giant Vesicles of Ternary Mixtures of Phospholipids and Cholesterol. *Biophys. J.* **85(5)**, 3074–3083.
12. Veatch, S. L., Polozov, I. V., Gawrisch, K., and Keller, S. L. (2004) Liquid domains in vesicles investigated by NMR and fluorescence Microscopy. *Biophys. J.* **86(5)**, 2910–2922.
13. Kahya, N., Scherfeld, D., Bacia, K., Poolman, B., and Schwille, P. (2003) Probing lipid mobility of raft-exhibiting model membranes by fluorescence correlation spectroscopy. *J. Biol. Chem.* **278(30)**, 28,109–28,115.
14. Baumgart, T., Das, S., Webb, W. W., and Jenkins, J. T. (2005) Membrane elasticity in giant vesicles with fluid phase coexistence. *Biophys. J.* **89(2)**, 1067–1080.
15. Kahya, N., Brown, D. A., and Schwille, P. (2005) Raft partitioning and dynamic behaviour of human placental alkaline phosphatase in giant unilamellar vesicles. *Biochemistry* **44(20)**, 7479–7489.
16. Bacia, K., Schuette, C. G., Kahya, N., Jahn, R., and Schwille, P. (2004) SNAREs prefer liquid-disodered over "raft" (liquid-ordered) domains when reconstituted into giant unilamellar vesicles. *J. Biol. Chem.* **279(36)**, 37,951–37,955.
17. Hammond, A. T., Heberle, F. A., Baumgart, T., Holowka, D., Baird, B., and Feigenson, G. W. (2005) Crosslinking a lipid raft component triggers liquid ordered-liquid disordered phase separation in model plasma membranes. *Proc. Natl. Acad. Sci. USA* **102(18)**, 6320–6325.
18. Liu, A. P. C. Personal communication.
19. Bernardino, S. J., Perez-Gil, J., Simonsen, A. C., and Bagatalli, L. A. (2004) Cholesterol rules: direct observation of the coexistence of two fluid phases in native pulmonary surfactant membranes at physiological temperatures. *J. Biol. Chem.* **279(39)**, 40,715–40,722.

20. Edidin, M. (2003) The State of Lipid Rafts: From Model Membranes to Cells. *Annu. Rev. Biophys. Biomol. Struct.* **32**, 257–283.
21. Munro, S. (2003) Lipid rafts: elusive or illusive? *Cell* **115(4)**, 377–388.
22. Veatch, S. L., Gawrisch, K., and Keller, S. L. (2006) Closed-loop miscibility gap and quantitative tie-lines in ternary membranes containing diphytanoyl PC. *Biophys. J.* **90(12)**, 4428–4436.
23. Angelova, M. I., Soleau, S., Mekeard, P., Faucon, J. F., and Bothorel, P. (1992) Preparation of giant vesicles by external AC electric fields. Kinetics and application. *Progr. Colloid Polym. Sci.* **89**, 127–131.
24. Mathivet, L., Cribier, S., and Devaux, F. (1996) Shape change and physical properties of giant phospholipid vesicles prepared in the presence of an AC electric field. *Biophys. J.* **70**, 1112–1121.
25. Veatch, S. L. (2004) Liquid Immiscibility in Model Bilayer Lipid Membranes, University of Washington, Seattle, WA, 172 p.
26. Veatch. S. L. and Keller, S. L. (2005) Miscibility phase diagrams of giant vesicles containing sphingomyelin. *Phys. Rev. Lett.* **94(14)**, 148,101.
27. Eckfeldt, E. L. and Lucasse, W. W. (1943) The Liquid–Liquid Phase Equilibria of the System Cyclohexane–ethyl Alcohol in the Presence of Various Salts as Third Components. *Phys. Chem.* **47(2)**, 164–183.
28. Ambroggio, E. E., Kim, D. H., Separovic, F., et al. (2005) Surface behavior and lipid interaction of Alzheimer beta-amyloid peptide 1-42: a membrane-distupting peptide. *Biophys. J.* **88(4)**, 2706–2713.
29. Evans, E., Hwinrich, V., Ludwig, F., and Rawicz, W. (2003) Dynamic tension spectroscopy and strength of biomembranes. *Biophys. J.* **85(4)**, 2342–2350.
30. Bacia, K., Schwille, P., and Kurzchalia, T. (2005) Sterol structure determines the separation of phases and the curvature of the liquid-ordered phase in model membranes. *Proc. Natl. Acad. Sci. USA* **102(9)**, 3272–3277.
31. Bagatolli, L. A. (2003) Direct observation of lipid domains in free standing bilayers: from simple t complex lipid mixtures. *Chem. Phys. Lipids* **122(1–2)**, 137–145.
32. Jin, L., Millard, A. C., Wuskell, J. P., et al. (2006) Characterzation and application of a new optical probe for membrane lipid domains. *Biophys. J.* **90(7)**, 2563–2575.
33. Huang, J., Bubolts, J. T., and Feigenson, G. W. (1999) Maximum solubility of cholesterol in phosphatidylcholine and phosphatidylethanolamine bilayers. *Biochim. Biophys. Acta* **1417(1)**, 89–100.

7

Fluorescence Correlation Spectroscopy

Kirsten Bacia and Petra Schwille

Summary

Fluorescence correlation spectroscopy (FCS) is a technique that allows for an extremely sensitive determination of molecular diffusion properties, down to the level of single molecules. It thus provides an attractive alternative to FRAP, requiring much less laser power and lower concentrations of fluorophores. FCS has recently been applied on live cells, and in comparison on domain-forming model membrane systems, to systematically study the influence of cholesterol on local membrane structure by investigating the mobility of selected lipid probes. The findings demonstrate the ability of FCS to sensitively distinguish between different local lipid structures, and emphasize the value of model systems for understanding membrane dynamics in general.

Key Words: Confocal microscopy; diffusion; fluorescence fluctuations; giant unilamellar vesicles; liquid-disordered; liquid-ordered; single molecules.

1. Introduction

Fluorescence correlation spectroscopy (FCS) was first devised in the early 1970s as a fluctuation analysis to be applied on small ensembles of fluorescently labeled molecules *(1)*. Its underlying idea is to confine the observation volume by focusing light down to the optical resolution limit, and simultaneously work at concentrations well below micromolar. In this way, molecular dynamics such as diffusion, structural relaxation, and interaction between different species can be observed with extremely high accuracy and at minimal interference with the molecular system. In the past years, the instrumental similarity between FCS, applied in confocal geometries, and laser scanning microscopy has triggered a series of intracellular FCS studies, which first yielded exact mobility analysis in cellular compartments (e.g., **refs.** *2–4*) and have recently been aiming at quantitative studies of enzymatic activity *(5)* and protein–protein interactions *in situ (6–8)*.

With respect to the analysis within cellular substructures, the plasma membrane, probably together with the nucleus, is one of the best environments to apply

From: *Methods in Molecular Biology, vol. 398: Lipid Rafts*
Edited by: T. J. McIntosh © Humana Press Inc., Totowa, NJ

FCS to, because their sizes are comfortably larger than the focal measurement volume of roughly a femtoliter (μm^3). Studies of membrane diffusion were among the first applications of FCS *(9–11)*. Recently, the raft hypothesis has triggered a large number of studies devoted to the possibility of distinguishing different modes of molecular mobility, i.e., diffusion coefficients, on heterogeneous membranes using various optical techniques *(12–14)*. It is widely assumed that on association with lipid rafts, proteins and lipids undergo considerable changes in their diffusion characteristics, in the extreme case a full immobilization. FCS would in principle be a perfectly suited technique to address this phenomenon, with the only limitation that rafts are assumed to be well below the optical resolution limit, and that a standard FCS measurement on a fragmented plasma membrane always spatially averages over rafts and the surrounding "non-raft" membrane patches.

To shed light on the principal difference in local fluidity between membranes of different lipid compositions, e.g., more fluid (liquid-disordered) domains of unsaturated glycerophospholipids, and liquid-ordered domains containing sphingolipids and cholesterol, model membrane systems have recently been used to help the understanding of cellular FCS data, and to establish a framework of reference for quantitative measurements on membrane systems in general. The most prominent of these model systems are giant unilamellar vesicles (GUVs), which have turned out to be invaluable not only as a reference system for FCS diffusion measurements on free-standing membranes, but also to understand the phenomenon of domain formation from a fundamental physico-chemical standpoint. In GUVs with sizes up to tens of micrometers, the so-called "canonical raft mixture" of cholesterol, sphingomyelin (SM), and 1,2-dioleoyl-*sn*-glycero-3-phosphocholine (DOPC) *(15)* yields at room temperature easily distinguishable domains, which can be assigned with a raft-like, liquid-ordered (L_o), and a more fluid liquid-disordered (L_d) membrane state. Parking the FCS measurement spot on either of the domains, a difference in local mobility of selected probes—1,1-dioctadecyl-3,3,3′,3′-tetramethylindocarbo cyanine dye diI as L_d marker, cholera toxin bound to GM1 as L_o marker—of one to two orders of magnitude can be easily observed **(Fig. 2)**. Cholesterol depletion by methyl-β-cyclodextrin (MβCD), which results in a clear dissolution of the domains on macroscopic scale, induces a shift of molecular diffusion such that the two probes now show similar mobility. The observation of reduced mobility of diI after MβCD treatment correlates with similar results on live cells treated in the same way **(Fig. 3)** *(14)*. Owing to the importance of the GUV model as a reference system for domain identification by FCS, this chapter covers the preparation of both GUVs and cells for FCS as well as the essential steps for FCS analysis on membranes in general.

2. Materials
2.1. Cell Culture and Labeling

1. Adherent cell lines rat basophilic leukemia (RBL-2H3) and human embryonic kidney (HEK293), obtained from the American Type Culture Collection.
2. Minimum essential medium (with Earle's salts, with L-glutamine), Dulbecco's modified Eagle's medium, and L-glutamine from Gibco/Invitrogen (Carlsbad, CA). Mycoplex fetal calf serum from PAA Laboratories (Pasching, Austria).
3. A 16 mg/mL (w/v) solution of MβCD (from Sigma St. Louis, MO) in water is prepared fresh before use. For an average degree of substitution of 1.8 methyl groups per glucose, this corresponds to approx 12 mM.
4. Dialkylcarbocyanine fluorescent probes with different alkyl tails and spectral properties, e.g., "diI" (1,1′-dioctadecyl-3,3,3′,3′-tetramethylindocarbocyanine perchlorate), "diO" (3,3′-dioctadecyloxacarbocyanine perchlorate), and "diD" (diIC$_{18}$(5), 1,1′-dioctadecyl-3,3,3′,3′-tetramethylindodicarbocyanine perchlorate) from Molecular Probes/Invitrogen. Fluorescent lipid analogs are dissolved at approx 1 to 1.5 mM in ethanol and stored at 4°C. Cholera toxin B subunit (ctxB) labeled with Alexa Fluor dyes from Molecular Probes/Invitrogen (Carlsbad, CA) (*see* **Note 1**). Cholera toxin solutions (2 mg/mL in phosphate buffered saline [PBS] or water) are stored as single-use aliquots at –20°C to avoid repeated freeze–thaw cycles.
5. Albumin from bovine serum (BSA) (from Fluka St. Louis, MO).
6. PBS: 137 mM NaCl, 2.7 mM KCl, 8.1 mM Na$_2$HPO$_4$, and 1.5 mM KH$_2$PO$_4$, (pH 7.2).
7. For optimal stability, cover slips (no.1 thickness, 25 mm diameter, from Menzel-Gläser, Braunschweig, Germany) and a custom-made holder are used (*see* **Note 2**).

2.2. Giant Unilamellar Vesicles

1. Lipids, such as DOPC, N-stearoyl-D-*erythro*-sphingosylphosphorylcholine (stearoyl SM), and cholesterol are obtained from Avanti Polar Lipids (Alabaster, AL). Ganglioside GM1 from Calbiochem/Merck (EMD) (Darmstadt, Germany). Lipids are stored in chloroform in glass vials with teflon-lined caps under argon or nitrogen at –20°C (*see* **Note 3**).
2. Chloroform (Fluka) and methanol (Merck) are used as solvents. To preserve lipids and dyes, it is important that chloroform decomposition into reactive products is prevented (chloroform stabilized with 1% ethanol and stored in the dark). Follow safe procedures for solvent handling and disposal.
3. Electroformation requires the aqueous solution to be of low ionic strength. Sucrose solution may be used and later exchanged by a buffer. Herein, a 12 mM sucrose solution is used, matching the osmolarity of the 12 mM MβCD solution.
4. A commercial flow-chamber (RC-21, Warner Instruments, Hamden, CT) or specially designed chamber, anodized for electrical insulation. Round cover slips of no.1 thickness and 25-mm diameter from Menzel-Gläser, custom-coated on one side by Gesim (Großerkmannsdorf, Germany) with indium-tin-oxide (ITO).
5. 10-µL Hamilton syringe (Hamilton, Reno, NV), vacuum grease (Glisseal, Borer, Switzerland), polyethylene tubing (Warner Instruments), disposable syringes, and

hypodermic needles. Copper conducting tape (3M, St. Paul, MN), a pulse generator, and a multimeter (Conrad Electronic, Hirschau, Germany).

2.3. Confocal Microscopy and FCS

Confocal imaging greatly facilitates performing FCS measurements in selected locations on phase-separated GUVs. Commercial setups combining confocal microscopy and FCS are manufactured by different companies. The protocol described herein was devised for the ConfoCor2 system (Carl Zeiss, Jena, Germany).

3. Methods
3.1. Cell Culture and Labeling

1. RBL cells are cultured in minimum essential medium, 15% fetal calf serum, nonessential amino acids, 2 mM glutamine, and 1 mM sodium pyruvate; human embryonic kidney cells in Dulbecco's modified Eagle medium, 10% fetal calf serum, 2 mM glutamine, 100 U/mL penicillin, and 100 µg/mL streptomycin, in a humidified incubator at 37°C and 5% CO_2. Cells are seeded onto cover slips in six well plates (Nalge-NUNC, Rochester, NY) with phenol-red free media 24–72 h before measurements.
2. Cells can be treated to interfere with membrane composition, topology, or cytoskeleton. For example, MβCD is commonly used to deplete cholesterol, although it may have secondary effects. For cholesterol depletion, cells are washed three times with PBS and treated with 12 mM MβCD in serum-free media in the incubator for 30 min.
3. Immediately before measurements, cells adhering to the cover slip are washed three times with PBS and labeled with a fluorescent lipid analog. For labeling, a small volume of lipid analog stock and 1 mL of a 1 mg/mL BSA in PBS solution in a glass vial are preheated to 37°C. While vortexing the BSA/PBS solution, a few microliters of lipid analog stock are injected, resulting in a clear labeling solution, which is applied fresh to the cells for approx 1 min at room temperature. Concentrations of the labeling solution and labeling times are adjusted to produce a low degree of labeling for FCS or stronger labeling for confocal imaging (*see* **Note 4**).
4. Stained cells are washed three times with PBS and mounted in a custom-made holder with phenol-red free media, supplemented with 10 mM HEPES buffer. For cholesterol-depleted cells, 12 mM MβCD is added to serum-free media to avoid reversal of the treatment.

3.2. Giant Unilamellar Vesicles

Electroformation can be performed with platinum wires as electrodes. However, GUVs adhering to wires are typically too far above the cover slip for using objectives of high numerical aperture with short working distance. ITO-coated cover slips are electrically conductive and optically transparent, allowing GUV electroformation close to the objective directly in the observation chamber

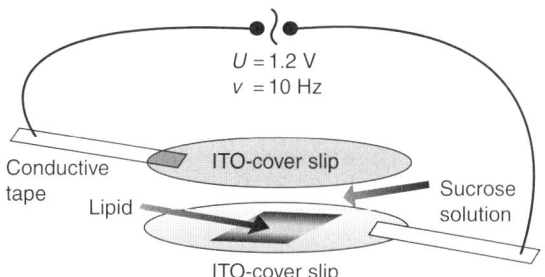

Fig. 1. Configuration for GUV electroformation. Schematic view of ITO-cover slip configuration. The flow chamber is not shown.

(*see* **Note 5**). For lipid mixtures involving lipids with high-melting temperature, electroformation is performed above the melting temperature of the highest melting lipid in an oven or on a heating block (here 65°C).

1. Pieces of copper tape are attached to the conductive side of the ITO-cover slips (*see* **Note 6**).
2. The insert of the chamber is prepared with vacuum grease. In the case of high-melting lipids, the sucrose solution, ITO-cover slips, and the metal parts of the flow chamber are preheated.
3. The lipids are mixed at the desired molar ratio, e.g., DOPC:SM:Chol (1:1:1). A fluorescent lipid analog is added at 0.001–0.01% (for FCS) or 0.1% (for imaging). GM1 is added at 0.1%. The final total lipid concentration is around 10 mg/mL in chloroform:methanol (2:1) (*see* **Note 7**).
4. Using a clean Hamilton syringe, approx 5 µL of lipid mixture are spread on the conductive side of a cover slip. If cover slips are not heated, they are placed in a desiccator for 30 min to ensure complete evaporation of the solvent.
5. The flow chamber and the ITO-cover slips are assembled with the conductive sides facing the void (**Fig. 1**).
6. Using a syringe and polyethylene tubing, the chamber is carefully filled with the sucrose solution in one step, avoiding air bubbles.
7. The copper tape is connected to the pulse generator, set at a sinosoidal alternating voltage of 10 Hz frequency and 1.1 to 1.3 V (*see* **Note 8**).
8. GUV electroformation is allowed to proceed for 1 to 3 h.
9. Imaging and FCS measurements are performed on GUVs directly in the flow chamber. The additional port is used for injecting MβCD solution (12 mM) or ctxB (5 µg/mL in 12 mM sucrose).

3.3. Confocal Microscopy and FCS

In GUVs, phase separation into microscale domains that become selectively enriched with a fluorescent lipid analog can be visualized with confocal microscopy. In many, but not all cases, diI prefers the L_d over the L_o phase and ctxB is bound to GM1 the L_o phase. Phase assignment is therefore clarified by FCS diffusion measurements in both phases. In GUVs, the focus is selectively placed in one type of domain. In cells, the focus presumably covers both rafts and surrounding bilayer. All measurements are performed at room temperature and cells are used for at most 1 h to avoid interference from internalized fluorescent lipid probes.

3.3.1. Fluorescence Correlation Spectroscopy

1. *Adjustment of the FCS setup:* using the same type of cover slip as for the intended GUV or cell measurements, the position of the pinhole and the correction colar setting of the objective are optimized.
2. *Calibration measurement:* a quick calibration measurement with free dye in water is performed to check the performance of the setup. Important parameters are:
   a. The particle brightness η obtained at a defined excitation power or at saturation (counts-per-molecule). It is determined by dividing the mean fluorescence intensity by the number of particles (N_{eff}) (*see* **Note 9**).
   b. The structure parameter (S), reporting on the shape of the detection volume (*see* **Note 9**).
   c. The diffusion time (τ_{diff}) of the dye in the lateral dimension (*see* **Note 9**).
3. *Focus positioning on the GUV or cell membrane:* the focus is centered on the lower or upper flat membrane of the cell or GUV. To this end, the time-averaged fluorescence intensity is recorded while scanning the membrane slowly through the focus in axial direction. The membrane is then positioned at the location of maximum fluorescence count-rate (*see* **Note 10**).
4. *Laser excitation power:* a series of FCS measurements at different laser powers directly on the membrane system is used to determine the power at which maximum brightness (counts-per-molecule) is obtained, but no bleaching artifacts are yet incurred (*see* **Note 11**). Laser power values (in watts) are determined with a powermeter (Newport, Irvine, CA).
5. *Pinhole size:* a pinhole size of 90 µm is chosen, which is well in the plateau region of the counts-per-molecule curve and still yields membrane FCS curves that fit reasonably well to the model equation for two-dimensional diffusion (*see* **Note 12**).
6. *Measurement duration, number of acquisitions, averaging:* acquisitions of 100 s are typically used for membrane measurements and raw data (photon arrival times) is additionally recorded. Raw data enables to repeat the correlation offline, excluding singular events, such as bright-dye aggregates. For measurements of fast diffusion in solution, several short (e.g., 10 s) measurements may be acquired, overlaid to identify singular events, and averaged (*see* **Note 13**).
7. *Fluorophore density, background:* samples labeled with diI to yield count-rates around 200 kHz have been found to work well. At very high count-rates, avalanche photodiode detection is nonlinear. At very low fluorophore densities, it is difficult to position the focus accurately on the membrane (*see also* **Note 10**). Finally, the count-rate of the labeled sample should be far above the autofluorescence level, which is determined from an unlabeled sample.

8. *Data processing:* the following model equation is fitted to the experimental FCS curves to obtain the τ_{diff} of a freely diffusing lipid in a membrane (two-dimensional case) (**Eq. 1**). It contains an exponential term to account for a photophysical blinking effect of the fluorophore.

$$G_{diff}(\tau) = N_{eff}^{-1} \frac{1 - F_{blink} + F_{blink} \cdot \exp(-\tau/\tau_{blink})}{1 - F_{blink}} \cdot (1 + \tau/\tau_{diff})^{-1} + \text{const} \quad (1)$$

where N_{eff} is the average number of particles in the effective detection volume; F_{blink} and τ_{blink} are the dark fraction and relaxation time of the blinking process, respectively. Depending on the range of τ values included in the fitting process, another exponential term may be necessary to account for both triplet and isomerization blinking encountered in cyanine dyes *(16)*. An anomalous diffusion model may be more appropriate for describing diffusion in native cell membranes *(11)*. Fitting is performed with a Levenberg–Marquardt algorithm implemented in the ConfoCor2 instrument software or in a scientific data analysis software (Origin, Originlab, Northampton, MA). In order to calculate the diffusion coefficient (D) from τ_{diff} according to

$$\omega_o^2 = 4D\tau_{diff} \quad (2)$$

the detection volume lateral radius (ω_o) has to be determined by calibration (*see* **Note 9**). For measurements performed on a setup with a stable detection volume, FCS curves can be summarized in a single plot for visualization (**Figs. 2** and **3**) after amplitude-normalization (*see* **Note 9**).

4. Notes

1. Alternatively, ctxB (from Calbiochem/Invitrogen) is labeled with an amino-reactive dye of choice and purified on a gel-filtration column (10 DG, Bio-Rad Hercules, CA).
2. As an alternative to a special holder, Labtek chambered coverglass (Nalge-NUNC) or glass bottom culture Petri dishes (MatTek, Ashland, MA) can be used.
3. Thin-layer chromatography is used to assess lipid purity. Samples are spotted on silica gel 60 plates (Merck), developed with an appropriate solvent system and visualized by spraying with 20% sulfuric acid and charring at 180°C.
4. Mixing the fluorescent lipid analog with the BSA/PBS at elevated temperature greatly reduces the occurrence of dye aggregates and improves homogeneity of the cell labeling. Cells are incubated with the labeling solution only shortly at room temperature to minimize staining of internal membranes.
5. This electroformation protocol produces closely and stably packed GUVs above a fluorescent lipid layer on the ITO-cover slip. It is well suited for confocal techniques. For applications requiring low out-of-focus background, GUVs are transfered into a new chamber after electroformation.

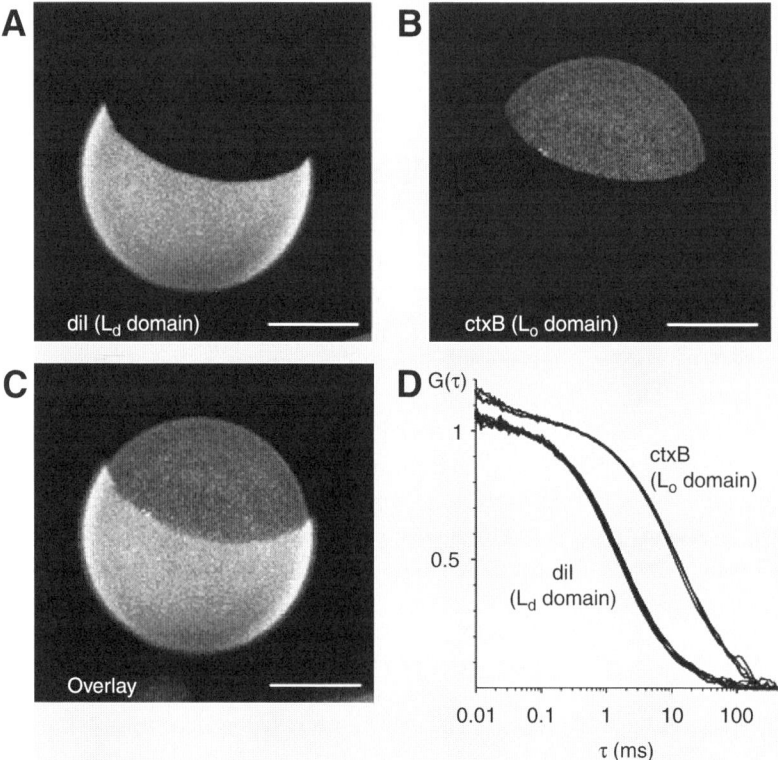

Fig. 2. Three-dimensional confocal reconstruction of a GUV made up of DOPC:SM: Chol, 1:1:1, with 0.1% GM1 and 0.1% diI. Bar = 10 μm. **(A)** Red channel showing diI fluorescence enriched in one phase. **(B)** Green channel showing fluorescence from Alexa-488-labeled ctxB bound to GM1, enriched in the other phase. **(C)** Overlay. **(D)** FCS measurements of lipid analog (diI) and ctxB (ctxB-488) show distinct diffusion characteristics and allow assignment of diI to the more fluid L_d phase (τ_{diff} = 1.5 ms, $D = 5 \times 10^{-8}$ cm^2/s) and ctxB-GM1 to the more ordered L_o phase (τ_{diff} = 12 ms, $D = 5 \times 10^{-9}$ cm^2/s).

6. An ohmmeter (multimeter, *see* **Subheading 2.2.**) is used to identify the conductive side of the cover slip and to check conductivity after attaching the copper tape.
7. Note that concentrations of lipid solutions in organic solvent change owing to solvent evaporation. Small volumes of highly concentrated lipids in large vials are to be avoided and wherein possible, aliquots of defined total lipid amount instead of defined concentration are used. Variability between GUV lipid composition in samples is estimated to be approx 2% *(15)*. Some lipids may undergo reactions during GUV production and laser microscopy.
8. A voltage meter (multimeter, *see* **Subheading 2.2.**) is used to check the applied voltage and exclude a short circuit.

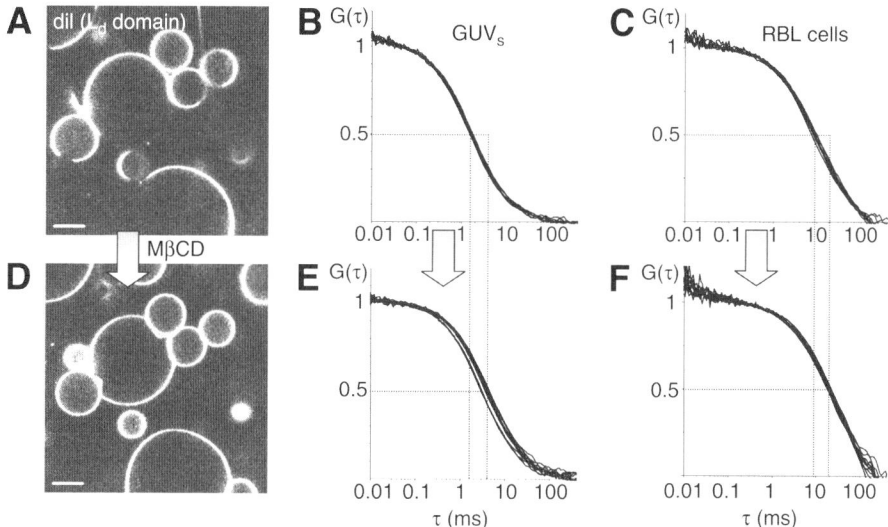

Fig. 3. Effect of cholesterol depletion on GUVs and cells. Upper panels (**A–C**) are before cholesterol depletion, lower panels (**D–F**) are after cholesterol depletion with MβCD. (**A**) Confocal fluorescence image of GUVs from DOPC:SM:Chol, 1:1:1, labeled with diI. Cholesterol depletion with MβCD (**panel D**) removes visible phase separation. Bar = 10 μm. (**B**) FCS measurements performed on diI in the phase of enrichment (L_d phase) in the same type of GUVs. Cholesterol depletion (**panel E**) significantly slows down diffusion from τ_{diff} = 1.5 ms to τ_{diff} = 4 ms ($D = 5 \times 10^{-8}$ cm²/s to $D = 2 \times 10^{-8}$ cm²/s). (**C**) FCS measurements performed on RBL cells, labeled with diI. Cholesterol depletion (**panel F**) significantly slows down diffusion from τ_{diff} = 9 ms to τ_{diff} = 19 ms ($D = 8 \times 10^{-9}$ cm²/s to $D = 4 \times 10^{-9}$ cm²/s).

9. The following model equation (three-dimensional diffusion) is used for fitting the dye calibration measurement:

$$G_{diff}(\tau) = N_{eff}^{-1} \frac{1 - F_{blink} + F_{blink} \cdot \exp(-\tau / \tau_{blink})}{1 - F_{blink}} \cdot (1 + \tau / \tau_{diff})^{-1}$$
$$\times \left[1 + \tau/(S^2 \tau_{diff})\right]^{-1/2} \quad (3)$$

where $S = z_o/\omega_o$ is the ratio of the axial (z_o) and lateral dimension (ω_o) of the detection volume, called structure parameter, and the other parameters have the same meanings as in **Eq. 1**. The τ_{diff} of the dye is related to the lateral radius of the detection volume (ω_o) according to **Eq. 2** and typically used for detection volume calibration. Rhodamine dyes have been commonly assumed to have a diffusion coefficient of $D = 2.8 \times 10^{-6}$ cm²/s at room temperature in water. If the diffusion time of the free dye is stable throughout all experiments, amplitude-normalized correlation curves and diffusion times from different membrane measurements (for instance obtained in different phases) can be compared directly. (For amplitude-normalization of FCS curves, $G(\tau)$ values are multiplied with N_{eff}, obtained from the fit.) If a setup with considerable day-to-day variability of the detection volume size and shape is used, ω_0 has to be accounted for, and only diffusion coefficients can be compared. For small detection volumes, routine recalibration with free dye can be difficult because of the interference of triplet blinking and nonideal detection volume shapes. In this case, a membrane diffusion sample (e.g., diI in DLPC [1,2-Dilauroyl-*sn*-Glycero-3-Phosphocholine]–GUVs) may be more suitable as a standard. Note also that the detection volume size has to be determined separately for different wavelengths.

10. Measuring FCS on a membrane means measuring on a nonisotropic sample. If the focus is not exactly centered or if the membrane undulates strongly, the diffusion time will be biased towards larger values. Undulations may be recognized from the concentration-dependent nature of their effect on the FCS curves: When the membrane is labeled with high marker concentrations and monitored at very low laser intensities, single fluorophore diffusion transits contribute less to the FCS curve and undulations are more prominent. In contrast, labeling in the single molecule regime minimizes the contribution from undulation-derived fluorescence fluctuations. However, working in the single molecule regime does not eliminate the bias in the diffusing time that arises from the temporary mispositioning. Slow membrane undulations can make it impossible to derive conclusions about anomalous diffusion or to assess very slow membrane diffusion processes. New approaches are being developed to deal with undulating membranes *(17)*.

11. Photobleaching affects FCS curves in different ways:
    a. Artifactual reduction in measured diffusion times. To avoid erroneous diffusion times, measurements have to be taken at laser powers below the bleaching threshold, which depends on the setup (focal volume), photostability of the dye, and dye mobility. The diffusion-time reducing effect of photobleaching is independent of fluorophore concentration.
    b. Photobleaching causes a continuous reduction of the fluorescent count-rate, which leads to an additional decay in the correlation curve. This decay may be misinterpreted as another diffusion decay, or, if its characteristic time is close to that of the diffusion, may lead to an artifactually increased fit value for the diffusion time. The count-rate decay affects measurements more seriously for higher particle concentrations, and the exact effect on the correlation curve depends on the normalization scheme used by the correlator.
    c. Note that if there is more than one species in the sample with different mobilities (e.g., markers in different lipid environments), bleaching (also using a software "prebleach" function) preferentially depletes the less mobile fraction.

12. Excessively large pinholes cause the Gaussian approximation of the detection volume to fail and FCS curves to deviate from the model equation. On the other hand, molecular brightness (counts-per-molecule) and therefore signal-to-noise become reduced at small pinhole sizes. The optimal pinhole size determined by Hess and

Webb *(18)* for the ConfoCor2 setup, 1.2 NA C-Apochromat objective, and 550 nm wavelength is 53 μm. Here, 90 μm pinhole is used for GUV and cell measurements in favor of approx 50% increased counts-per-molecule.

13. Short measurements provide less information about the long time-scale tail of the FCS curve. Therefore, it is preferred to record long FCS measurements (acquisition time ≥1000 × τ_{diff}). Rather than excluding additional slow fluctuations (e.g., from undulations) from the FCS curve by taking short acquisitions, the resulting additional decay in the FCS curve is accounted for by an offset in the fitting procedure.

Acknowledgments

The authors acknowledge support by the Europäische Fonds für Regionale Entwicklung (4212/04-01).

References

1. Magde, D., Elson, E. L., and Webb, W. W. (1974) Fluorescence correlation spectroscopy. II. An experimental realization. *Biopolymers* **13,** 29–61.
2. Politz, J. C., Browne, E. S., Wolf, D. E., and Pederson, T. (1998) Intranuclear diffusion and hybridization state of oligonucleotides measured by fluorescence correlation spectroscopy in living cells. *Proc. Natl. Acad. Sci. USA* **95,** 6043–6048.
3. Brock, R., Vamosi, G., Vereb, G., and Jovin, T. M. (1999) Rapid characterization of green fluorescent protein fusion proteins on the molecular and cellular level by fluorescence correlation microscopy. *Proc. Natl. Acad. Sci. USA* **96,** 10,123–10,128.
4. Schwille, P. (2001) Fluorescence Correlation Spectroscopy and Its Potential for Intracellular Applications. *Cell Biochem. Biophys.* **34,** 383–408.
5. Kohl, T., Haustein, E., and Schwille, P. (2005) Determining protease activity in vivo by fluorescence cross-correlation analysis. *Biophys. J.*, **89(4),** 2770–2782.
6. Baudendistel, N., Muller, G., Waldeck, W., Angel, P., and Langowski, J. (2005) Two-hybrid fluorescence cross-correlation spectroscopy detects protein-protein interactions in vivo. *Chemphyschemistry* **6,** 984–990.
7. Kim, S. A., Heinze, K. G., Waxham, M. N., and Schwille, P. (2004) Intracellular calmodulin availability accessed with two-photon cross-correlation. *Proc. Natl. Acad. Sci. USA* **101,** 105–110.
8. Kim, S. A., Heinze, K. G., Bacia, K., Waxham, M. N., and Schwille, P. (2005) Two-photon cross-correlation analysis of intracellular reactions with variable stoichiometry. *Biophys. J.* **88,** 4319–4336.
9. Fahey, P. F., Koppel, D. E., Barak, L. S., Wolf, D. E., Elson, E. L., and Webb, W. W. (1977) Lateral diffusion in planar lipid bilayers. *Science* **195,** 305–306.
10. Korlach, J., Schwille, P., Webb, W. W., and Feigenson, G. W. (1999) Characterization of lipid bilayer phases by confocal microscopy and fluorescence correlation spectroscopy. *Proc. Natl. Acad. Sci. USA* **96,** 8461–8466.
11. Schwille, P., Korlach, J., and Webb, W. W. (1999) Fluorescence correlation spectroscopy with single-molecule sensitivity on cell and model membranes. *Cytometry* **36,** 176–182.

12. Schutz, G. J., Kada, G., Pastushenko, V. P., and Schindler, H. (2000) Properties of lipid microdomains in a muscle cell membrane visualized by single molecule microscopy *EMBO J.* **19,** 892–901.
13. Fujiwara, T., Ritchie, K., Murakoshi, H., Jacobson, K., and Kusumi, A. (2002) Phospholipids undergo hop diffusion in compartmentalized cell membrane. *J. Cell Biol.* **157,** 1071–1081.
14. Bacia, K., Scherfeld, D., Kahya, N., and Schwille, P. (2004) Fluorescence correlation spectroscopy relates rafts in model and native membranes. *Biophys. J.* **87,** 1034–1043.
15. Veatch, S. L. and Keller, S. L. (2003) A closer look at the canonical 'Raft Mixture' in model membrane studies. *Biophys. J.* **84,** 725–726.
16. Widengren, J. and Schwille, P. (2000) Characterization of photoinduced isomerization and back-isomerization of the cyanine dye Cy5 by fluorescence correlation spectroscopy. *J. Phys. Chem.* **104,** 6416–6428.
17. Ries, J. and Schwille, P. (2006) Studying Slow Membrane Dynamics with Continuous Wave Scanning Fluorescence Correlation Spectroscopy. *Biophys. J.*, **91(5),** 1915–1924.
18. Hess, S. T. and Webb, W. W. (2002) Focal volume optics and experimental artifacts in confocal fluorescence correlation spectroscopy. *Biophys. J.* **83,** 2300–2317.

8

Multiphoton Laser-Scanning Microscopy and Spatial Analysis of Dehydroergosterol Distributions on Plasma Membrane of Living Cells

Avery L. McIntosh, Barbara P. Atshaves, Huan Huang, Adalberto M. Gallegos, Ann B. Kier, Friedhelm Schroeder, Hai Xu, Weimin Zhang, Suojin Wang, and Jyh-Charn Liu

Summary

Multiphoton laser-scanning microscopy (MPLSM) imaging in combination with advanced image analysis techniques provides unique opportunities to visualize the arrangement of cholesterol in the plasma membrane (PM) of living cells. MPLSM makes possible the use of a naturally occurring sterol, dehydroergosterol (DHE), for observing sterol-enriched areas of the PM. Pure DHE has properties similar to cholesterol as observed in model and cellular membranes but with a conjugated double-bond system that fluoresces at ultraviolet wavelengths. MPLSM enables the excitation of DHE at infrared wavelengths that many laser-scanning microscopy systems are able to transmit effectively and that are less harmful to the cell. Thus, with the incorporation of DHE into living cells and the advent of MPLSM, real-time images of the cellular distribution of DHE can be obtained. In juxtaposition, notably the application of newly advanced techniques in image analysis, aids not only the identification and segmentation of sterol-rich regions of the PM of cells, but also the elucidation of the statistical nature of the observed patterns. In studies involving murine L-cell (Larpt-+K-) fibroblasts, DHE is shown to exhibit strong cluster patterns within the PM.

Key Words: Caveolae; cholesterol; cluster, complete spatial randomness; dehydroergosterol; domains; ergosterol; multiphoton; plasma membrane; raft.

1. Introduction

Since the inception of the basic lipid bilayer hypothesis of biological membrane structure nearly 35 yr ago *(1)*, exciting new insights have come from new findings, membrane lipids and proteins are nonrandomly organized not only across but also within the lateral plane of the bilayer *(2–9)*. Increasing evidence, largely obtained by subfractionation and isolation in vitro, indicates that lipids such as cholesterol spontaneously form cholesterol-rich lipid domains *(2,4,10,11)*. Such cholesterol-rich microdomains, also called lipid rafts, appear to reside in all cell plasma membranes (PMs) examined *(2)*.

Such cholesterol-rich domains are thought to be highly important in both normal and pathophysiological functioning. Regarding normal roles, cholesterol-rich domains appear to function in transbilayer signaling (insulin and Endothelial nitric oxide synthetase eNOS), reverse cholesterol transport, immune recognition, cell–cell interaction, and transcytosis *(8,12–14)*. Cholesterol depletion or disruption abolishes many functions associated with cholesterol-rich lipid rafts *(15)*. Regarding pathophysiological roles, increasing data now suggest that cholesterol-rich domains are also the entry portals for multiple bioterror toxins (anthrax toxin, cholera toxin, shiga toxin, enterohemorrhagic *Escherichia coli* shiga-like toxin, rotaviral, and retroviral enterotoxic peptides, and ricin) *(16–20)*, parasites (malaria) *(21)*, *(22)*, and viruses (Ebola, Marburg, Echovirus, and Influenza) *(23–28)*.

However, despite the importance of cholesterol-rich lipid rafts in biology, there is actually very little evidence demonstrating their existence in PMs of living cells. This is because of the fact that, with rare exception *(29,30)*, nearly all other studies of cholesterol-rich rafts are based on subcellular fractionation and biochemical characterization of cholesterol-enriched fractions isolated from the PM or cell *(31–34)*.

The major problem limiting demonstration of cholesterol-rich domains in the PMs of living cells has been the lack of a suitable, nonperturbing 'tag' to detect cholesterol in living cells. This significant problem was recently overcome through use of a naturally occurring fluorescent sterol (dehydroergosterol [DHE]) and three-photon laser-scanning microscopy in living cells *(29,30)*. DHE is a useful probe for cholesterol because it: (1) is a naturally occurring fluorescent sterol, (2) is a close structural analog of cholesterol, (3) exhibits the same exchange kinetics as cholesterol in both model membranes and biological membranes, (4) is taken up by cultured L-cells such that more than 80% of endogenous sterol is replaced by DHE without altering membrane lipid composition or sterol-sensitive enzymes, (5) codistributes with cholesterol in model and biological membranes, and (6) is nontoxic to cultured cells or animals (reviewed in **refs.** *2,35–37)*. The focus of this chapter is to describe in detail the use of DHE and multiphoton imaging to visualize and analyze the distribution of cholesterol-rich rafts in the PM of living cells. This powerful technology will now allow studies of cholesterol-domain dynamics in real-time, in living cells.

2. Materials
2.1. DHE Synthesis

1. Acid-wash all glassware including amber vials.
2. Ergosterol (99+%pure) and cholesterol (99+%pure) from Steraloids (Wilmington, NH).
3. Burdick-Jackson (Muskegon, MI), "Purified Plus" solvents (methanol, chloroform, and ethyl ether), and JT Baker (Mallinckrodt Baker, Phillipsburg, NJ). "Baker Analyzed" glacial acetic acid (99+% pure) and acetic anhydride (99+% pure) from VWR Scientific (West Chester, PA).
4. Mercuric acetate (anhydrous) was purchased from Fisher Scientific (Pittsburgh, PA).
5. Filtered milli-Q/deionized water (Millipore, Billerica, MA) was used for all aqueous solutions.

2.2. DHE Characterization

2.2.1. Determination of DHE Purity: High-Performance Liquid Chromatography

1. Because caveolae are disrupted by small amounts of oxidized sterols *(38)*, it was essential to determine the purity of the synthesized DHE.
2. DHE was chromatographically resolved from ergosterol and other sterols using reversed phase high-performance liquid chromatography (HPLC) as described *(39)*.
3. The HPLC analysis was performed through use of a Perkin–Elmer Series 4 Liquid Chromatograph (Perkin–Elmer Inc., Wellesley, MA) equipped with a Shimadzu RF 535 Fluorescence HPLC monitor (Shimadzu, Kyoto, Japan) and a Perkin–Elmer LC-95 ultraviolet (UV)/Visible spectrophotometer detector (Perkin–Elmer Inc.). Runs were performed using a solvent mixture of methanol:acetonitrile in a ratio of 3:1 v/v. An Adsorbosphere UHS 10 μm C18 column with inner diameter × length of 4.6 × 250 mm (Alltech Associates Deerfield, IL) was used.

2.2.2. Determination of DHE Purity: Absorbance and Fluorescence Spectroscopy

1. Absorbance spectra and extinction coefficients of DHE in ethanol were ontained using a Lamoda 2 UV/Visible spectrometer Perkin-Elmer Inc.
2. Steady-state fluorescence excitation and emission spectra of DHE in ethanol were determined with an ISS PC1 Photon counting spectrofluorometer (ISS Instruments Inc., Champaign, IL) configured with the provided 300 W xenon arc source and emission monochromator in the standard L-format (right angle). Excitation spectra were obtained with the emission monochromator set at 375 nm with 8-nm spectral slit width. Emission spectra were obtained with the excitation monochromator (8-nm spectral slit width) set at 324 nm.
3. Quartz fluorometer and spectrophotometer cuvets (VWR Scientific) were used.

2.2.3. Determination of DHE Purity: Atmospheric Pressure Chemical Ionization Mass Spectrometry

1. Mass spectra of commercially available DHE, DHE synthesized as described previously, and commercially available ergosterol starting material were determined through use of a Thermo Finnigan LCQ Deca ion trap Liquid Chromatograph mass spectrometer (Thermo Finnigan, San Jose, CA) equipped with an orthogonal atmospheric pressure chemical ionization (APCI) ion source. Samples were 1 mM in a MeOH:H2O (equal volume MeOH, equal volume H2O) solution, with a sample flow rate of 200 mL/min at the mass analyzer.
2. Both methanol and H_2O were HPLC grade Burdick and Jackson (Muskegon, MI) purchased from VWR Scientific. The instrument utilized a quadrupole ion trap mass analyzer with ultra-high-purity helium gas (99.998%) as the damping gas in order to reduce chemical noise.

2.3. Preparation of Large Unilamellar Vesicles (LUVs)

1. Acid-wash all glassware including amber vials (VWR Scientific).
2. Store all fluorescently labeled compounds in amber vials.
3. Prepare concentrated (e.g., 20 mg/mL) stock solutions of 1-Palmitoyl-2-oleoyl-*sn*-glycero-3-phosphocholine (POPC) (Avanti Polar Lipids, Alabaster, AL) and DHE (Δ5,7,9[11],22-ergostatetraen-3β-ol) (Sigma, St. Louis, MO) in new chloroform (Sigma, St. Louis, MO). These should be stored at –80°C.
4. Prefilter autoclaved solutions of 10 m*M* phosphate-buffered saline (PBS) buffer with a pH 7.4 through a 0.1–0.2-μm filter. Store at room temperature.

2.4. Probes and Cell Culture

1. Higuchi medium Higuchi, K, (1970). An improved chemically defined culture medium for strain L mouse cells based on growth responses to graded levels of nutrients including iron and Zinc. *J. Cell Physiol.* 75, 65–72 supplemented with 10% fetal bovine serum (Hyclone, Logan, VT).
2. Stock solutions of Nile red (Molecular Probes, Eugene, OR) and any other PM probes such as Vybrant DiO (3′-dioctadecyloxacarbocyanine perchlorate) (Molecular Probes). Prepare dilute aliquots as necessary for cellular labeling.

2.5. Multiphoton Laser-Scanning Microscopy

1. Lab-Tek two-well-chambered cover glass (Nalge Nunc, Naperville, IL)
2. Emission filters that cover the region 360–430, 485–515, and 525–650 nm ranges. (Chroma Technology Corp., Rockingham, VT).

2.6. Statistical Image Analysis

1. Application routines written in Microsoft Visual C++, .net framework, R, and Matlab (The Mathworks, Inc., Natick, MA). R (available at URL:http://www.R-project.org; R Foundation for statistical Computing, Vienna, AU).

3. Methods

Because cholesterol is not readily detectable in cell membranes either in vitro or in living cells, the availability of a fluorescent sterol with properties closely mimicking those of cholesterol represents a major technological advance for studies of cholesterol structure, dynamics, and protein interactions. The discovery of DHE in membranes of eukaryotes such as yeast *Candida tropicalis (40)* and Red Sea sponge *Biemna fortis (41)* provided for the first time, a naturally occurring fluorescent sterol. Subsequent studies showed that DHE is an excellent probe molecule whose structural and functional properties closely mimic those of cholesterol in model membranes *(29,42,43)*, biological membranes *(3,6,32,44,45)*, lipoproteins *(46,47)*, protein–sterol interactions (reviewed in **refs.** *6,48–50*), and fluorescence resonance energy transfer (FRET) studies to determine intermolecular distance *(48,51–53)*. Most recently, DHE has proven useful in beginning to resolve the structure and dynamics of cholesterol in lipid rafts/caveolae both in vitro *(2,31,32)* and in living cells *(30,32)*. The following describes the methodology for incorporating DHE into cell membranes without inducing formation of crystalline DHE, which does not exist in PMs of living cells *(29)*.

3.1. Synthesis of DHE

The successful application of the fluorescent cholesterol analog DHE *(2,31)* to investigation of the function and organization of sterols in membranes, especially lipid raft/caveolae microdomains of living cells, requires the use of highly purified DHE. This is because of the fact that lipid rafts/caveolae are highly sensitive not only to cholesterol content *(2,8,12,15,54–57)*, but also to sterol structure *(36,37)* and sterol oxidation *(38)*. Unfortunately, DHE obtained from commercial sources varies in purity from batch to batch; therefore, DHE was synthesized from ergosterol obtained from Steraloids (*see* **Subheading 2.**). The DHE synthesis procedure was based on the method of Ruyle *(58)* as described earlier *(39,59)*.

3.2. DHE Characterization

3.2.1. Spectroscopic Analysis of Synthetic DHE

1. Perform an absorbance spectrum of more than 99% pure DHE. For the DHE synthesized for use in spectroscopic and multiphoton laser-scanning microscopy (MPLSM) imaging investigations, it revealed maxima at 310, 325, and 342 nm. No other peaks appeared (*see* **Note 1**).
2. Calculate a difference spectrum to resolve any potentially small peaks. No peaks appeared (not shown). The spectra indicated a purity more than 99%.

3.2.2. HPLC Analysis of Synthetic DHE

1. Monitor the HPLC elution profile (not shown) of the synthesized DHE and pure ergosterol at 205 nm. Only a single peak was detectable with a retention time of 11.7 min.
2. Monitor the elution of DHE at 242 nm. No peak has been observed (data not shown).
3. Monitor the elution of DHE at 281 nm. No significant peak was observed showing that there was no significant amount of starting material remained and calculations showed a purity ≥99%.
4. Monitor the absorbance at 325 nm while monitoring the fluorescence emission at 375 nm. In each case only a single peak was detectable, again with retention time of 11.7 min. Quantitative analysis of such HPLC chromatograms indicated that the synthesized DHE used in the MPLSM imaging investigations was ≥99% pure.

3.2.3. Examination of the Synthesized Product by Mass Spectrometry

Examine by APCI mass spectroscopy as described in **Subheading 2.2.3.** to determine the mass of the synthesized product including any impurities. The mass spectra of the synthesized product should detect DHE at $(M + 1)^+$ of 395.3 and $(M–OH)^+$ of 377.5, as expected. Any other masses detected can be used to determine possible impurities. The APCI mass spectroscopy detected only a single peak with more than 99% (data not shown).

3.3. Preparation of LUVs

1. Mix POPC and DHE in a small acid-washed amber vial at the mole ratio of 65:35 (POPC:DHE) by using the proportioned amounts taken from the stock solutions of POPC and DHE, and vortex.
2. Evaporate the liquid chloroform by first drying slowly under N_2 to form a thin film that is layered on the inside surface of the vial. Then fully remove any residual chloroform under a vacuum desiccation for at least 4 h.
3. Rehydrate at 1–10 mM using PBS buffer (10 mM, pH 7.4) that has been prefiltered through a 0.1-µm filter. First, vortex the solution for 1 min and then sonicate for approx 10 min.
4. Extrude the aqueous mixture using a handheld mini-extruder (Avanti Polar Lipids) through two 0.1-µm polycarbonate membranes with membrane supports. The mixture should be cycled back and forth for 10.5 full cycles or for a total of 21 transfers through the membrane. Empty the solution into an amber vial that is sterile.
5. Use immediately or store under N_2 in an amber vial at 4°C for up to a week typically.

3.4. Probe labeling and Cell Culture

1. Seed L-cell fibroblasts onto Lab-Tek II chambered cover glass at a density of 2–5 × 10^{-7} cells per chamber. Layer cells with Higuchi medium containing 10% fetal bovine serum so that there is 1 mL of media.

2. Maintain for a period of time up to overnight in a CO_2 incubator at 37°C so that the cells can attach the cover glass.
3. Supplement healthy cells with DHE-containing LUVs so that the concentration of the DHE is between 10 and 20 μg/mL in media.
4. Grow cells in the DHE-containing media in a temperature-, humidity-, and CO_2-controlled incubator for 1–2 d before imaging. This will enable the cells to replace a majority of the cholesterol with DHE.
5. Wash cells immediately before imaging using 10 mM PBS (pH 7.4) or similar buffer.
6. Label cells with 100 nM Nile red in ethanol solution and/or third fluorescent probe such as DiO ($DiOC_{18}$[3]) (Molecular Probes) according to manufacturer's instructions. Do not use anything that will extract the fluorescent sterol analog.

3.5. Multiphoton Laser-Scanning Microscopy

1. Prepare and turn on the Bio-Rad MRC-1024MP MPLSM system (Zeiss, Thornwood, NY) and Spectra-Physics Millennia X (Spectra-Physics, Mountain View, CA), and Coherent Mira 900F (Coherent, Mira 900F titanium: Sapphire laser (Coherent, Palo Alto, CA) for operation at least 1 h before use so that the lasers and system stabilize.
2. Tune wavelength of the Coherent Mira 900 F titanium: Sapphire laser (Coherent, Palo Alto, CA) to 920 nm with modelocked output power more than 700 mW. Typical transmission through the scan head not including objective will be approx 40–60% depending on the excitation wavelengths (*see* **Notes 1** and **2**).
3. Place the two-well-chambered cover glass containing labeled cells onto the microscope stage. Using widefield illumination change the *z*-focus in order to locate the plane containing the cells, and finally focus the objective on a focal plane inside the cells.
4. Initiate scanning of the cells using the Bio-Rad LaserSharp acquisition software, (Zeiss, Thornwood, NY) starting with low multiphoton excitation power by attenuating the beam. Adjust the gain and black level of the photomultiplier tubes in order to detect the sample or at least the autofluorescence. Probes that use two-photon excitation should appear to be detected at lower power, typically. Use fast scan speeds for reduced dwell time and thereby reduce photodamage to cells (*see* **Notes 3** and **4**).
5. Readjust focus so that the confocal plane of excitation and emission detection is about midway between the upper and lower cell surface. The contour of the PM will appear around the cytoplasmic area of the cell when the DHE is imaged.
6. Image the Nile red and any two-photon excited probes first, typically at about 10–50% of the power needed for three-photon excitation of DHE. The amount of power will depend on the multiphoton cross-section absorption of the other probes (e.g., Nile red) as well as their localized cellular environment.
7. Increase power to enable detection of DHE in the 360–430-nm filtered photomultiplier and scan the cells. Adjust black level of this photomultiplier tube alone.
8. Remove the chambered cover glass with labeled cells and replace it with a chambered cover glass that has unlabeled cells. Using the set conditions determine the

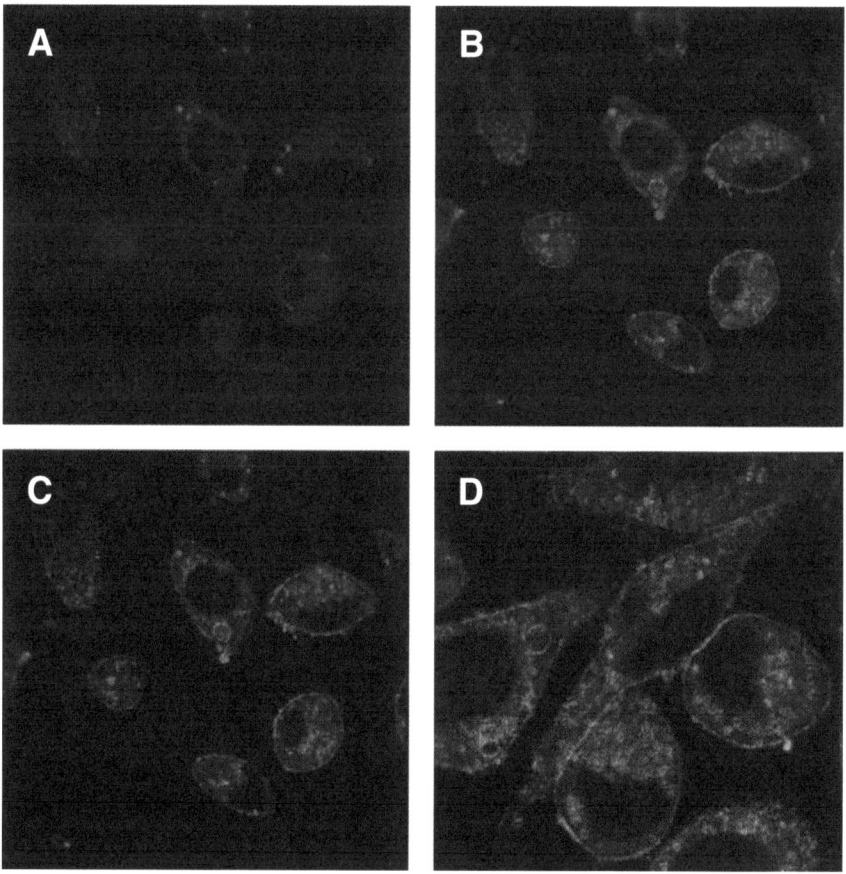

Fig. 1. MPLSM of DHE and Nile red in murine L-cell fibroblasts. Excitation at 920 nm. **(Panel A)** Image of L-cells labeled with DHE and Nile red at low power (25% attenuation). **(Panel B)** Image of L-cells labeled with DHE and Nile red at high power (40% attenuation). **(Panel C)** Image with high-power DHE (green channel, **panel B**) merged with the low-power Nile red (red channel, **panel A**). **(Panel D)** Image acquired at high power with zoom = 2.

background autofluorescence levels and readjust to minimize autofluorescence levels while trying to maintain as much detection sensitivity as possible for all probes at the power levels determined previously.
9. Replace the previous chambered slide and check sensitivity at the previously determined power levels and with the changes in gain and black level. If necessary increase power slightly in order to enhance sensitivity and maximize dynamic range. Recheck background autofluorescence levels again.
10. Prepare a new chambered cover glass with labeled cells and acquire the Nile red emission first at the predetermined lower excitation power (**Fig. 1A** red channel)

and then increase the predetermined upper excitation power to acquire the image of DHE emission (**Fig. 1B** green). Keep averaging to a minimum and acquire images as rapidly as possible.
11. Merge the Nile red channel (red) of the low-power-acquired image with the DHE channel (green) of the high-power-acquired image (**Fig. 1C**).
12. Scan with increased zoom as necessary (**Fig. 1D**) (*see* **Notes 3** and **5**).

3.6. Statistical Image Analysis

1. *PM extraction:* the PM is separated from interior cellular compartments by simple subtraction of the intensity of the Nile red emission from the DHE emission. Negative results from the subtraction are set to zero. This step exploits the property that Nile red mostly stains internal compartments, but not the PM *(30)*. The flowchart of the process is given in **Fig. 2A** (*see* **Notes 6–8**).
2. *Noise suppression:* if the computed signal-to-noise ratio in an analysis window is less than a threshold value, all pixels in the window will be set to zero. An analysis window is defined as adjacent pixels arranged in a particular shape (usually square or rectangle), at a fixed size *(60)*. Scattered spots whose sizes are smaller than a threshold are removed. Window size, signal-to-noise ratio, and intensity threshold are used to optimize the effectiveness of the denoising process. **Figure 3A,B** denote an original image of living L-cell fibroblasts and the image of the PM after noise suppression, respectively.
3. *Estimate PM trajectory:* use an intensity moments function *(61–63)* to reduce the edge effects caused by jigsaw boundary of DHE pixels caused by the digital imaging process. The PM areas are transformed into smooth planar structures; *see* the example given in **Fig. 4**. The geometric trajectory of PM follows the smooth contour.
4. *Selection of segments for analysis:* lamellipodia portions of adjacent cells tended to overlap more with each other, making it potentially unreliable for assessment of the DHE intensity distributions. To avoid this problem, only nonoverlapped PM regions should be chosen for analysis, and five regions of interest (**Fig. 5A**) were randomly selected for analysis.
5. *Estimate the PM boundary:* PM of the cell sample does not rest on a flat plane. To control the potential anomalies caused by the boundary of the extracted PM, a statistical technique is used to define the boundary from the ridge line. A smaller confidence interval implies a smaller area being analyzed; and therefore, more reliable assessment on the properties of the DHE distributions. For instance, when two red lines forming the boundaries of a (very small) 65% confidence interval of a segment are plotted on the segment (**Fig. 6**); DHE pixels represent about 42% of the area within the two red lines.
6. *Derive the ridge line of a PM segment:* using a polynomial fitting routine in Matlab to reduce a (multipixel thick) segment into a single-pixel line along the contour of the PM. In the routine, a third-order polynomial fitting routine was used to shrink a segment into a single-pixel line, **Fig. 5B**. The general third-order polynomial *(64)* can be expressed as:

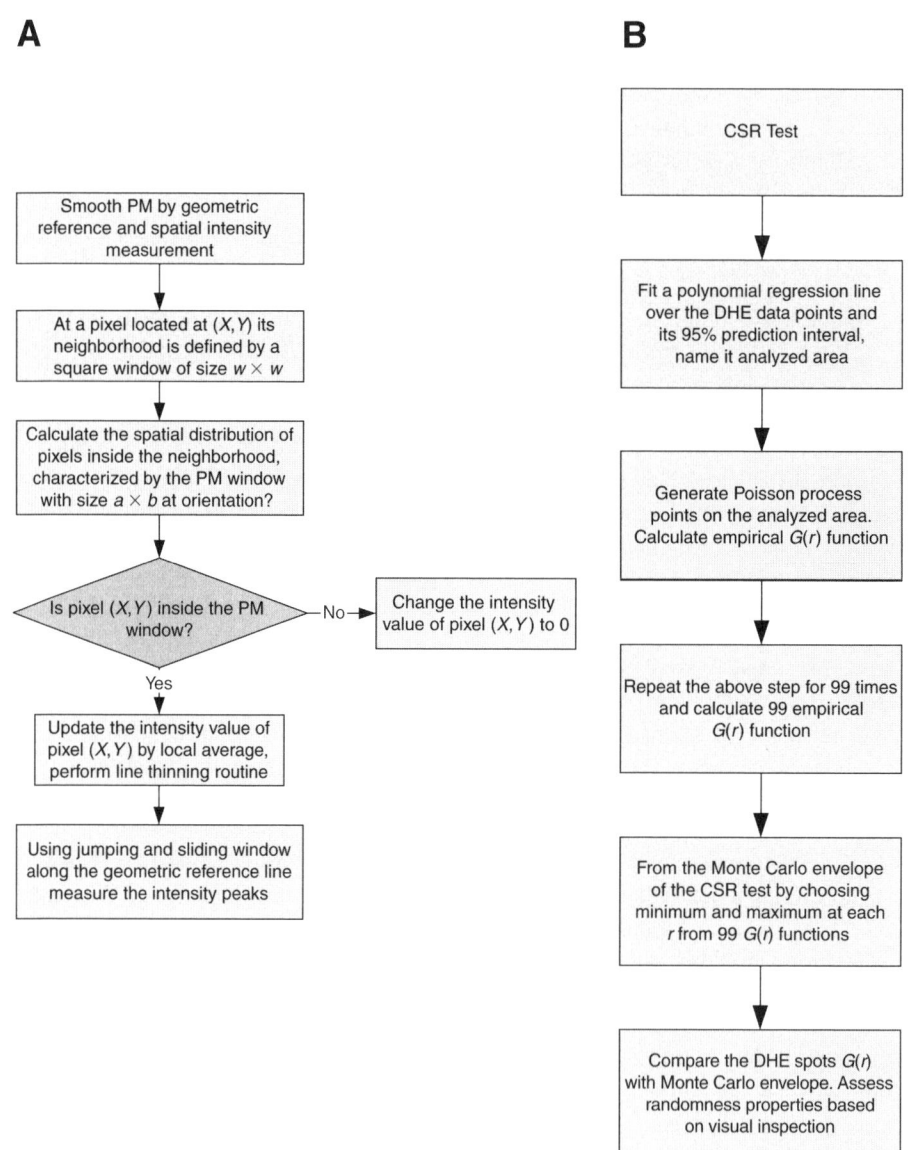

Fig. 2. System flowchart. (**Panel A**) Smoothing and geometric reference with spatial intensity measurement. (**Panel B**) CSR test.

$$y_i = \beta_0 + \beta_1 x_i + \beta_2 x_i^2 + \beta_1 x_i^3 + \varepsilon_i$$

where $i = 1, 2, \ldots, n$, n is the number of pixels, and the relative regression matrix form is $Y = \beta X + \varepsilon,$ where

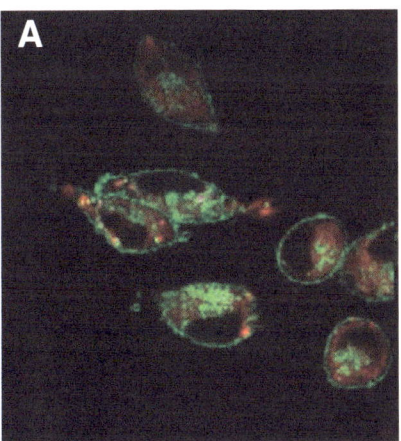 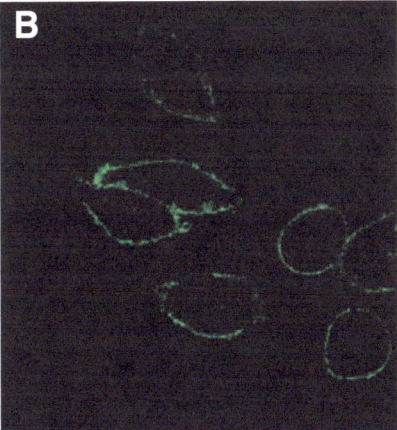

Fig. 3. PM extraction results. (**Panel A**) A merged image showing L-cell fibroblasts supplement with DHE (green) in the form of LUVs and labeled with Nile red (red). (**Panel B**) PM extraction result using the intensity subtraction method with noise removal.

$$Y_{1\times n} = [y_1, y_2, ..., y_n], \quad \beta_{1\times n} = [\beta_0, \beta_1, \beta_2, ..., \beta_3],$$

$$X_{4\times n} = \begin{bmatrix} 1 & 1 & 1 & 1 \\ x_1 & x_2 & ... & x_n \\ x_1^2 & x_2^2 & ... & x_n^2 \\ x_1^3 & x_2^3 & ... & x_n^3 \end{bmatrix}, \quad \varepsilon_{1\times n} = [\varepsilon_0, \varepsilon_1, \varepsilon_2, ..., \varepsilon_n]$$

To estimate the regression coefficient matrix, the least-square optimization criterion is used:

$$Q = \sum_{i=1}^{n_i} \left(y_i - \beta_0 - \beta_1 x_i - \beta_2 x_i^2 - \beta_3 x_i^3 \right)^2$$

and the estimator of β is $\underset{1\times 4}{b} = \begin{bmatrix} b_0 & b_1 & b_2 & b_3 \end{bmatrix}$.

The least-square normal equation for the general regression model is: $X'Xb = X'Y$. The least-square estimator matrix is: $b = (X'X)^{-1}(X'Y)$. With computed estimator matrix b, one is able to obtain the estimated Y, where $Y = bX$.

By applying the aforementioned third-order polynomial equations to the location matrix (X,Y) of each of the five segments S1–S5, one can obtain their skeleton lines plotted in **Fig. 5B**.

7. *DHE intensity measurement:* following the PM trajectory, the DHE intensity distribution can be measured along their perpendicular directions, and the results for

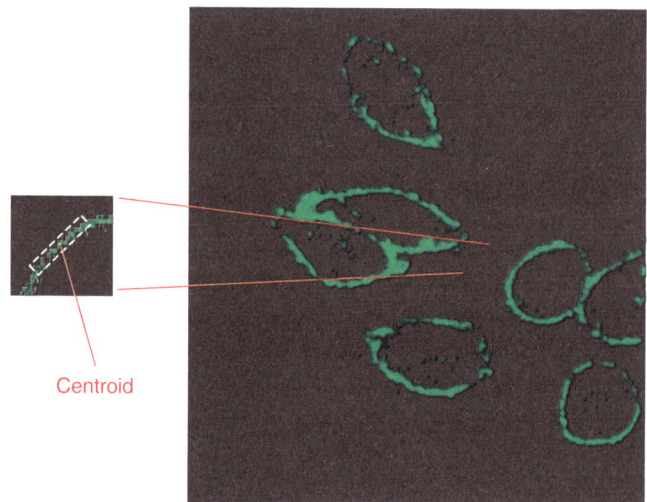

Fig. 4. Image smoothing using a nonlinear median window filter based on a moments functions. The smoothed image shows the result after application of a continuous planar model of the DHE image using window size 20 × 20 and neighborhood 5 × 5. Each pixel in the smoothed image represents the intensity centroid calculated from the application of this filter to the raw data. A blowup of a particular PM region from the raw data was shown to illustrate the window and its intensity centroid.

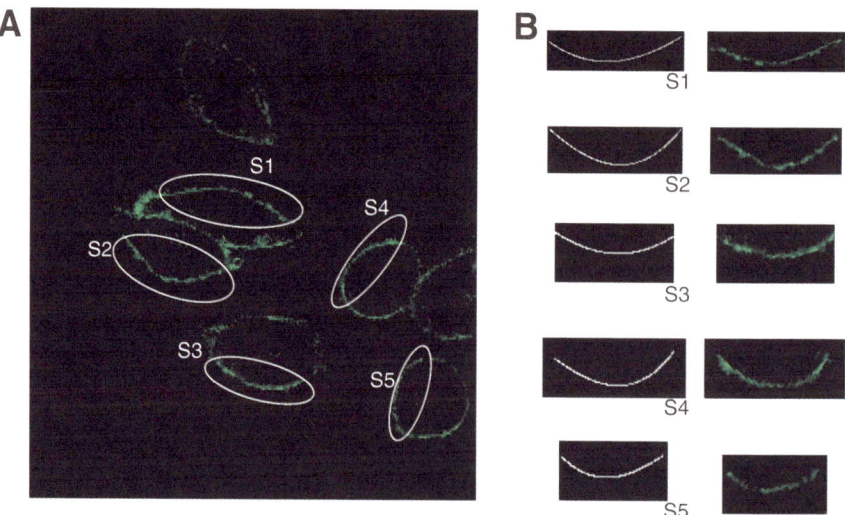

Fig. 5. Five segments of the DHE emission in extracted PMs from MPLSM images of living fibroblasts. **(Panel A)** The monomeric DHE along the PM with the five segments marked that were used for spatial measurement study. **(Panel B)** The geometric reference lines from a third-order polynomial fitting routine on the PM and the corresponding raw data values for each segment.

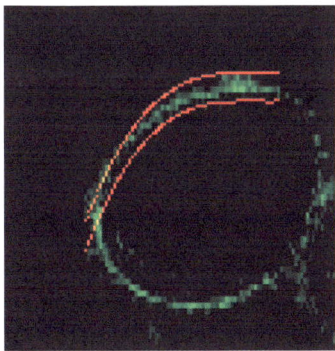

Fig. 6. Segmented PM area of DHE, superimposed with the 65% confidence interval from Nile red marker PM area.

the five selected PM segments S1–S5 are given in **Fig. 7A** (using sliding window) and **Fig. 7B** (using jumping window).
8. *Measure normality of DHE emissions*: use normality probability plots and Anderson–Darling method to ascertain the normality for the DHE emission. This step ensures uniformity of the scanning steps, and detect anomaly of energy distribution.
9. *Spatial pattern randomness test*: analyze the chosen data sets using statistical two-dimensional spatial analysis to perform a test of complete spatial randomness (CSR) and then perform Monte Carlo simulations of the DHE data sets. Statistical analysis on the spatial properties of the DHE locations can be performed using the R system; the flowchart is as **Fig. 2B**.

3.6.1. Spatial Pattern Analysis

A configuration of a statistical spatial point pattern, such as location of DHE pixels on the PM is said to exhibit CSR, if with high probabilities it is a realization of a homogeneous Poisson process *(65)*. If the CSR test is rejected, the data set would be considered having either a *regular* or *cluster* pattern. There are several summary statistics to capture the spatial properties of a spatial pattern. One of these statistics is the nearest neighbor function $G(y)$. If the nearest neighbor function values from a spatial data set are larger (smaller) than those from a CSR point pattern, then the data set has a cluster (regular) pattern. Analysis of the spatial properties of DHE locations on the PM follows the following steps.

1. *Select data samples for analysis*. A simple technique is to use the intensity as a selection criterion of DHE pixels for analysis. When the threshold value is zero, all pixels are included for analysis. When the 75% value is set as the threshold, only high-intensity pixels (called peak value pixels) are retained for analysis. Typical PM segments before **(Fig. 8A)** and after **(Fig. 8B)** removal of DHE pixels with values below the 75% were analyzed to generate a spatial map of the peak DHE pixels, whose intensities represent strong DHE emission. This simple technique

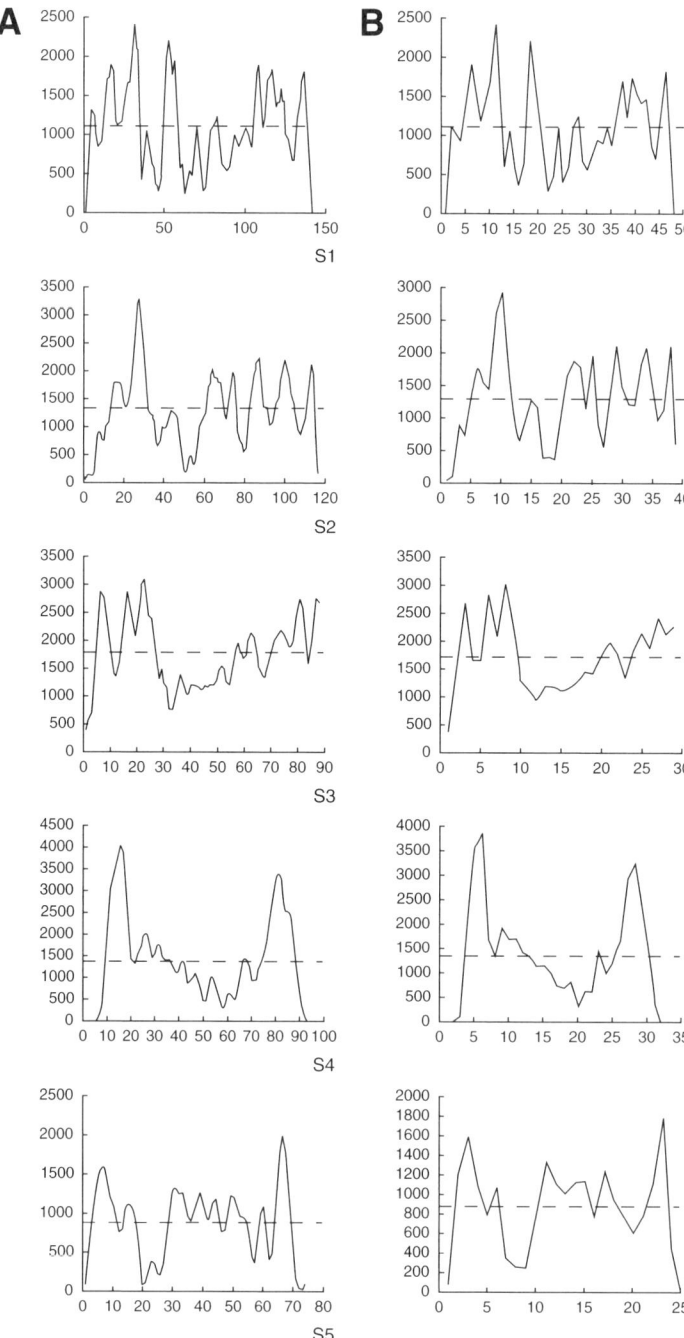

Fig. 8. Zoomed images comparing thresholded (peak) data to the full data set of a typical PM segment of DHE emission detected over the wavelength region 350–400 nm. **(Panel A)** Image before 75% thresholding was applied. **(Panel B)** Image after 75% thresholding was applied.

   allows one to study any potential differences in spatial properties of strong and weak DHE signals on the PM.
2. Calculate the *empirical distribution function (EDF)* of one of the summary statistics, such as *nearest neighbor $G_1(y)$ function,* for the peak DHE pixels. The algorithm used is the spatial Kaplan–Meier estimator of $G(y)$.
3. *Generate s – 1 sets of Monte Carlo simulations **(65)** of CSR pattern.* All simulations are generated from an enlarged area and cropped by an area whose shape is as that of peak DHE pixels. Each set of simulation has the same number of points as that of peak DHE pixels. Simulation of CSR pattern is relatively simple: generate points that are independently and uniformly distributed on an area. The area shape on which peak DHE pixels reside is formed from fitting a polynomial regression line, where *y*-coordinates are the dependent variable, *x*-coordinates and their exponentials are the independent variables, and then a 95% predict interval over DHE spots is predicted.
4. *Estimate the EDF of one of summary statistics, such as $G_i(y)$ function, for each simulated points set, and then create the Monte Carlo envelopes.* The upper and lower bounds of Monte Carlo envelopes obtained from the simulation are defined as **(63)**

$$U(y) = \max_{i=2,\,...,\,s} \{G_i(y)\} \qquad L(y) = \min_{i=2,\,...,\,s} \{G_i(y)\}$$

5. A data set is compatible with CSR if its EDF line falls inside the envelope. The data set is called a cluster (regular) pattern if its <EQN>line rises above (falls below) the upper part (low bound) of the Monte Carlo *G* function envelope **(65)** **(Fig. 7)**.

Fig. 7. *(Opposite page)* Intensity measurement of DHE along PM segments of different cells. **(Panel A)** DHE intensity measurements from segments 1–5 using a sliding window. **(Panel B)** DHE intensity measurements from segments 1–5 using a jumping window. Window size is 10 and bin size is 3. The vertical axis represents the sum of the intensities within the window; whereas the horizontal axis represents the window sequence number. The dashed line represents the mean intensity over the segment sequences.

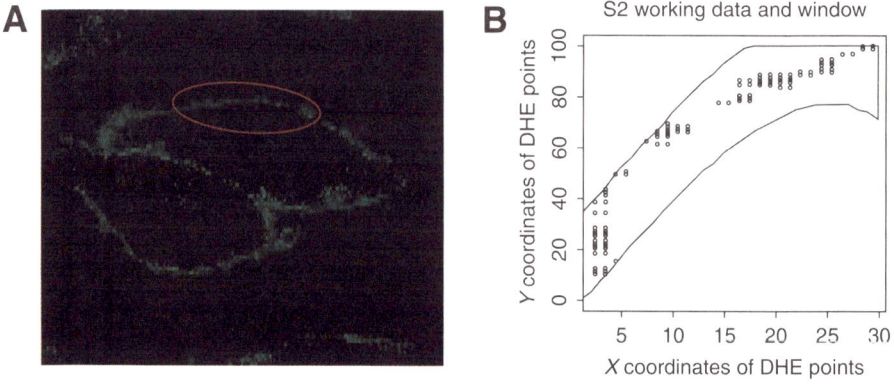

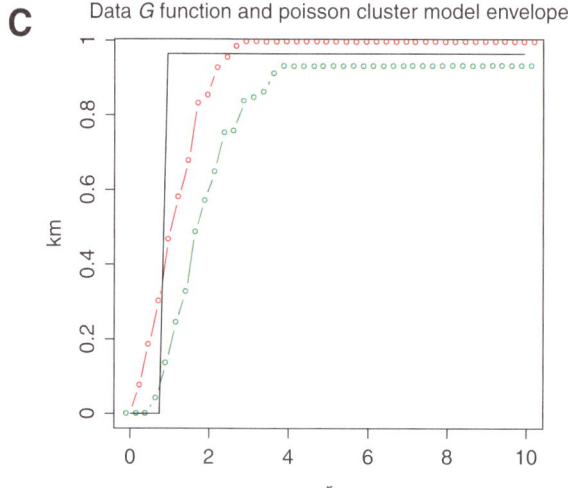

Fig. 9. CSR test result of a selected PM segment. **(Panel A)** The PM segment being used for analysis. **(Panel B)** The spatial locations of "peak" DHE pixels by setting the mean as the threshold. **(Panel C)** The CSR testing outcome.

Figure 9 gives the CSR testing outcome for a selected PM segment. In **Fig. 9C**, the *x*-axis is for the nearest neighbor distance, and *y*-axis for the *G* function. The data *G* function for the data is smaller than the Monte Carlo lower envelope at distance $y < 1$. This means that below the distance of 1, the data present a regular pattern than that of a CSR. This is reasonable because the nearest neighbor distance for pair of pixels is 1. It is also noticed that *G* function for this data set is greater than the Monte Carlo upper envelope when distance $y > 1$. It is concluded that the DHE points follow a spatial clus-ter pattern, which deviates significantly from the null hypothesis in such a way that the points of the pattern tend to form local concentrations. Herein in the empirical G function estimation, the Kaplan–Meier estimator *(67)* was used, which was implemented in the R-statistical computing language *spatstat* package (available at URL:http://www.R-project.org; R Foundation for Statistical Computing, Vienna, AU) Baddeley, A. and Turner, R. (2005) Spatstat: An R package for Analyzing Spatial Point Patterns *J. Stat. Software* **12**, 1–4.

4. Notes

1. Using a probe like DHE that emits fluorescence in UV wavelengths requires optics that allows transmission from the visible down to near 350 nm.
2. Owing to the intensity needed to effectively excite DHE at the level of sensitivity, caution must be exercised in managing length of the imaging times used so that the cells are not damaged because of heat or photodegradation.
3. Zoom levels enhance resolution but also produce more photobleaching especially of probes with higher cross-sectional absorption such as Nile red (**Fig.1D**).
4. Block the entrance of all ambient light into the detection path in order to enhance the sensitivity of detection.
5. Consistency and validation of the technique should be verified with several probes (e.g., DiO or enhanced cyan fluorescent protein labeled marker ECFP-Mem (Clontech, Palo Alto, CA), *(30)* that localize within the PM and with another image segmentation technique (e.g., rank-statistic-based method using rank-based comparisons).
6. Careful choices of the segmented PM regions are necessary but are based on simple criteria based on regions of clean membrane contours with no overlapping or interfacing regions (can easily be seen in central portion of **Fig. 1D**), such as where cells or cell debris are overlapping, or the places of low z distance between PM surfaces such as where lamellipodia exist.
7. Selection of sparsely vs densely populated areas did not appear to have significant effects on the analysis outcomes.
8. Different segmentation methods can be adopted to test consistency of the extraction results of the PM areas.

Acknowledgments

This work was supported in part by the National Institute of Health GM31651 (F.S.), an Administrative Supplement for the Study of Complex Biological Systems to GM31651 (J.C.S.L.), and National Institute of Health P20 grant "Fluorescence Probes for Multiplexed Intracellular Imaging" GM72041 (Project 2, A.K. and F.S; Project 4 to J.C.S.L).

References

1. Singer, S. J. and Nicolson, G. L. (1972) The fluid mosaic model of the structure of cell membranes. *Science* **175**, 720–731.

2. Schroeder, F., Atshaves, B. P., Gallegos, A. M., et al. (2005) Lipid rafts and caveolae organization, in *Advances in Molecular and Cell Biology*, (Frank, P. G. and Lisanti, M. P., eds.), Elsevier, Amsterdam, pp. 3–36.
3. Schroeder, F., Frolov, A. A., Murphy, E. J., et al. (1996) Recent advances in membrane cholesterol domain dynamics and intracellular cholesterol trafficking. *Proc. Soc. Exp. Biol. Med.* **213,** 150–177.
4. Brown, D. A. and London, E. (1998) Structure and origin of ordered lipid domains in biological membranes. *J. Membr. Biol.* **164,** 103–114.
5. Hooper, N. M. (1999) Detergent-insoluble glycosphingolipid/cholesterol rich membrane domains, lipid rafts, and caveolae. *Mol. Membr. Biol.* **16,** 145–156.
6. Schroeder, F., Gallegos, A. M., Atshaves, B. P., et al. (2001) Recent advances in membrane microdomains: rafts, caveolae and intracellular cholesterol trafficking. *Exp. Biol. Med.* **226,** 873–890.
7. Edidin, M. (2001) Shrinking patches and slippery rafts: scales of domains in the plasma membrane. *Trends Cell Biol.* **11,** 492–496.
8. Anderson, R. G. W. and Jacobson, K. (2002) A role for lipid shells in targeting proteins to caveolae, rafts, and other lipid domains. *Science* **296,** 1821–1825.
9. Lin, S. L. and Tian, P. (2003) Detailed computational analysis of a comprehensive set of group A rotavirus NSP4 proteins. *Virus Genes* **26,** 271–282.
10. Bretscher, M. S. and Munro, S. (1993) Cholesterol and the Golgi apparatus. *Science* **261,** 1280–1281.
11. Brown, R. E. (1998) Sphingolipid organization in biomembranes: what physical studies of model membranes reveal. *J. Cell Sci.* **111,** 1–9.
12. Lavie, Y. and Liscovitch, M. (2000) Changes in lipid and protein constituents of rafts and caveolae in multidrug resistant cancer cells and their functional consequences. *Glycoconjugate J.* **17,** 253–259.
13. Everson, W. V. and Smart, E. J. (2005) Caveolae and the regulation of cellular cholesterol homeostasis, in *Caveolae and Lipid Rafts: Roles in Signal Transduction and the Pathogenesis of Human Disease*, (Lisanti, M. P. and Frank, P. G. eds.), Elsevier Academic Press, San Diego, pp. 37–55.
14. Smart, E. J. (2005) Caveolae and the regulation of cellular cholesterol homeostasis, in *Advances in Molecular and Cell Biology*, (Lisanti, M. P. and Frank, P. G., eds.), Elsevier B. V., Amsterdam, 35p.
15. Smart, E. J. and van der Westhuyzen, D. R. (1998) Scavenger receptors, caveolae, caveolin, and cholesterol trafficking, in *Intracellular Cholesterol Trafficking*, (Chang, T. Y. and Freeman, D. A., eds.), Kluwer Academic Publishers, Boston, pp. 253–272.
16. Huang, H., Schroeder, F., Zeng, C., Estes, M. K., Schoer, J., and Ball, J. A. (2001) Membrane interactions of a novel viral enterotoxin: rotavirus nonstructural glycoprotein NSP4. *Biochemistry* **40,** 4169–4180.
17. Huang, H., Schroeder, F., Estes, M. K., McPherson, T., and Ball, J. M. (2004) The interactions of rotavirus NSP4 C-terminal peptides with model membranes. *Biochem. J.* **380,** 723–733.

18. Swaggerty, C. L., Huang, H., Lim, W. S., Schroeder, F., and Ball, J. A. (2004) Comparison of SIVmac239(352-382) and SIVsmmPBjJ41(360-390) enterotoxic synthetic peptides. *Virology* **320,** 243–257.
19. Abrami, L., Liu, S., Cosson, P., Leppla, S. H., and van der Groot, F. G. (2003) Anthrax toxin triggers endocytosis of its receptor via a lipid raft-mediated clathrin-independent process. *J. Cell Biol.* **160,** 321–328.
20. Sandvig, K. and van Deurs, B. (1999) Endocytosis and intracellular transport of ricin: recent discoveries. *FEBS Lett.* **452,** 67–70.
21. Shin, J.-S. and Abraham, S. N. (2002) Caveolae as portals of entry for microbes. *Microbes Infect.* **3,** 755–761.
22. Norkin, L. C. (2001) Caveolae in the uptake and targeting of infectious agents and secreted toxins. *Adv. Drug Delivery Rev.* **49,** 301–315.
23. Bavari, S., Bosio, C. M., Wiegand, E., et al. (2002) Lipid raft microdomains. A gateway for comportmentalized trafficking of Ebola and Marburg viruses. *J. Exp. Med.* **195,** 593–602.
24. Empig, C. J. and Goldsmith, M. A. (2002) Association of the caveola vesicular system with cellular entry by filoviruses. *J. Virol.* **76,** 5266–5270.
25. Marjomaki, V., Pietiainen, V., Upla, P., et al. (2002) Internalization of Echovirus 1 in caveolae. *J. Virol.* **76,** 1856–1865.
26. Scheiffele, P., Roth, M. G., and Simons, K. (1997) Interaction of influenza virus haemagglutinin with sphingolipid-cholesterol membrane domains via its transmembrane domain. *EMBO J.* **16,** 5501–5508.
27. Marsh, M. and Pelchen-Matthews, A. (2000) Endocytosis in viral replication. *Traffic* **1,** 525–532.
28. Sieczkarski, S. B. and Whittaker, G. R. (2002) Dissecting virus entry via endocytosis. *J. Gen. Virol.* **83,** 1535–1545.
29. McIntosh, A., Gallegos, A., Atshaves, B. P., Storey, S., Kannoju, D., and Schroeder, F. (2003) Fluorescence and multiphoton imaging resolve unique structural forms of sterol in membranes of living cells. *J. Biol. Chem.* **278,** 6384–6403.
30. Zhang, W., McIntosh, A., Xu, H., et al. (2005) Structural analysis of sterol distribution in the plasma membrane of living cells. *Biochemistry* **44,** 2864–2984.
31. Gallegos, A. M., McIntosh, A. L., Atshaves, B. P., and Schroeder, F. (2004) Structure and cholesterol domain dynamics of an enriched caveolae/raft isolate. *Biochem. J.* **382,** 451–461.
32. Atshaves, B. P., Gallegos, A., McIntosh, A. L., Kier, A. B., and Schroeder, F. (2003) Sterol carrier protein-2 selectively alters lipid composition and cholesterol dynamics of caveolae/lipid raft vs non-raft domains in L-cell fibroblast plasma membranes. *Biochemistry* **42,** 14,583–14,598.
33. Eckert, G. P., Igbavboa, U., Muller, W., and Wood, W. G. (2003) Lipid rafts of purified mouse brain synaptosomes prepared with or without detergent reveal different lipid and protein domains. *Brain Res.* **962,** 144–150.
34. Pike, L. J., Han, X., Chung, K.-N., and Gross, R. W. (2002) Lipid rafts are enriched in arachidonic acid and plasmenylethanolamine and their composition is

independent of caveolin-1 expression: a quantitative electrospray ionization/mass spectrometric analysis. *Biochemistry* **41,** 2075–2088.
35. Gallegos, A. M., Atshaves, B. P., Storey, S. M., et al. (2001) Gene structure, intracellular localization, and functional roles of sterol carrier protein-2. *Prog. Lipid Res.* **40,** 498–563.
36. Gimpl, G. and Fahrenholz, F. (2000) Human oxytocin receptors in cholesterol-rich vs cholesterol-poor microdomains of the plasma membrane. *Eur. J. Biochem.* **267,** 2483–2497.
37. Burger, K., Gimpl, G., and Fahrenholz, F. (2000) Regulation of receptor function by cholesterol. *Cell Mol. Life Sci.* **57,** 1577–1592.
38. Smart, E. J., Ying, Y., Conrad, P. A., and Anderson, R. G. W. (1994) Caveolin moves from caveolae to the golgi apparatus in response to cholesterol oxidation. *J. Cell Biol.* **127,** 1185–1197.
39. Fischer, R. T., Stephenson, F. A., Shafiee, A., and Schroeder, F. (1985) Structure and dynamic properties of dehydroergosterol, delta 5,7,9(11),22-ergostatetraen-3 beta-ol. *J. Biol. Phys.* **13,** 13–24.
40. Sica, D., Boniforti, L., and DiGiacomo, G. (1982) Sterols of *Candida tropicalis* grown on n-alkanes. *Phytochemistry* **21,** 234–236.
41. Delseth, C., Kashman, Y., and Djerassi, C. (1979) Ergosta-5,7,9(11),22-tetraen-3beta-ol and its 24epsilon-ethyl homolog, two new marine sterols from the red sea sponge *Biemna fortis*. *Helv. Chim. Acta* **62,** 2037–2045.
42. Nemecz, G., Fontaine, R. N., and Schroeder, F. (1988) A fluorescence and radiolabel study of sterol exchange between membranes. *Biochim. Biophys. Acta* **943,** 511–521.
43. Schroeder, F. (1984) Fluorescent sterols: probe molecules of membrane structure and function. *Prog. Lipid Res.* **23,** 97–113.
44. Schroeder, F., Jefferson, J. R., Kier, A. B., et al. (1991) Membrane cholesterol dynamics: cholesterol domains and kinetic pools. *Proc. Soc. Exp. Biol. Med.* **196,** 235–252.
45. Gallegos, A. M., Atshaves, B. P., Storey, S. M., Schoer, J., Kier, A. B., and Schroeder, F. (2002) Molecular and fluorescent sterol approaches to probing lysosomal membrane lipid dynamics. *Chem. Phys. Lipids* **116,** 19–38.
46. Schroeder, F., Goh, E. H., and Heimberg, M. (1979) Regulation of the surface physical properties of the very low density lipoprotein. *J. Biol. Chem.* **254,** 2456–2463.
47. Bergeron, R. J. and Scott, J. (1982) Cholestatriene and ergostatetraene as in vivo and in vitro membrane and lipoprotein probes. *J. Lipid Res.* **23,** 391–404.
48. Fischer, R. T., Cowlen, M. S., Dempsey, M. E., and Schroeder, F. (1985) Fluorescence of delta 5,7,9(11),22-ergostatetraen-3 beta-ol in micelles, sterol carrier protein complexes, and plasma membranes. *Biochemistry* **24,** 3322–3331.
49. Schroeder, F., Frolov, A. Schoer, J., et al. (1998) Intracellular sterol binding proteins, cholesterol transport and membrane domains, in *Intracellular Cholesterol Trafficking*. (Chang, T. Y. and Freeman, D. A., eds.), Kluwer Academic Publishers, Boston, pp. 213–234.

50. Stolowich, N. J., Petrescu, A. D., Huang, H., Martin, G., Scott, A. I., and Schroeder, F. (2002) Sterol carrier protein-2: structure reveals function. *Cell Mol. Life Sci.* **59,** 193–212.
51. Loura, L. M. S. and Prieto, M. (1997) Dehydroergosterol structural organization in aqueous medium and in a model system of membranes. *Biophys. J.* **72,** 2226–2236.
52. John, K., Kubelt, J., Muller, P., Wustner, D., and Hermann, A. (2002) Rapid transbilayer movement of the fluorescent sterol dehydroergosterol in lipid membranes. *Biophys. J.* **83,** 1525–1534.
53. Schroeder, F., Butko, P., Nemecz, G., and Scallen, T. J. (1990) Interaction of fluorescent delta 5,7,9(11),22-ergostatetraen-3β-ol with sterol carrier protein-2. *J. Biol. Chem.* **265,** 151–157.
54. Anderson, R. (1998) The caveolae membrane system. *Annu. Rev. Biochem.* **67,** 199–225.
55. Everson, W. V. and Smart, E. J. (2001) Influence of caveolin, cholesterol, and lipoproteins on nitric oxide synthase. *TCM* **11,** 246–250.
56. Brown, D. A. and London, E. (2000) Structure and function of sphingolipid- and cholesterol-rich membrane rafts. *J. Biol. Chem.* **275,** 17,221–17,224.
57. Kirsch, C., Eckert, G. P., and Mueller, W. E. (2002) Statins affect cholesterol micro-domains in brain plasma membranes. *Biochem. Pharmacol.* **65,** 843–856.
58. Ruyle, W. V., Jacob, T. A., Chemerda, J. M., et al. (1953) The preparation of delta7,9(11)-allo-steroids by the action of mercuric acetate on delta7-allo steroids. *J. Am. Chem. Soc.* **75,** 2604–2609.
59. Fischer, R. T., Stephenson, F. A., Shafiee, A., and Schroeder, F. (1984) Delta 5,7,9(11)-Cholestatrien-3 beta-ol: a fluorescent cholesterol analogue. *Chem. Phys. Lipids* **36,** 1–14.
60. Gonzalez, R. C., and Woods, R. E. (2002) *Digital Image Processing, 2nd ed.*, Prentice Hall, NJ, 222–275.
61. Mukundan, R. and Ramakrishnan, K. R. (1998) *Moment Function in Image Analysis: Theory and Applications*. World Scientific, River Edge, NJ.
62. Sluzek, A. (1995) Identification and Inspection of 2–D Objects Using New Moment-Based Shape Descriptors. *Patt. Recog. Lett.* **16,** 687–697.
63. Hu, M. (1962) Visual-Pattern Recognition by Moment Invariants. *IRE Trans. Info. Theory* **8,** 179–187.
64. Neter, J., Kutner, M. H., Wasserman, W., and Nachtsheim, C. J. (1996) *Applied Linear Statistical Models, 4th ed.*, McGraw-Hill/Irwin, Chicago, IL, 188–210.
65. Diggle, P. J. (2003) *Statistical Analysis of Spatial Point Patterns, 2nd ed.*, A Hodder Arnold Publication, London, 6–7, 9–10, 30–100.
66. Barnard, G. A. (1963) Contribution to the discussion of Professor Bartlett's paper. *J. R. Stat. Soc.* **B25,** 294.
67. Baddeley, A. and Gill, R. D. (1997) Kaplan–Meier estimators of distance distributions for spatial point processes. *Ann. Stat.* **25,** 263–292.

9

NMR Detection of Lipid Domains

Ivan V. Polozov and Klaus Gawrisch

Summary

Methods for detection of lateral domains by solid-state ^{2}H nuclear magnetic resonance (NMR) and ^{1}H magic angle spinning (MAS)-NMR in model- and biomembranes are presented. ^{2}H NMR has been used for decades to distinguish between liquid-ordered and solid-ordered lamellar phases of phospholipids with deuterated hydrocarbon chains. More recently, it was shown that superposition of liquid-ordered and -disordered phases is detected as well, taking advantage of the large differences in chain order parameters between them. Experiments require preparation of samples with deuterated lipids. In contrast, ^{1}H MAS-NMR utilizes the natural proton NMR signals of lipids in model- and biomembranes. Very good resolution of resonances according to their chemical shifts is achieved by rapid spinning of samples at the "magic angle" (54.7°) to the main magnetic field. Phase transitions to ordered states are detected as broadening of resonances. The method distinguishes liquid-disordered, liquid-ordered, and solid-ordered phases, has much higher sensitivity than ^{2}H NMR, and does not require labeling. In combination with pulsed magnetic field gradients, ^{1}H MAS-NMR yields diffusion rates that may report confinement of lipids to domains with submicrometer dimensions.

Key Words: 1H NMR; 2H NMR; cholesterol; diffusion; domain; liquid-disordered; liquid-ordered; magic angle spinning; membrane; pulsed magnetic field gradient; raft; solid-ordered; solid-state NMR.

1. Introduction

Solid-state ^{2}H nuclear magnetic resonance (NMR) experiments conducted on phospholipid/cholesterol mixtures were instrumental for developing the concept of the liquid-ordered (L_o) phase *(1)*. The large differences in spectral appearance between solid-ordered (S_o), and L_o phases has been widely used to construct phase diagrams of cholesterol-containing phospholipid mixtures *(1–3)*. Recently, it was shown that coexistence of large L_o and liquid-disordered (L_d) domains is detected as signal superposition, taking advantage of the higher cholesterol content in

L_o phases and the related increase of chain order parameters *(3)*. Unfortunately, sensitivity of solid-state ^{2}H NMR experiments is rather low, requiring milligram quantities of chain perdeuterated lipid per sample and acquisition times on the order of 1 h. Low sensitivity is the result of the low gyromagnetic ratio of deuterium nuclei and the large bandwidth of spectra. Whereas the requirement for deuteration and sample size are easily met for model membranes, preparation of deuterated cell membranes is challenging. But on the positive side, ^{2}H-labeling is sterically non-perturbing and its influence on the strength of molecular interactions is negligible.

^{1}H magic angle spinning (MAS)-NMR is much more sensitive and does not require labeling. The acquisition of spectra from a few milligram of membrane material takes less than one minute, permitting to record the phase state of membranes at small temperature increments. However, calibration of temperature inside the spinning MAS rotor is somewhat difficult. Proton resonances of phospholipids and cholesterol are well resolved. Membrane proteins contribute to a broad background signal only. Recently, it has been shown that ^{1}H MAS-NMR spectra of L_d-, L_o-, and S_o phases are distinctly different *(4,5)*. Experiments require the use of a MAS probe with a low-proton background signal, and sample spinning as well as the spectrometer electronics must have high stability to enable reproducible acquisition of spectra that are free of distortions. This is particularly important for diffusion measurements by ^{1}H MAS-NMR with application of pulsed magnetic field gradients *(4,6)*. Formation of L_o phases in membranes is reflected by a significant reduction of diffusion coefficients *(7,8)*. Confinement of lipids to fluid domains with submicrometer dimensions is seen as dependence of apparent diffusion rates on diffusion time. At temperatures below the percolation threshold, the fluid domains are discountinuous and lipid diffusion is confined to domain boundaries *(4)*.

2. Materials

2.1. Deuterium NMR Experiments

1. Deuterated lipids as stock solution in organic solvent or as dry powder.
2. The solvent of choice for cholesterol-containing lipid mixtures is high-performance liquid chromatography grade, stabilized chloroform, or chloroform/methanol mixtures (*see* **Note 1**).
3. Water for ^{2}H NMR experiments must be deuterium depleted (2–4 ppm residual ^{2}H content).
4. Cylinder(s) with pure argon or nitrogen gas. For preparation of a large number of samples, the use of liquid-nitrogen tanks for gas use is recommended.
5. Vacuum stand for removal of organic solvents and for lyophilization of water-containing samples.
6. Cylindrical sample containers that fit into the solenoid coil of the NMR probe

(typi-cally 5 mm outer diameter). Thin-walled containers yield higher signal intensity owing to a higher filling factor but break more easily if sample preparation requires centrifugation. Containers must be sealed with a tightly fitting cap that prevents loss of water during experiments. This is particularly critical for experiments conducted at temperatures above ambient over extended periods of time. The use of miniature glass vials with small ground glass stoppers is recommended.
7. Desktop centrifuge capable of spinning sample containers at up to 5000g.
8. Solid-state NMR spectrometer equipped with a static ^{2}H NMR probe with a solenoidal sample coil, e.g. Bruker Biospin Avance (Bruker Biospin Corp., Billerica, MA) (*see* **Note 2**). The homogeneity of the main magnetic field should be sufficient to yield a resonance width of 25 Hz or better at 50% intensity and 100 Hz or better at 5% intensity. The probe and transmitter must yield a $\pi/2$ pulse length of 5 µs or less (2–3 µs desirable). A longer pulse length results in spectral distortions. The radiofrequency channels must have sufficient bandwidth to enable detection of signals that are free of phase- and amplitude distortions over a spectral width of at least 200 kHz. The instrument must be equipped with a variable temperature control unit that permits temperature adjustment to ±0.5 K or better.

2.2. Proton MAS-NMR Experiments

1. Lipids, cell suspensions, or tissue samples.
2. Water for ^{1}H MAS-NMR experiments should be at least 98% deuterated.
3. Ultracentrifuge capable of spinning samples with a volume of 1–5 mL at 100,000g or higher.
4. Desktop centrifuge with swinging buckets capable of spinning samples with a volume of 15 mL at 2000g, e.g. IEC Centra CL2 (Thermo Fisher Scientific, Inc. Waltham, MA).
5. MAS rotors with Kel-F inserts for liquid samples (4-mm rotor outer diameter, inserts with a spherical volume of 10–20 µL recommended, Bruker Biospin, Inc., Billerica, MA).
6. Solid-state NMR spectrometer equipped with a ^{1}H MAS probe capable of spinning liquid samples at a frequency of 10 kHz or higher. The instrument should have a resonance frequency of at least 300 MHz to separate lipid resonances by their chemical shift. A higher-resonance frequency improves resolution and sensitivity. The instrument must allow shimming of resonances better than 10 Hz at 50% intensity and better than 50 Hz at 5% intensity. The temperature controller must allow adjustment of sample temperature to ±0.5 K or better. Temperature gradients over the spinning sample must be less than 3 K. For diffusion experiments a ^{1}H MAS probe with a gradient coil that generates a magnetic field gradient parallel to the rotor axis is required. The maximal gradient strength must be at least 0.3 T/m. Application of stronger gradients permits detection of smaller domains, but is technically more challenging.

3. Methods
3.1. Sample Preparation
3.1.1. ^{2}H NMR Samples

Samples should contain 2–10 mg of deuterated lipid. The signal acquisition time increases quadratically with decreasing lipid content (*see* **Note 3**). Liposome size must be in the micrometer range. Smaller liposomes are undesirable because their small radius of curvature, in combination with lateral diffusion, results in lipid reorientation that averages out quadrupolar splittings. Samples containing polyunsaturated lipids are very prone to oxidation and must be prepared in a glove box filled with an oxygen-free atmosphere. Also, addition of an antioxidant like butylated hydroxytoluene at a lipid/butylated hydroxytoluene molar ratio of 200/1 is recommended.

1. Mix the lipids in organic solvent. Remove the solvent in a stream of argon or nitrogen gas while rotating the glass vial to form a thin lipid film on the wall. The time for complete removal of solvents depends on lipid layer thickness, chemical composition of lipids, and solvent properties. ^{1}H MAS-NMR is a convenient tool for detection of trace amounts of remaining solvents. If necessary, residual solvent is removed by exposing samples to vacuum (0.1 mB).
2. Hydrate the lipids by addition of 50 wt% deuterium depleted water or buffer solution prepared with deuterium depleted water (*see* **Note 4**), and collect the sample at the bottom of the tube by centrifugation.
3. Transfer the membrane pellet to the sample container for solid-state NMR experiments. The efficient transfer of milligram-size samples is challenging. It is conveniently achieved by inserting the tube upside down into the wider end of a plastic pipet tip. The thin end of the tip is cut such that it fits snuggly into the end of the NMR sample tube. This assembly is placed into the rotor of a desktop centrifuge with swinging buckets and the membrane pellet transferred to the NMR container by mild centrifugation. Please note that the sample must be in a liquid crystalline phase for efficient transfer. After transfer, mechanical stirring of the sample with a plastic tooth pick, freeze–thaw cycles, as well as annealing at a temperature well above the main phase-transition temperature of lipids may be applied to improve homogeneity. Annealing of samples as well as lower levels of hydration increase liposome size. However, addition of less than 20 water molecules per lipid is not desirable because insufficient hydration may alter the phase state.
4. ^{2}H NMR experiments on cell membranes requires at least partial deuteration of lipids. This is achieved by growing cells on fully or partially deuterated growth substrates and/or deuterated water (*see* **Note 5**). Cell membranes were also successfully doped with deuterated lipids by fusion with liposomes made up of deuterated lipids or by transfer of deuterated lipids to cell membranes from micellar solution, from cyclodextrin, from bovine serum albumin, or from lipid exchange proteins. However, such transfer may disturb the natural composition of lipids including their lateral distribution, as well as the asymmetry of transbilayer distribution.

3.1.2. Samples for ^{1}H MAS-NMR

The recommended size of ^{1}H MAS-NMR samples is 1–5 mg of lipid in a volume of 10–20 µL. Whereas this does not pose a problem for model membranes, cell membrane preparations may require ultracentrifugation to achieve such densities. For diffusion measurements by ^{1}H MAS-NMR, average liposome size must be 1 µm or larger.

1. Prepare lipid mixtures as described in **Subheading 3.1.1.**, but transfer them to a MAS rotor equipped with an insert for liquid samples. Concentrate cell membranes before the transfer by ultracentrifugation, resuspend the pellet in ^{2}H$_2$O or deuterated buffer, and pellet it again. This exchanges most of the H$_2$O for ^{2}H$_2$O and deuterates exchangeable protons. For a higher level of deuteration, samples may be repeatedly dispersed in deuterated water/buffer with subsequent pelleting. If sample density is insufficient, the H$_2$O/^{2}H$_2$O exchange can be combined with concentrating the cell material by osmosis. Cells and membrane vesicles are deflated and aggregated by addition of strong osmolytes like a solution of polyethylene glycol (PEG) in ^{2}H$_2$O or concentrated salt solutions, with subsequent separation of sample and osmolyte by centrifugation. High molecular weight PEGs also dehydrate samples through a dialysis bag, which avoids contamination of samples with PEG. Rates of water removal are greatly enhanced by continuously stirring the viscous PEG solution.
2. Transfer the membrane pellet to the MAS rotor using a plastic pipet tip of proper diameter and mild centrifugation as described in **Subheading 3.1.1., step 3** (*see* **Note 6**).

3.2. Solid-State NMR

3.2.1. ^{2}H NMR

3.2.1.1. ACQUISITION AND PROCESSING PARAMETERS

1. Set up the solid-state NMR instrument for detection of the ^{2}H NMR frequency. Resolution of ^{2}H resonances from deuterated lipid segments that have protonated groups as next neighbors, for example, a single deuterated methylene group in protonated hydrocarbon chain, may benefit from application of proton decoupling during signal acquisition. The latter requires a dual channel ^{1}H/^{2}H solids probe and spectrometer.
2. Select a quadrupolar echo pulse sequence (d_1-90°$_x$–τ-90°$_y$-acquire) *(9)*. The first pulse turns magnetization from the *z*-axis to the *x*–*y* plane. Magnetization spreads rapidly in the plane from the interaction between the electric quadrupole moment of the ^{2}H nucleus and the electric field gradients in C–D bonds. The second phase shifted 90°$_y$ pulse refocuses the signal to form an echo that has a maximal amplitude after the delay time τ. Acquisition of the free induction decay (FID) must begin well before the echo maximum is reached.

3. The ^{2}H carrier frequency is placed exactly at the center of the spectrum. Typical acquisition parameters are:
   a. Spectral band width 200 kHz or higher.
   b. Filter bandwidth of 500 kHz or higher.
   c. Dwell time between data points in the FID equal to 2.5 µs or less.
   d. Delay time τ between the π/2-pulses of 30–100 µs, depending on the ring-down time of the probe.
   e. 8192 data points per FID or higher.
   f. Recycle delay of 250 ms or longer.
   g. The π/2-pulses must have duration of less than 5 µs to ensure homogeneous excitation of resonances.
4. Left shift the data points of the FID at increments of 1/10th of a dwell time unit such that the first data point coincides precisely with the echo maximum *(10)*. This avoids the need for a first-order phase correction of the spectra after Fourier transformation and the related distortions of the spectral baseline.
5. Multiply the FID with an exponential window function equivalent to a line broadening of 50–200 Hz to reduce noise, convert the signal to the frequency domain by a Fourier transformation, and phase correct it using zero-order phase correction only. A computer program written for Mathcad (Mathsoft) (Parametric Technology Corp., Needham, MA) with tools for computer-aided adjustment of data points to the echo maximum, Fourier transformation, automatic phase adjustment, as well as calculation of smoothed chain order parameter profiles is available on request.

3.2.1.2. Interpretation of ^{2}H NMR Spectra

The ^{2}H NMR experiment yields so-called "powder pattern spectra" that are a superposition of resonances from bilayers oriented at random to the magnetic field **(Fig. 1)**. Every methylene group of a hydrocarbon chain yields a doublet with an orientation-dependent splitting. The two maxima in the powder pattern stem from lipid bilayers that are oriented with their normal perpendicular to the magnetic field. The separation between these maxima is called quadrupole splitting ($\Delta\nu_Q$). For methylene groups, it is converted to a bond-order parameter by dividing it by 125 kHz. Order parameters are a very sensitive measure of average bond orientation to the bilayer normal and of motional freedom of ^{2}H–C bonds *(11)*.

In the fluid L_d- and L_o phases, chain isomerization and lipid reorientation about the bilayer normal are sufficiently rapid to yield 100–300 Hz wide, superimposed resonances that appear as symmetric peaks in the spectra. The terminal methyl group of chains has lowest order parameters, and therefore, the smallest quadrupole splitting. Order of saturated hydrocarbon chains increases rapidly from the methyl end to the middle of the chain and remains constant up to the lipid carbonyl group (order parameter plateau). Quadrupole splittings from the region of the order parameter plateau are heavily superimposed. The L_o phase has higher cholesterol content as compared with L_d, which greatly

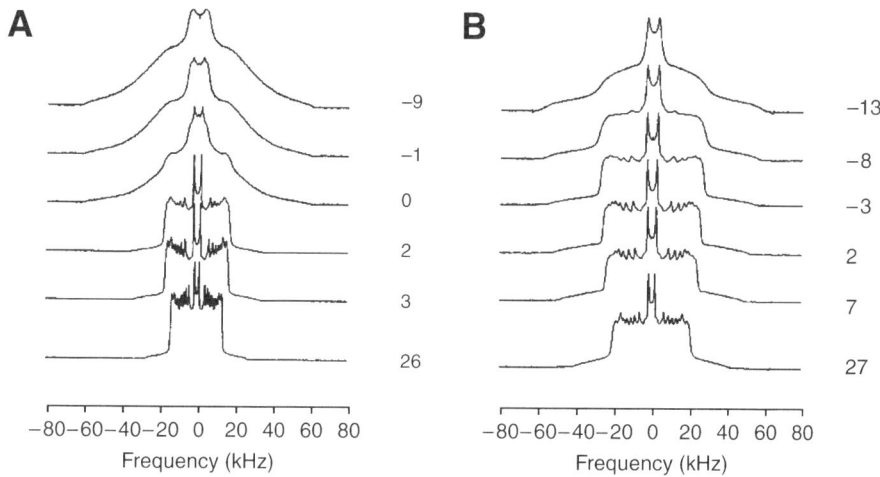

Fig. 1. (**A**) Deuterium NMR spectra of 1-stearoyl(d35)-2-oleoyl-*sn*-glycero-3-phosphocholine (SOPC-d35) recorded at −9, −1, 0, 2, 3, and 26°C. (**B**) ^{2}H NMR spectra SOPC-d35/cholesterol (7/3, mol/mol) recorded at −13, −8, −3, 2, 7, and 27°C.

increases quadrupolar splittings and makes spectra of both phases easily distinguishable, in particular in spectra with phase coexistence (**Fig. 2**).

In contrast to the fluid L_d- and L_o phases, S_o-phase spectra are broader and poorly resolved (**Fig. 1**). In the S_o phase, lipid hydrocarbon chains are packed in a crystalline lattice, which reduces rates of chain isomerization and lipid orientation by orders of magnitude. Therefore, order parameters of lipids in S_o are much higher and spectral width is increased. But slow motions of lipids in S_o also increase the resonance linewidth to values in the kilohertz range, resulting in mostly featureless spectra, except for the somewhat resolved quadrupolar splitting of terminal methyl groups. The rates of chain motions decrease further with decreasing temperature, resulting in a strong temperature dependence of S_o-phase spectra. Furthermore, at sufficiently low temperature, slow transitions between crystalline phases may take place that are reflected in the spectra as well.

For convenience, spectral changes as a function of temperature are often plotted as spectral first moment, M_1, calculated according to $M_1 = \int_0^\infty \omega f(\omega)d\omega / \int_0^\infty f(\omega)d\omega$, where $\omega = 2\pi\nu$, and ν is the resonance frequency in Hz (**Fig. 3**). By convention $\nu = 0$ corresponds to the center of the symmetric spectra. The first spectral moment is proportional to the averaged chain-order parameters of all deuterated lipid segments of the sample. It increases in the sequence $L_d < L_o < S_o$. Phase transitions are seen as discontinuity or change of slope in the plot of M_1 vs temperature.

The spectral differences between L_d-, L_o-, and S_o phases enable determination of phase state as a function of composition and temperature. For an

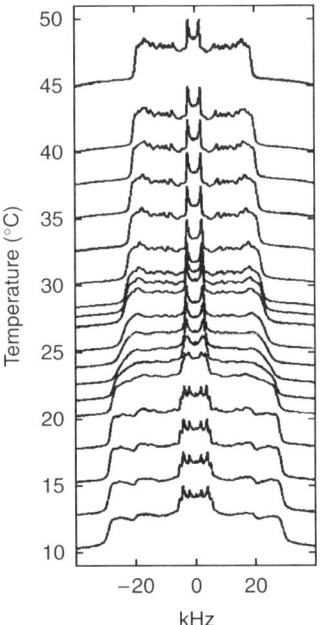

Fig. 2. Temperature dependence of first moments, M_1, of the ^{2}H NMR spectra of SOPC-d35 (△), and SOPC-d35/cholesterol (7/3, mol/mol) (▲) shown in **Fig. 1**.

unambiguous interpretation, the spectra of pure L_d-, L_o-, and S_o phases must be generated first. This is achieved by recording spectra at sample compositions and temperatures, which yield enrichment of a particular phase. The task of assigning a phase state(s) to a particular spectrum is complicated by the temperature and concentration dependence of spectra. For L_d- and L_o phases, small spectral changes from a change of temperature are reasonably well approximated by increa-sing/decreasing all quadrupolar splittings by the same factor. Small differences from a change of cholesterol content can be approximated by a scaling factor as well. Correction for the temperature dependence of S_o-phase spectra is challenging and should be avoided.

Spectra of pure phases are either directly recorded or generated by spectral subtraction. For example, a judiciously determined fraction of a spectrum obtained at the higher cholesterol concentration, which is enriched in L_o, is subtracted from the spectrum at lower cholesterol content to yield the spectrum of a pure S_o phase. The phase composition of other samples is then determined by computer-aided fitting, using a superimposition of L_d-, L_o-, and S_o-phase spectra at proper proportions (*see* **Note 7**). If lipids exchange between L_d-, L_o-, and S_o phases on a time-scale in the range of 10^{-5}–10^{-4} s then additional spectral changes

NMR of Lipid Domains

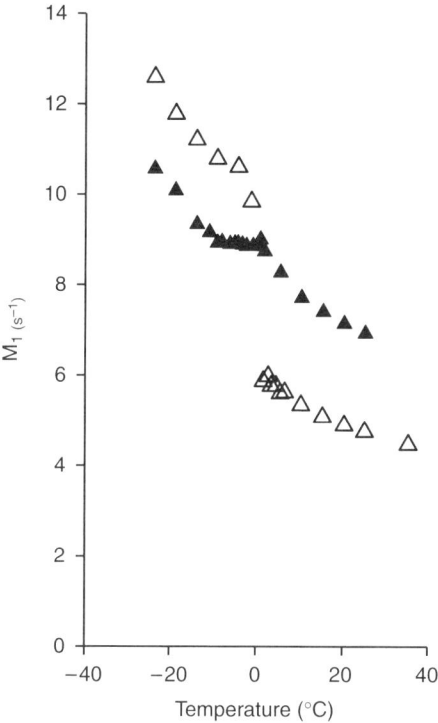

Fig. 3. ^{2}H NMR spectra of a DPPC-d62, 1,2-dileoyl-*sn*-glycero-3-phosphocholine (DOPC), cholesterol mixture, DPPC-d62/DOPC (1/1, mol/mol), containing 30 mol% cholesterol, recorded as a function of temperature. At temperatures more than 30°C this lipid mixture is in a L_d phase. Below 20°C coexisting L_d- and L_o phases are observed (reproduced from **ref. 3** with permission).

occur, which are not reproduced by a superposition of spectra of the corresponding phases. Exchange is recognized by severe broadening of resonances.

3.2.2. ^{1}H MAS-NMR

3.2.2.1. ACQUISITION AND PROCESSING PARAMETERS

1. Spin the sample at a MAS frequency of 10 kHz and adjust the sample temperature as needed (*see* **Note 8**).
2. Adjust the spectral bandwidth of the ^{1}H channel to cover all resonances of the MAS spinning centerband as well as the two first spinning side bands. Acquire the FID with a single π/2 hard pulse. Select a delay time between data acquisitions equal to 3–5 times the longest spin–lattice relaxation time of any resonance that will be used for data analysis. Chose a proper delay time between the hard pulse and the acquisition of the first data point of the FID to eliminate the need for a

first-order phase correction of spectra after Fourier transformation (*see* **Note 9**). Multiply the FID with an exponential window function equivalent to a line broadening of 1–20 Hz before Fourier transformation.

3. Record the spectrum of the MAS rotor filled with deuterated buffer at identical conditions and compare its intensity with the spectrum from the membrane-containing sample. Subtract the spectrum of the buffer-filled rotor from the membrane spectrum if needed. This eliminates residual proton signals from the MAS probe hardware. However, because such subtraction may also introduce spectral distortions, the use of a MAS probe with a low-proton background signal is highly desirable.

3.2.2.2. INTERPRETATION OF ^{1}H NMR SPECTRA

Information about the phase state of lipids in the membrane is extracted from chemical shifts, resonance intensities, and the linewidth of resonances (**Fig. 4**). The proton chemical shifts of lipids in organic solvent serve as guide for an initial assignment. Verification of assignments is achieved by two-dimensional homo- and heteronuclear chemical shift correlation experiments, conducted on the membrane sample in the spinning rotor *(12)*. The peak of highest integral intensity in membrane spectra is usually the hydrocarbon chain methylene resonance at 1.3 ppm. Information about the phase state of lipids is conveniently extracted from the linewidth of this peak as well as its spinning sideband/centerband intensity ratio (**Fig. 5**). The linewidth increases in the sequence $L_d < L_o < S_o$. At a MAS-spinning frequency of 10 kHz it is 25–100 Hz for L_d, 500–1000 Hz for L_o, and larger than 1.5 kHz for S_o. The 10-fold increase of linewidth and corresponding decrease of resonance intensity on a transition from L_d to L_o or S_o is easily detected. The S_o phase is distinguishable from L_o by its much higher spinning sideband/centerband intensity ratio. Typical sideband/centerband ratios at a MAS frequency of 10 kHz are 0.02 for L_d, 0.05 for L_o, and 0.2 for S_o *(5)*.

The onset and completion of phase transitions in membranes are detected as discontinuities in plots of peak intensity vs temperature (**Fig. 6**). Formation of L_o or S_o at lower temperature is reflected as break points in the curves. The plot of spinning sideband/centerband intensity ratio is most sensitive to S_o formation. Also, the much higher intensity of spinning sidebands of ordered phases results in discernible spectral changes between spinning center and -sidebands. In spectra of coexisting disordered and ordered phases the centerband intensity has stronger contributions from L_d whereas, signal intensity of sidebands is dominated by S_o and L_o. Therefore, the linewidth of resonances in the spinning sidebands tends to be larger than the corresponding linewidth of the spinning centerband. The magnitude of differences is a qualitative measure for the amount of S_o and L_o in the sample.

The strong linebroadening of chain resonances on transition of lipids into L_o or S_o makes ^{1}H MAS-NMR an uniquely sensitive tool for detection of ordered

NMR of Lipid Domains

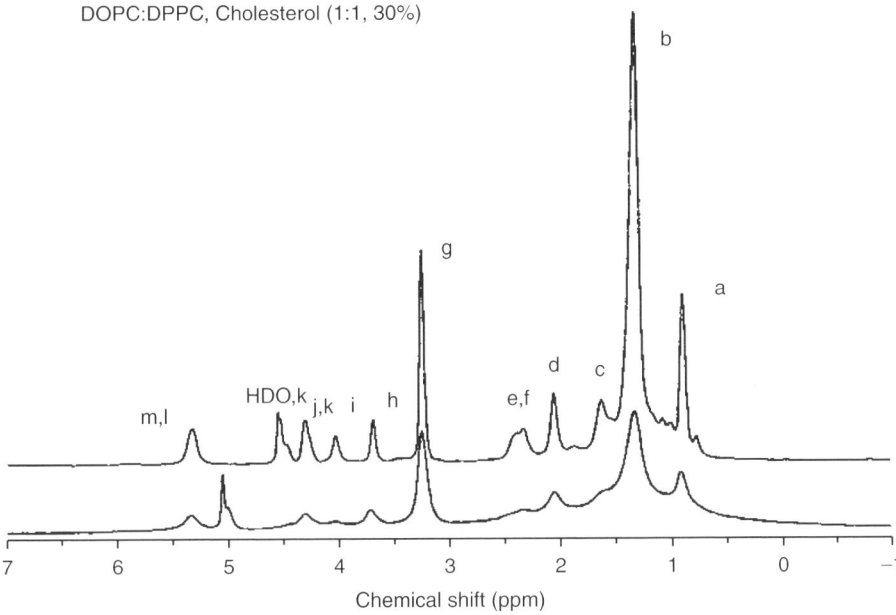

Fig. 4. ^{1}H MAS-NMR spectra of multilamellar DPPC/DOPC/cholesterol liposomes. The top spectrum (L_d phase) was acquired at 45°C and the bottom one (L_d- and L_o-phase coexistence) at 8°C. Lipid resonances are assigned by labels as shown on the DOPC molecule. The DPPC spectrum is similar to DOPC, except for the lack of doubled bond-related resonances (d, m) (reproduced from **ref. 3** with permission).

lipid domains, irrespective of their size. However, appearance of spectra depends on the rate of lipid exchange between domains. If fluid domains are sufficiently small, lipids exchange rapidly between them on the time-scale of 10^{-4} s. Resonance lines have Lorentzian shape with a linewidth, $\Delta v_{1/2}$, which reflects the fractional contributions of ordered and disordered phases

$$\Delta v_{1/2} = p_o \Delta v^o_{1/2} + p_d \Delta v^d_{1/2}$$

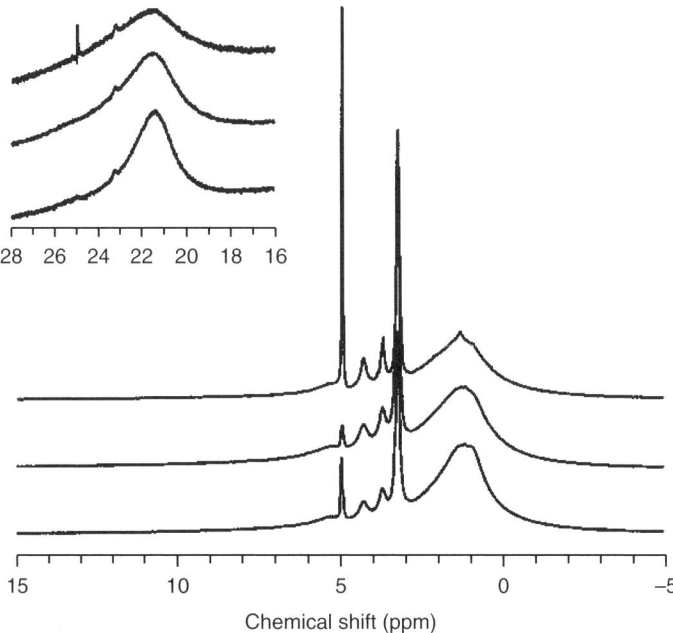

Fig. 5. ^{1}H MAS-NMR spectra (from top to bottom) of SOPC, SOPC with 15 mol% cholesterol, and SOPC with 30 mol% cholesterol at 4, −2, and −4°C, respectively, corresponding to the S_o phase (SOPC), S_o–L_o coexistence (15 mol% cholesterol), and mostly L_o (30 mol% cholesterol). The spinning sidebands of first-order of the chain methylene resonance at 1.3 ppm are shown at 20-fold magnification in the inset (reproduced from **ref. 5** with permission).

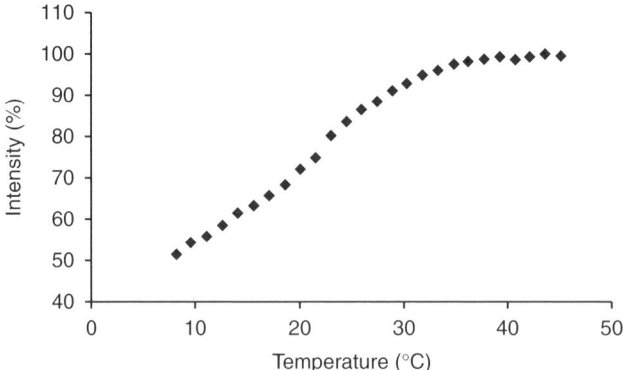

Fig. 6. Peak height of the methylene resonance (1.3 ppm) of the spectrum shown in **Fig. 3** plotted as a function of temperature. The decrease of peak height at temperatures less than 30°C reflects formation of an ordered lipid phase (reproduced from **ref. 3** with permission).

where p_o and p_d are the mole fraction of L_o- and disordered phases, respectively, and $\Delta v_{1/2}^o$ and $\Delta v_{1/2}^d$ are the linewidths of resonances in those phases *(5)*. Because the linewidth of chain resonances in L_o is more than one order of magnitude larger compared with L_d, even a small fraction of an L_o phase in exchange with L_d increases $\Delta v_{1/2}$ significantly and reduces signal height.

If coexisting L_d- and L_o domains are large, spectra are a superposition of resonances from the two phases. Coexistence with an S_o-phase results almost always in signal superposition because lipid exchange rates between fluid and solid phases are usually low, irrespective of their size. Signal superposition from phase coexistence is detected as isosbestic points in superimposed spectra recorded as a function of temperature. The fraction of lipids in the phases is obtained from a fit of the experimental spectrum as superposition of the spectra from the contributing phases (*see* **ref. 5**).

Spinning of samples in MAS rotors at 10 kHz generates centrifugal forces of up to $6 \times 10^5 g$, which may separate membranes from excess water of hydration *(13)*. However, even at such high centrifugal forces, severe dehydration, which alters membrane lateral organization is unlikely to occur. If necessary, dehydration is prevented by adjusting the density of the water phase to the density of lipid membranes, for example, by hydrating the sample with a proper mixture of H_2O and ^2H_2O *(14)*.

3.2.3. Diffusion Measurements by ^1H PFG MAS-NMR

3.2.3.1. ACQUISITION AND PROCESSING PARAMETERS

1. Spin the sample prepared according to **Subheading 3.1.2.** in a MAS probe equipped with a coil for pulsed magnetic field gradients that are oriented parallel to the axis of the spinning rotor. Choose a MAS-spinning frequency in the range of 3–10 kHz at which spinning is very stable (*see* **Note 10**).
2. Select a stimulated echo sequence with shaped bipolar gradient pulses for data acquisition (*see* **Fig. 7**). This sequence reduces spectral baseline distortions from a perturbation of the main magnetic field *(4,6,15)* (*see* **Note 11**).
3. Set the delay time between data acquisitions to three to five times the longest spin–lattice relaxation time of any resonance that will be used for data analysis. Carefully determine the length of the π/2- and π-pulses. Select the maximal gradient strength, gradient length, and diffusion time to achieve a significant reduction of signal intensity at the highest gradient strength (*see* **Note 12**). Increase the length of the recovery time (*T*), until spectral perturbations are tolerable.
4. Acquire sets of spectra as a function of gradient strength. After Fourier transformation, carefully correct the spectral phase and baseline before measuring signal amplitudes. Follow instructions in *(6,12)* to calculate diffusion rates from the dependence of signal intensity on gradient strength.
5. Measure apparent diffusion rates as a function of diffusion time.

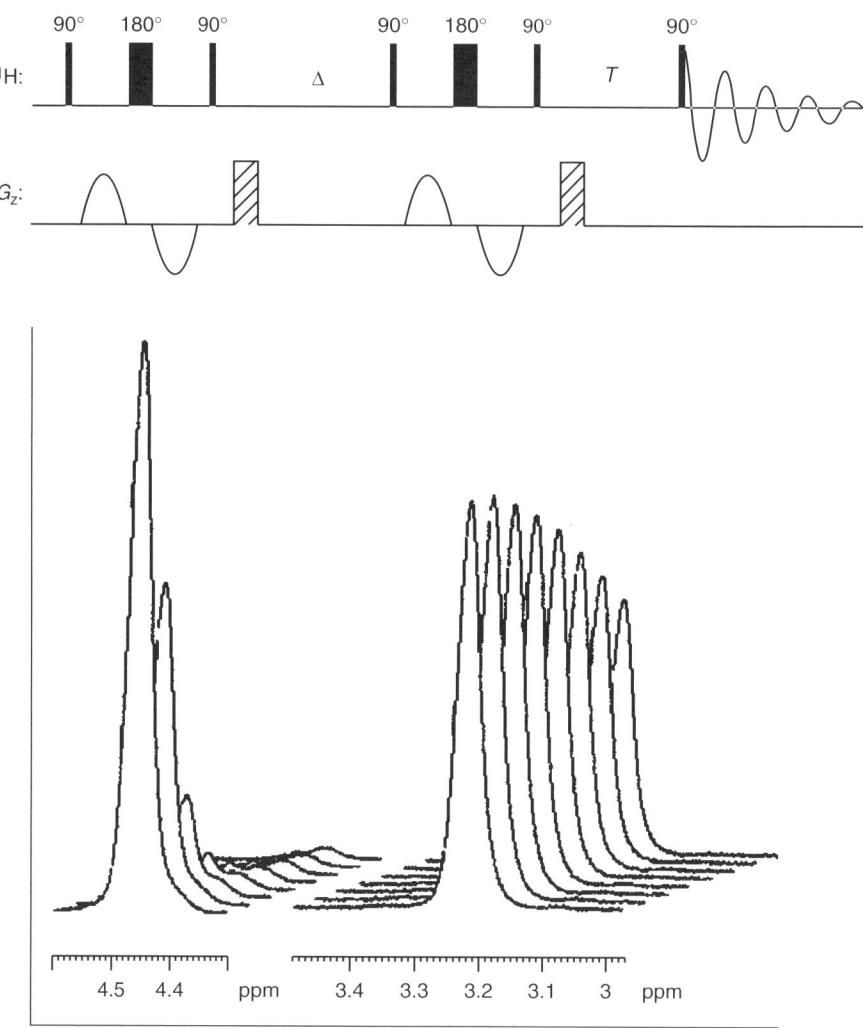

Fig. 7. Stimulated echo sequence with shaped bipolar gradient pulses for diffusion measurements by ^{1}H MAS-NMR. The spectra show water **(left)** and lipid resonances **(right)** of multilamellar liposomes, recorded at increasing strength of pulsed magnetic field gradients. The water signal decays much faster with increasing gradient strength, indicating higher rates of lateral diffusion of water molecules.

3.2.3.2. Data Analysis—Determination of Domain Size

The diffusion of lipids and of interlamellar water is confined to layers. If samples are prepared as described under **Subheading 3.1.2.**, those layers are oriented at random to the applied magnetic field gradient. The diffusion coefficient (D) of such randomly oriented layers is determined from the dependence of peak

NMR of Lipid Domains

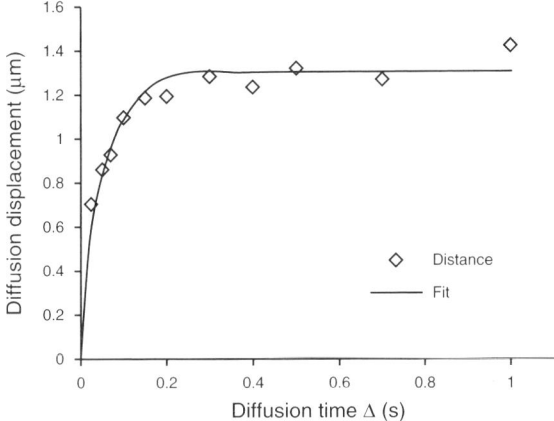

Fig. 8. Average distance that lipids move during the diffusion time (Δ). Liposomes are mixtures of SOPC/1-palmitoyl-2-oleolyl-*sn*-glycero-3-ethanolamine (3/7, mol/mol). The experiments were conducted at a temperature above the L_d–S_o phase transition range (33.4°C). The solid line is a fit to the data assuming an average liposome radius of 1.3 ± 0.1 μm. The fit yields a curvature corrected lateral diffusion constant $D = 0.8 \pm 0.1 \times 10^{-11}$ m²/s (reproduced from **ref. 4** with permission).

intensity on gradient strength according to *(5)*: $\ln\left(\dfrac{I}{I_0}\right) = -\dfrac{2}{3}kD + \dfrac{2}{45}(kD)^2$, where I/I_0 is the normalized intensity of the lipid peak. For square-shaped pulses $k = 4\gamma^2 g^2 \delta^2 \left(\Delta - \dfrac{d_2}{16} - \dfrac{\delta}{3}\right)$ where γ is the gyromagnetic ratio of protons, g is the gradient strength, δ is the gradient pulse length, and d_2 the time between the gradient pulses that sandwich the π-pulse. Please note that the aforementioned formula is an approximation, which is only valid for a gradient-induced signal attenuation of up to 90%.

The data analysis yields an apparent diffusion constant D_{app} that is equal or smaller than the true lateral diffusion constant in bilayers depending on membrane curvature or on spatial restrictions to lipid diffusion from formation of domains. Curvature effects and size restriction are reflected as a reduction of D_{app} with increasing diffusion time. For analysis, plot the average distance (x), lipids that move during the diffusion time (Δ), calculated according to $x(\Delta) = (2D_{app}\Delta)^{1/2}$ (**Fig. 8**). Diffusion is spatially restricted if $x(\Delta) = constant$ for sufficiently large Δ. A reduction of *constant* with decreasing temperature indicates constraints to lipid diffusion from domain formation. The value of *constant* is a measure of average domain size. Such spatial restrictions to lipid lateral diffusion in a binary lipid mixture were observed in the region of a L_d–S_o

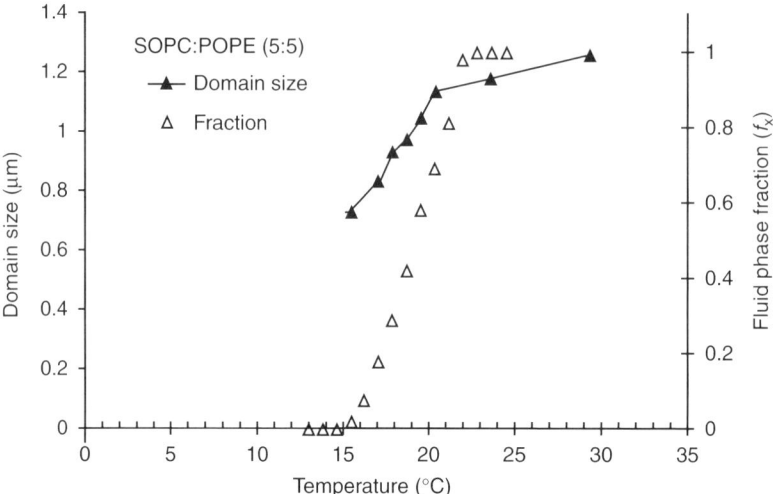

Fig. 9. Fraction of lipids in the L_d phase in SOPC/1-palmitoyl-2-oleolyl-*sn*-glycero-3-ethanolamine (3/7, mol/mol), liposomes (open triangles), and diffusion distances (filled triangles) as function of temperature. Above the onset of the L_d–S_o phase transition, average lipid displacement reports the size of liposomes. Below the percolation threshold, the distance of lipid movement reflects the size of L_d-phase domains (reproduced from **ref. 4** with permission).

phase transition *(4)* (**Fig. 9**). Lipid diffusion experiments by ^{1}H PFG MAS-NMR can be conducted on both model and biological membranes.

4. Notes

1. Purchase only high-grade, stabilized chloroform in small bottles that are used up within weeks after being opened. Old chloroform may contain trace amounts of hydrochloric acid that is harmful to lipids.
2. The resonance frequency of the spectrometer is of secondary importance because the spectral width of ^{2}H NMR spectra is entirely determined by quadrupolar interactions that do not depend on magnetic field strength. However, sensitivity of signal detection increases in a nonlinear fashion with increasing field strength, making the use of higher fields desirable. Unfortunately, at higher field strength membrane samples tend to slowly gain a preferred orientation in the magnetic field, which may perturb a quantitative data analysis. Protein-free lipid bilayers tend to orient with their normal perpendicular to the field, whereas samples with a high content of proteins with transmembrane helices tend to orient parallel to the field. In most cases, a field of 7 T is an acceptable compromise between decent sensitivity and a sufficiently low tendency for a field-induced bilayer orientation.
3. The low sensitivity of ^{2}H-NMR usually necessitates acquisition of a large number of FID for signal detection. The signal-to-noise ratio of NMR spectra increases

with the square root of the number of acquisitions. Therefore, the acquisition time increases with the square of decreasing sample size, setting practical lower limits for sample size. For example, a sample of 1/4th the size requires a 16-fold increase of acquisition time to yield the same signal-to-noise ratio.

4. Expenses for deuterium-depleted water are reduced by preparation of small quantities of deuterium-depleted buffer, for example, 1 mL. This is conveniently achieved by starting out from a stock solution of the buffer prepared with regular, deionized water. A proper amount of this stock solution is lyophylized and the residue dissolved in an equal amount of deuterium-depleted water in the same vial.

5. Deuterated water inhibits cell growth, might alter membrane composition, and is lethal for most living organisms at high concentrations. The tolerable limits of deuterated water in growth solutions must be determined individually. In contrast, feeding of deuterated growth substrates like fatty acids, glucose, and so on, is usually well tolerated. ^{2}H labeling of lipids does not change lipid conformation and motional properties. However, there is a small effect from deuteration on the strength of hydrogen bonds and on phase transition temperatures. For example, chain perdeuterated 1,2-dipalmitoyl(d62)-sn-glycero-3-phosphocholine DPPC-d62) has a gel–fluid phase transition at T_m = 310 K whereas protonated DPPC has a T_m = 314.5 K. As a rule of the thumb, the main phase-transition temperature of lipids is lowered by 2 K per deuterated hydrocarbon chain.

6. The transfer of protein-free liposomes that contain an excess of ^{2}H$_2$O by centrifugation is challenging because the deuterated water has usually a higher density than the membranes. Therefore, any ^{2}H$_2$O that was added in excess will fill the 10–20 µL volume of the MAS insert first, although the membrane material forms an upper layer that may not reach the NMR active volume. Therefore, excess ^{2}H$_2$O that may separate from the membranes during the transfer must be removed, for example, by centrifugation followed by removal of water with a syringe or by blotting with filter paper.

7. The quadrupolar echo pulse sequence attenuates spectral intensities according to differences in the spin–spin relaxation times T_2 in the sequence $S_o > L_o > L_d$. As a result, S_o- and L_o-spectral contributions have somewhat lower intensity compared with L_d. To correct for the differences in T_2, spectra are recorded at several delay times τ (*see* pulse sequence in **Subheading 3.2.1.**), and the $L_d/L_o/S_o$ intensity ratio extrapolated to τ = 0.

8. The sample temperature in a MAS rotor depends on the temperature of the bearing gas. Some MAS probes control sample temperature additionally by a flow of temperature-controlled gas into the probe chamber of the MAS stator. The influence from the air temperature that controls MAS spinning is minor, but the spinning frequency itself is critical. At spinning frequencies up to 5 kHz, the rotor temperature is a few degrees lower than the temperature of the bearing gas because of the Joule–Thompson effect on gas expansion at the bearings. At spinning frequencies above 5 kHz the rotor temperature increases in a nonlinear fashion with increasing spinning frequency resulting from frictional heating at the bearings. The temperature inside the spinning rotor is conveniently measured through the chemical shift

difference between water and choline in a micellar solution of 1,2-dicaproyl-*sn*-glycero-3-phosphocholine (Avanti Polar Lipids; Alabaster, AL) loaded into a MAS rotor with insert for liquid samples. The chemical shift to temperature calibration curve is determined in high-resolution NMR experiments using a temperature calibrated probe. Please note that every spinning frequency requires a separate calibration curve. The calibration is usually reproducible to within ±0.5 K if rotors and inserts are of the same type, spinning frequencies are equal or less than 10 kHz, and samples spin without hesitation. The temperature inside the MAS rotor can be verified by recording spectra of lipids with known gel–fluid phase transition temperatures. Also, the chemical shift of water in the samples is a very sensitive measure of relative changes of temperature.

Most of the heating takes place at the two bearings at each end of the rotor, although the center of the rotor is chilled by the expanding gas. This may cause significant temperature gradients across the sample. Highly conductive samples may also heat because of induction of eddy currents. However, the latter is unlikely for biomembrane samples at physiological concentrations of saline. In the authors' experience, a spinning frequency of 10 kHz is an acceptable compromise between sufficient resolution of gel phase spectra, reproducibility of sample temperature, and tolerable temperature gradients over the sample of 3 K or less. A significant reduction of temperature gradients is achieved by using rotor materials with higher heat conductivity.

9. Proper adjustment of this delay time is critical for reproducibility of spectra with a large bandwidth and broad resonances. If first-order phase correction must be applied, such spectra may have a baseline roll, which complicates data analysis. The proper value of the delay time depends on the bandpass filters in the NMR receiver. Values in the range from zero to one dwell time unit (time between two data points in the FID) are typical.

10. Any movement of the rotor along the spinning axis during the diffusion time disturbs the experiment. Requirements for spinning stability increase with increasing gradient strength. The application of the gradient pulses exerts a force on the gradient coil that may trigger mechanical vibrations. Indicators for spinning instability and mechanical vibrations are:
    a. Modulation of tune and match of probe resonances from spinning and application of field gradients.
    b. Poor signal reproducibility between consecutively acquired spectra.
    c. Instability of the spinning frequency.
    d. Strong pinging sounds from the gradient coil on application of gradient pulses.

11. Trapezoidal-shaped gradient pulses with short raise and decay times yield almost the same effective gradient strength as square pulses but without the strong perturbations of the main field.

12. Do not exceed gradient current, gradient length, and gradient duty cycle beyond the limits specified by the probe manufacturer to avoid equipment damage. With increasing diffusion time, maximal signal intensity declines exponentially owing to spin–spin and spin–lattice relaxation. The intensity of resonances with shorter relaxation times declines more rapidly, which changes the appearance of spectra.

Also, long diffusion times result in magnetization transfer between resonances through the nuclear Overhauser effect that perturbs the experiment. For most samples, diffusion times of less than 200 ms are desirable. Application of stronger gradients enables conducting experiments at shorter diffusion times. The strength of pulsed field gradients as well as the linearity of gradient strength is measured by conducting experiments on a substance with a known diffusion constant, for example, water *(16)*.

Acknowledgments

This work was supported by the Intramural Research Programs of National Institute on Alcohol Abuse and Alcoholism (NIAAA) and National Institute of Child Health & Human Development (NICHD), NIH.

References

1. Vist, M. R. and Davis, J. H. (1990) Phase equilibria of cholesterol/dipalmitoylphosphatidylcholine mixtures: 2H nuclear magnetic resonance and differential scanning calorimetry. *Biochemistry* **29**, 451–464.
2. Thewalt, J. L. and Bloom, M. (1992) Phosphatidylcholine—cholesterol phase diagrams. *Biophys. J.* **63**, 1176–1181.
3. Veatch, S. L., Polozov, I. V., Gawrisch, K., and Keller, S. L. (2004) Liquid domains in vesicles investigated by NMR and fluorescence microscopy. *Biophys. J.* **86**, 2910–2922.
4. Polozov, I. V. and Gawrisch, K. (2004) Domains in binary SOPC/POPE lipid mixtures studied by pulsed field gradient H-1 MAS NMR. *Biophys. J.* **87**, 1741–1751.
5. Polozov, I. V. and Gawrisch, K. (2006) Characterization of the liquid-ordered state by proton MAS NMR. *Biophys. J.* **90**, 2051–2061.
6. Gaede, H. C. and Gawrisch, K. (2003) Lateral diffusion rates of lipid, water, and a hydrophobic drug in a multilamellar liposome. *Biophys. J.* **85**, 1734–1740.
7. Filippov, A., Orädd, G., and Lindblom, G. (2003) The effect of cholesterol on the lateral diffusion of phospholipids in oriented bilayers. *Biophys. J.* **84**, 3079–3086.
8. Scheidt, H. A., Huster, D., and Gawrisch, K. (2005) Diffusion of cholesterol and its precursors in lipid membranes studied by H-1 pulsed field gradient magic angle spinning NMR. *Biophys. J.* **89**, 2504–2512.
9. Davis, J. H., Jeffrey, K. R., Bloom, M., Valic, M. I., and Higgs, T. P. (1976) Quadrupolar echo deuteron magnetic resonance spectroscopy in ordered hydrocarbon chains. *Chem. Phys. Lett.* **42**, 390–394.
10. Davis, J. H. (1983) The description of membrane lipid conformation, order and dynamics by 2 H-NMR. *Biochim. Biophys. Acta* **737**, 117–171.
11. Seelig, J. (1977) Deuterium magnetic resonance: theory and application to lipid membranes. *Q. Rev. Biophys.* **10**, 353–418.
12. Gaede, H. C. and Gawrisch, K. (2004) Multi-dimensional pulsed field gradient magic angle spinning NMR experiments on membranes. *Magn. Reson. Chem.* **42**, 115–122.

13. Nagle, J. F., Liu, Y. F., Tristram-Nagle, S., Epand, R. M., and Stark, R. E. (1999) Re-analysis of magic angle spinning nuclear magnetic resonance determination of interlamellar waters in lipid bilayer dispersions. *Biophys. J.* **77,** 2062–2065.
14. Koenig, B. W. and Gawrisch, K. (2005) Specific volume of unsaturated phosphatidylcholines in the liquid crystalline phase. *Biochim. Biophys. Acta* **1715,** 65–70.
15. Gaede, H. C. and Gawrisch, K. (2004) Multidimensional PFG-MAS-NMR experiments on membranes. *Magn. Reson. Chem.* **42,** 115–122.
16. Mills, R. (1973) Self diffusion in normal and heavy water in range 1–45 degrees. *J. Phys. Chem.* **77,** 685–688.

10

Lateral Diffusion Coefficients of Raft Lipids From Pulsed Field Gradient NMR

Greger Orädd and Göran Lindblom

Summary

The pulsed field gradient-nuclear magnetic resonance diffusion technique has an appreciable potential for biophysical investigations in membrane biology, various lyotropic liquid crystals, and other complex fluid systems. In particular, topics like transport of molecules both across and within the plane of a lipid membrane can be successfully studied, as well as the formation of lipid domains and their intrinsic dynamics. The pulsed field gradient-nuclear magnetic resonance technique and the preparation of oriented samples for investigations of lipid lateral diffusion in macroscopically aligned bilayers, oriented by a goniometer probe in the main magnetic field, are described. Some recent results illustrating the potential of the method in detecting and characterizing domain formation are also presented.

Key Words: Lateral diffusion; PFG-NMR; phospholipids; cholesterol; lateral phase separation; domains; liquid-ordered phase; liquid-disordered phase.

1. Introduction

The nuclear magnetic resonance (NMR) methods with pulsed magnetic field gradients provide some of the most attractive techniques for studies of molecular transport and the applicability of the techniques has been growing fast because of many improvements of the NMR equipments for diffusion, measurements, and microscopy *(1)*. One of the most successful applications of pulsed magnetic field gradients (PFG)-NMR is its use in extracting structural information about heterogeneous systems such as complex liquids and liquid crystals *(2–4)*, and in particular, lipid lateral diffusion coefficients in a macroscopically oriented lamellar liquid-crystalline (L_α) phase that can be directly measured *(2,5,6)*.

1.1. Method Prerequisites

The PFG-NMR method, which is based on the refocusing of a spin echo, requires that the transverse relaxation time is sufficiently long to allow for insertion of PFG pulses in the defocusing and refocusing periods. This is usually not a problem for isotropic liquids and liquid crystals but for anisotropic phases, the dipolar coupling will cause a substantial line broadening that prevents the observation of the spin echo, unless special solid-state NMR techniques are used *(7,8)*. However, all the static NMR interactions have a common scaling term, namely, $3\cos^2\theta_{LD} - 1$, where θ_{LD} is the angle between the bilayer normal and the main magnetic field (B_0) *(9)*. For $\cos^2\theta = 1/3$, i.e., $\theta = 54.7°$, this scaling term becomes zero and the static interactions disappear. In order to achieve this, the lipid bilayers must be oriented in parallel on a flat solid support (e.g., glass plates) so that the bilayer normals are perpendicular to the glass plates and oriented at an angle of 54.7° with respect to B_0.

1.2. PFG-NMR and Domain Formation

It is generally accepted that macro- and microdomains exist within the plane and across the lipid bilayers, and that their presence in a biological membrane may be associated with important cell functions, such as signal transduction, lipid trafficking, transcytosis, protein sorting, and virus budding *(10–12)*. In the simplest form, such domains occur in systems of two-phase coexistence of a liquid-ordered (L_o) phase in which the lipids are more tightly packed, and a surrounding liquid-disordered (L_d) phase *(13,14)*. However, the formation of biologically functional domains, also referred to as rafts, requires additional components of lipids and/or proteins *(10)*. The influence of domains on the translational motion of lipids enables the PFG-NMR diffusion method to detect domain formation in lipid membranes. Lipids will either diffuse in and out of the domain and the "lipid sea" by an exchange mechanism, or provided that the border between different domains presents an obstacle to lipid diffusion, lipids will encounter restrictions in the diffusional motion. Both these processes will affect the observed diffusion coefficient and permit studies of both formation and size of the domains.

2. Materials

1. Organic solvents—PA quality.
2. Deuterated water—99.9%.
3. Lipids—the lipids are stored at −30°C, and when needed, care is taken to avoid exposure to strong light and oxygen.
4. Glass plates—microscope cover plates, thickness no. 00 (0.06–0.08 mm) are cut to specified sizes by Marienfeldt Laboratory Glassware (Lauda-Königshofen, Germany). The size depends on the dimensions of the sample holder and the

size of the rf-coil of the NMR probe. Two different sizes are used; 14 × 4.7 and 14 × 2.5 mm².
5. Sample holders—the glass plates are stacked into glass tubes with quadratic inner cross-section. These glass tubes are ordered from Wale Apparatus (Hellertown, PA) and have an inner side of 3 (S-103) and 5 (S-105) mm. The tubes are cut to specific lengths to fit into the NMR probes, 68 mm (S-103) and 18 mm (S-105) and after the sample loading the tubes are sealed with square caps specially made of polymethyl methacrylate (PMMA) (S-105) or by wax (S-103). A number of specially made sample holders with cylindrical outer dimensions were also manufactured in polyetheretherketon (PEEK) for the larger glass plates in order to facilitate the sealing of the ends. For these holders the ends were sealed with o-rings and screw caps.

3. Methods

3.1. Preparation of Macroscopically Oriented Bilayers

Lipids dissolved in a 1:4 mixture of methanol:1-propanol are deposited onto thoroughly cleaned, but otherwise untreated glass plates, to a concentration of about 5–15 µg/mm². The solvent is evaporated and the plates are placed into high vacuum for at least 4 h to remove traces of solvent (*see* **Note 1**). The plates are then stacked on top of each other and placed into a glass tube with square cross-section. Typically, 35 (5-mm tube) or 60 (10-mm tube) plates are used for one sample. The sample tube is placed for several days in a humid atmosphere above the gel to L_α-phase transition temperature. During this period of time hydrated and oriented bilayers are formed.

Finally, after attaining the desired water content (checked by weighing), the tube is sealed and the sample is left another day or two for final equilibration. This procedure results, in that, large dark areas are observed in the sample when it is watched perpendicular to the glass plates between crossed polarizers, as the bilayers are optically isotropic along the normal (the director) *(6)*. Mostly, more than 85% of the sample is found to be oriented as determined by ^{31}P NMR, whereby the area under the signal from the oriented part of the sample is compared with the total signal area *(6)* (*see* **Note 2**).

3.2. Orienting the Samples in the Magnetic Field

As mentioned in the Subheading 1., the static interactions that produce linewidths of several kilohertz must be removed by orienting the bilayer normals at the magic angle (θ_{LD} = 54.7°). To accomplish this, a specially built goniometer probe is used (Fig. 1). By turning the rod connected to the goniometer stage the sample can be rotated from outside of the magnet so that the bilayer normal makes an angle of 54.7° with respect to B_0, thereby canceling the dipole interactions and producing an isotropic-like spectrum. In practice this is achieved by maximizing the spin echo signal. In doing so the effective transverse

Fig. 1. The rf-coil of the 10-mm probe with a goniometer sample holder that enables one to turn the sample about an axis perpendicular to B_0. The glass tube that holds the sample (*see* lower enlarged picture) is turned by operating the angular gear observed in the upper enlarged picture. The anti-Helmholtz coils that produce the field gradients have been removed from the probe in order to show the goniometer stage.

relaxation time changes from submilliseconds to 20–50 ms, thereby enabling the spin echo experiment to be performed (Fig. 2).

3.3. PFG-NMR Diffusion Measurements

The use of NMR for diffusion measurements rests on the ability to create transverse magnetization with a precession rate that is dependent on the local magnetic field. The details of this method are beyond the scope of this chapter and the interested reader is referred to previously published reviews *(2–4,15)*. Herein, only a brief introduction will be given. Two basic spin echo experiments for diffusion measurements are illustrated in **Fig. 3**. The spin echo (SE) sequence *(16)* (**Fig. 3**, top) depends on the creation of transverse magnetization in a time interval in which the nuclear spins with different precession rates are allowed to dephase in the xy-plane. After a time τ the dephasing process is reversed by the application of a 180° rf-pulse and the nuclear spins begin to rephase and eventually they meet again to form a spin echo.

Lateral Diffusion Coefficients of Raft Lipids

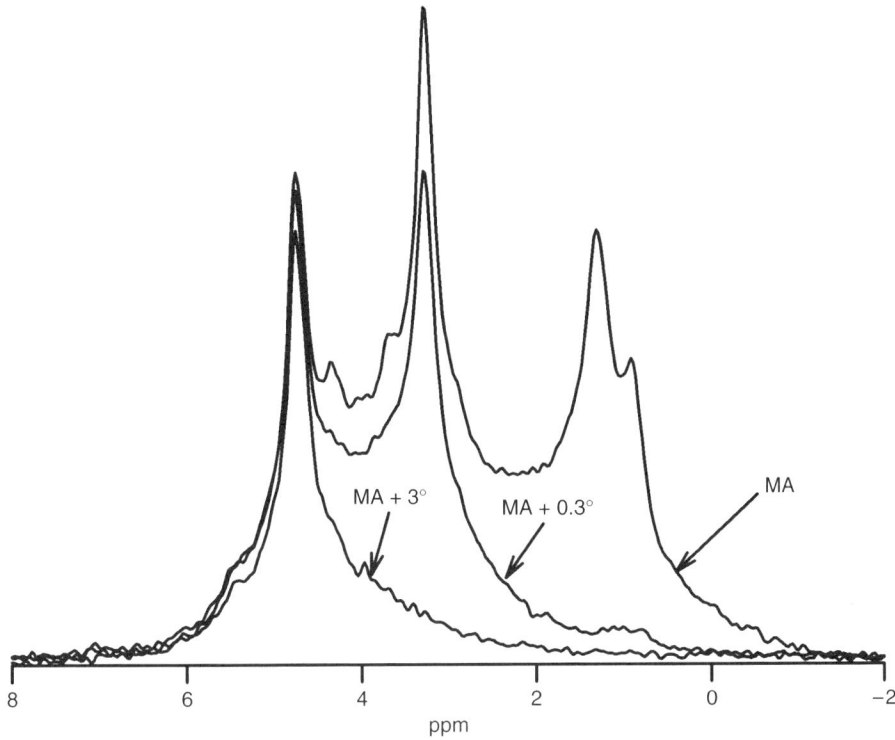

Fig. 2. The effect on the proton NMR lineshape in a STE experiment as θ is moved away from the MA, is illustrated for the lipid signals of (DMPC) dimyristoyl phosphatidylcholine in an oriented sample. The main oservable signals are from water (4.6 ppm), the choline head group (3.1 ppm) and the hydrocarbon chains (0.9 and 1.1 ppm). As the transverse relaxation owing to the static dipolar coupling increases the amplitudes are diminished and finally lost in the baseline noise. Note that this happens at deviations of less than 0.3° from the magic angle (MA) for the signal from the hydrocarbon chains.

Spins moving in an inhomogeneous magnetic field will experience a varying precession rate and the refocusing of the magnetization will in general not be complete after the application of a refocusing pulse. This effect is enhanced by the application of magnetic field gradients during the dephasing/rephasing periods, resulting in an attenuation of the spin echo amplitude according to *(16)*

$$A = \sum_i A_{0i} \exp\left[-\gamma^2 g^2 \delta^2 D_i (\Delta - \delta/3)\right] \quad (1)$$

where D is the self-diffusion coefficient of the molecules. In **Eq. 1** the summation goes over all diffusing components. A_{0i} is the echo amplitudes without applied gradients, γ is the gyromagnetic ratio, Δ is the time interval between gradient pulses, and δ and g are the duration and amplitude of the PFG, respectively

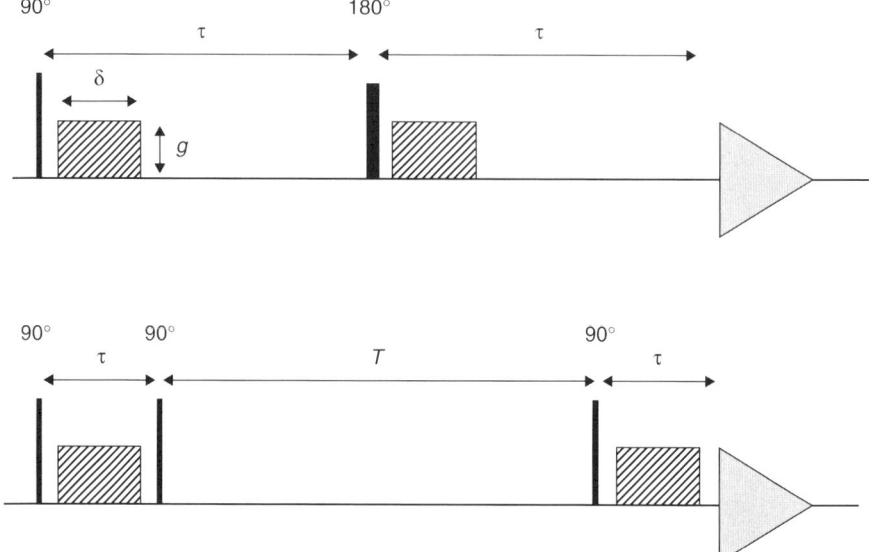

Fig. 3. The basic pulse sequences used in the PFG diffusion experiments. **(Top)** SE experiment. **(Bottom)** STE experiment. The magnetic field gradient pulses are shown as hatched rectangles, the rf-pulses as black rectangles and the refocused spin echo as a gray triangle.

(cf. **Fig. 3**). The initial echo amplitude A_{0i} is determined by the longitudinal and transverse NMR relaxation times. Usually, the second half of the spin echo is collected and Fourier transformed to obtain the spectrum in the frequency domain.

The stimulated spin echo (STE) *(17)* (**Fig. 3**, bottom) is a modification of the basic SE sequence in which the magnetization is stored along the z-axis during the time interval T. This makes it possible to extend Δ to longer times because of the difference between the transverse and longitudinal relaxation rates.

In the NMR experiment the translational diffusion is measured in the direction of the magnetic field gradient, which normally is directed along B_0. For a lipid membrane the observed diffusion coefficient, D, depends on the orientation and the two "local" diffusion coefficients, D_L and $D_\perp$, where the former one is the lateral diffusion coefficient representing translational diffusion in the plane of the membrane, and $D_\perp$ stands for the diffusion perpendicular to the bilayer. Then *(18)*,

$$D = D_L \sin^2 \theta_{LD} + D_\perp \cos^2 \theta_{LD} \qquad (2)$$

For a bilayer oriented at the magic angle $\sin^2 \theta_{LD} = \frac{2}{3}$ and, as it is reasonable to assume that $D_\perp$ is orders of magnitude smaller than D_L, the second term in **Eq. 2** can be neglected. Thus, $D_L = 1.5D$.

For the diffusion investigations there are two solid-state NMR spectrometers available in the laboratory, wherein one is equipped with a 10-mm ^{1}H probe for 100 MHz with a maximum gradient strength of 3 T/m (routinely used for lipid diffusion measurements in the order of 1–10 μm^2/s) and the other one with a 5-mm dual ^{1}H/X probe for a 400-MHz system that is capable of giving gradient strengths up to 10 T/m for slow diffusion of the order of 0.01–0.1 μm^2/s and for diffusion studies of isotopically enriched molecules, for example, with ^{2}H, ^{31}P, ^{19}F, and ^{13}C *(19)*. For measurements in which the gradient direction can be varied with respect to B$_0$, a microimaging system (Fraunhofer IBMT, St Ingbert Germany with a goniometer sample holder is used *(18)*.

3.4. Data Analysis

The analysis of the data is based on **Eq. 1**, in which the signal attenuation owing to diffusion can be studied by varying any of the parameters Δ, δ, or *g*. Usually δ or *g* is varied, as this leaves the diffusion time (Δ), constant during the experiment.

3.4.1. Peak Intensity or Integral Data

The most straightforward way of analyzing the spectral data is by using the peak intensity of some selected peak (*see* **Fig. 4A**). The decays can be analyzed by nonlinear least-square algorithms built into most spreadsheet software. The diffusion coefficients are obtained together with the relative contribution of each component to the chosen frequency in the spectra. If the peaks are well separated for different molecular species, this will provide the diffusion coefficients for each species in a mixture. This is a simple method that works well for spectra of good signal-to-noise (S/N) quality, but often fails if the diffusion coefficients are similar in magnitude and if the spectra are noisy. An improvement of this method is to use integrated intensities instead of single frequencies (**Fig. 4B**). This sometimes improves the S/N ratio and information about individual species can still be obtained for nonoverlapping peaks.

3.4.2. Component Resolved Spectroscopy

Studies of domain forming bilayer systems involve mixtures of lipids that are similar in structure; and therefore, the spectra usually contain severely overlapping signals. The situation is further deteriorated by the limited resolution that can be obtained with oriented samples. Thus, a method is needed to separate the experimental lineshapes into spectra corresponding to the different diffusion components. In order to achieve this, the component resolved spectroscopy (CORE) method for global analysis of the entire data set has proven to be very useful *(20)*. The CORE analysis provides values of the diffusion coefficients together with the individual amplitudes of the diffusing components for each frequency channel, i.e., the individual bandshapes of the diffusion components are obtained. The use

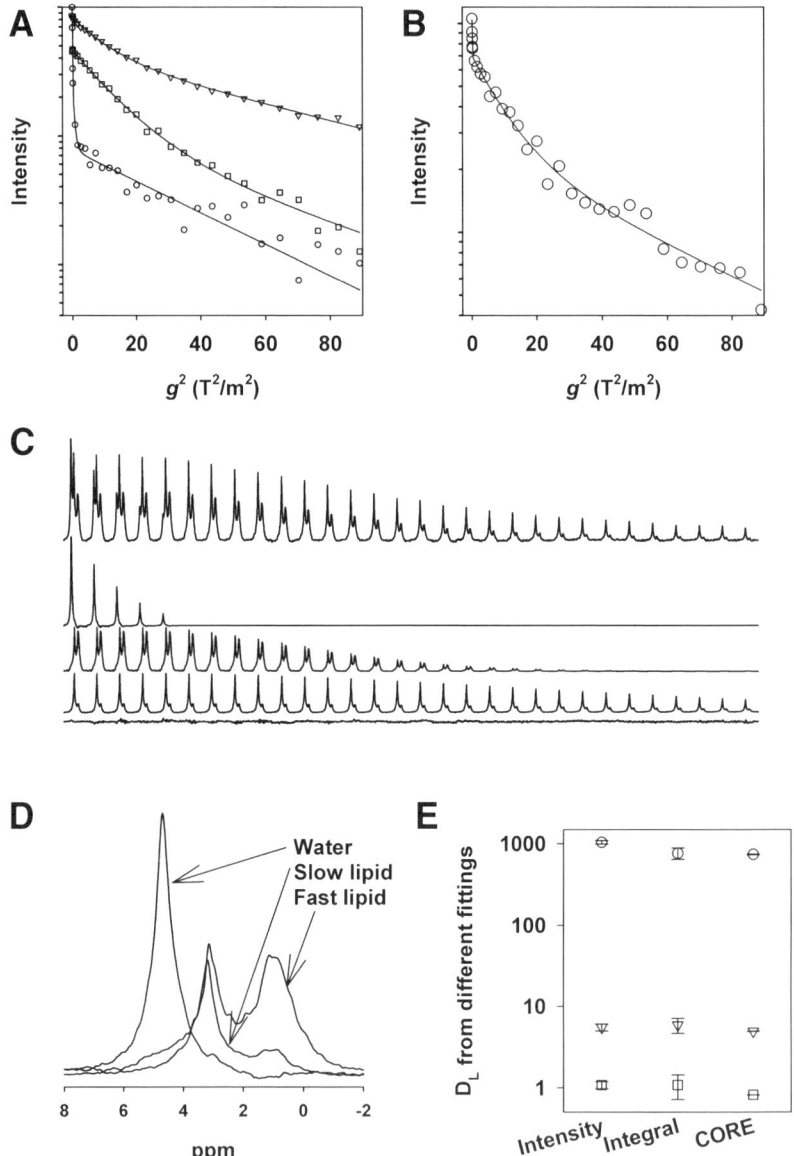

Fig. 4. A comparison of different methods to obtain D_L from the PFG-NMR data. Data is collected from a sample made up of DOPC/DPPC/CHOL, 35:35:30 mol% at 21°C. At this temperature the lipid bilayer is in a two-phase coexistence region of L_d and L_o phases. The STE pulse sequence was used with Δ = 15 ms and δ = 5 ms. In order to catch the decay of both the fast diffusing water and the slower diffusing lipids g was varied between 0.236–9.44 T/m, with a spacing of 0.059 T/m between the first five points and 0.354 T/m between the rest of the points. (**A**) Fit of peak intensity values. Two diffusion coefficients were used for the lipid peaks (triangles: 3.1 ppm,

of the entire data set for the fit makes this method quite robust and accurate even for data of low S/N quality. An example of the CORE analysis is presented in **Fig. 4C,D** and a comparison of the results obtained for the three different methods is shown in **Fig. 4E**. More practical details (data format, parameters, and so on) about the CORE method are given in **Note 3**.

3.4.3. Criteria for the Number of Diffusion Components in an Experiment

By applying the analysis on a system having a large number of components, one can normally decide on the definite number of components from the improvement in the normalized global error square sum (χ^2). This is illustrated in **Fig. 5A,B**.

3.5. Data Interpretation

In general, the experimental data will contain a spectral component of water that is easily distinguished from the lipids by the appearance of its corresponding peak at approx 4.6 ppm and by the large diffusion coefficient of typically 100–1000 µm^2/s. This component is of minor interest and will not be discussed further. The experimental parameters can be set up to eliminate the water peak from the spectra; and thus, concentrate on the lipid contribution to the signal. The observation of more than one lipid diffusion component can be explained in two ways. The simplest explanation is that each lipid species has a distinct diffusion coefficient so that in a mixture of lipids, separate diffusion coefficients will occur for each type of lipid. However, for liquid crystalline bilayers PFG-NMR measurements on isotopically labeled lipids have shown that all lipid species within the same domain have the same diffusion coefficient *(19)*. Rather, it is the bilayer properties, in particular the packing of the lipids in the bilayer that determine the lipid diffusion coefficient.

Fig. 4. (*Continued*) squares: 1.1 ppm) and three for the water peak (circles, 4.6 ppm). (**B**) Fit of integral values to three diffusion coefficients. The whole spectral region is included in the integration. (**C**) CORE fit of the entire data set. Stack plots of 30 experimental spectra as well as the three fitted components and the residuals. (**D**) Spectra for the three components obtained with the CORE method. The main three peaks correspond to water (4.6 ppm), choline CH$_3$ (3.1 ppm), and the chain CH$_2$ and CH$_3$ groups (1.1 and 0.9 ppm). Note the large overlap of the two lipid components and the diminished amplitude of the chain peak in the slow component. The latter is a consequence of faster relaxation in this region of the molecules in the more ordered L$_o$ phase. (**E**) Typical results and errors (reported as the standard error for the nonlinear least square fits and as the 80% confidence interval for the CORE analysis) from the three different fitting methods. The three diffusion coefficients correspond to water and lipid diffusion in the L$_d$ and L$_o$ phase, respectively.

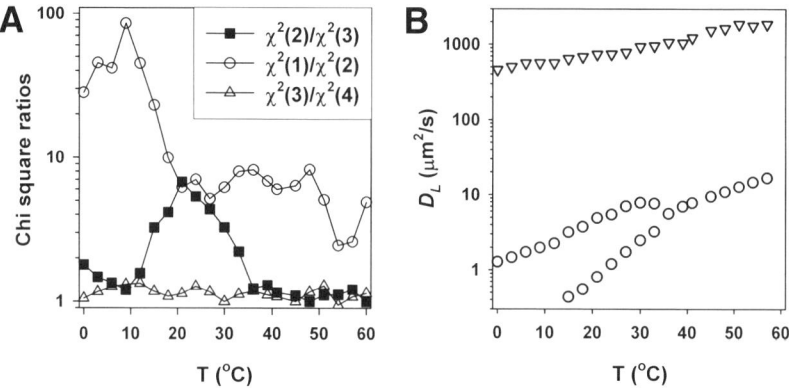

Fig. 5. The use of χ^2 ratios to determine the number of diffusion coefficients in the data set. CORE analyses have been performed for 1–4 diffusion coefficients in the temperature range 0–60°C for the same sample as in **Fig. 4**. **(A)** The ratio of the χ^2 values for i and $i + 1$ components is a measure of the improvement of the fit by adding another component. It can be observed that going from one to two components reduces the χ^2 by a factor of 10–100 over the whole temperature interval. Adding a third component improves the fit only in the temperature interval 15–33°C and adding a fourth component makes no further improvement. Thus, the lipid lateral diffusion is described by two diffusion coefficients for $T = 15$–33°C and by one diffusion coefficient for all other temperatures. **(B)** Results of the CORE fit with the number of components determined by the procedure of χ^2 ratios.

Therefore, the existence of two lipid diffusion components is caused by the formation of domains in which the diffusion in the L_o phase is slower than that in the L_d phase owing to the higher packing and the more ordered nature of the L_o phase. The observed diffusion coefficient will depend on the exchange rate of the lipid between the phases that in turn is related to the value on the lateral diffusion coefficient and on the size of the domains. The root mean square displacement, $\sqrt{\langle r^2 \rangle}$, is often used as a measure of the mean distance traveled by a diffusing particle in a specified time, and for Brownian two-dimensional motion this is equal to $\sqrt{4D_L \Delta}$. Inserting typical values for raft systems ($\Delta = 100$ ms, $D_L = 1$–10 μm²/s), this is about 1 μm. If the average size of the domains is much smaller than 1 μm the lipids will move in and out of different domains during the diffusion time, whereas if the domains are much larger than 1 μm, the lipids will remain in the same domain during the experiment. This has been treated theoretically *(21)* and four cases can be distinguished.

3.5.1. No Domains

For a homogeneous lipid bilayer only one diffusion coefficient will be detected, irrespective of what parameter is altered in the NMR-diffusion experiment,

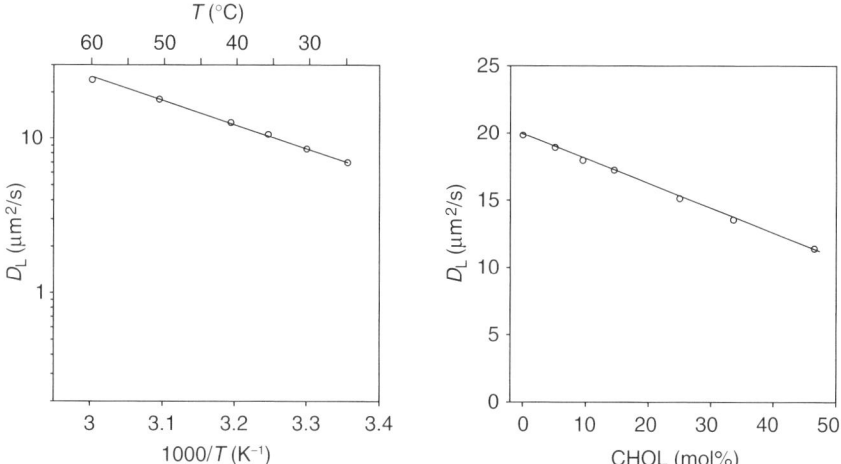

Fig. 6. Lipid lateral diffusion in the binary system of DOPC/CHOL. The left panel shows the temperature dependence for a CHOL concentration of 9.5 mol% and the right panel gives the dependence of D_L on the CHOL concentration at 50°C. The linear dependence in the Arrhenius plot and the monotonic decrease with increasing CHOL concentration is typical of a homogeneous membrane. Lines are merely a guide for the eye. From **ref. 24**.

i.e., Δ, δ, or g (**Fig. 6**). Furthermore, only one Arrhenius activation energy will be detected at all temperature regions.

3.5.2. Small Domains; Fast Exchange

If there is an exchange of lipids between two phases or domains that is much faster than the diffusion time, the observed diffusion coefficient, D_{OBS}, will be an average of the diffusion in the two phases, D_1 and D_2, weighted by the relative populations, p_1 and p_2:

$$D_{OBS} = p_1 D_1 + p_2 D_2 \tag{3}$$

This situation can be difficult to distinguish from that for a homogeneous membrane described in **Subheading 3.5.1**. However, as the relative populations of the two phases change with alterations in temperature or composition, D_{OBS} will deviate from that of the single phase in the two-phase area. This can be observed as kinks in the curves of D_{OBS} vs cholesterol (CHOL) and as a non-Arrhenius temperature dependence (**Fig. 7**). Thus, the nonlinear features of these curves can be taken as evidence of lateral separation into small domains.

3.5.3. Large Domains; Slow Exchange

When lipids are mainly residing in separate domains during the experiment, two lateral diffusion components will be observed, corresponding to the diffusion

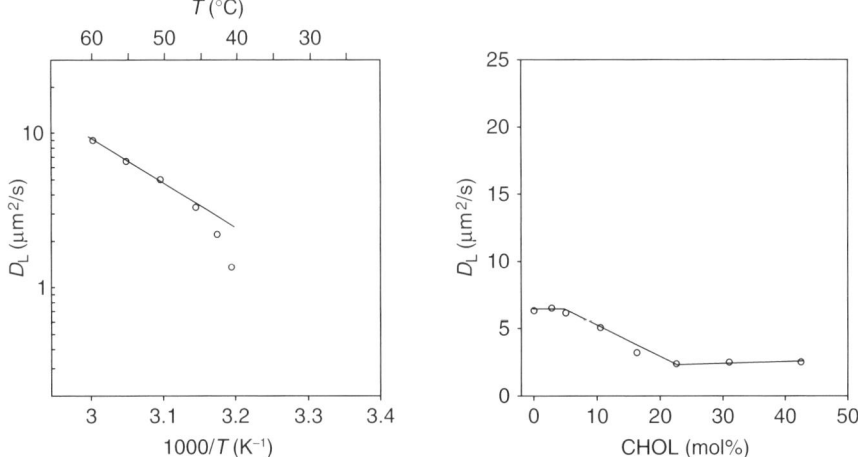

Fig. 7. Lipid lateral diffusion in the egg sphingomyelin eSM/CHOL system. The left panel shows the temperature dependence for a CHOL concentration of 10.5 mol% and the right panel gives the dependence of D_L on the CHOL concentration at 50°C. The curved feature of the Arrhenius plot and the kinks in the concentration plot are indicative of an inhomogeneous membrane in which the domains are sufficiently small so that fast exchange occurs. The lines are merely guides for the eye. From **ref. 24**.

in each one of the phases. The relative populations of the diffusion components are reflected in the fractions of lipids in the two domains (**Fig. 8**).

3.5.4. Domains With Intermediate Exchange

If the exchange between different domains is of the same order as the diffusion time, two diffusion components are observed. However, both the observed diffusion coefficients and the relative amplitudes will then be functions of Δ, the relative populations and the diffusion coefficients in the two phases. Thus, such a situation can be recognized by an alteration of Δ.

4. Notes

1. The solvent mixture should give a good adhesion to the glass surface and results in thin films covering the glass plates. The choice of solvent can be critical and in some cases the glass surface has been modified in order to obtain a suitable degree of hydrophobicity for the lipids to adhere to it *(22)*. The standard solvent mixture used by the authors is a 1:4 (v:v) mixture of methanol and 1-propanol. Sometimes it is easier to dissolve the lipids in only methanol and then add the propanol gradually. Some sterols and ceramides have a limited solubility in this mixture and for such cases a 3:2 (v:v) mixture of hexane and 2-propanol has been useful. Mixtures of methanol or ethanol with benzene have also been used with varying success in receiving good aligned bilayers *(23)*. Usually, the surface tension is low enough to

Lateral Diffusion Coefficients of Raft Lipids

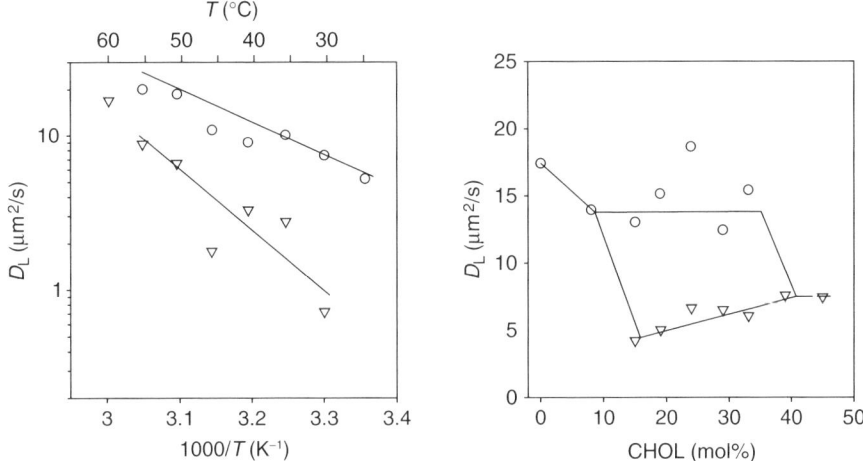

Fig. 8. Lipid lateral diffusion in the DOPC/eSM/CHOL system. The left panel shows the temperature dependence for a composition of 38:38:24 mol%. The right panel gives the dependence of D_L on the CHOL concentration when the other two lipids are kept at equimolar amounts at 50°C. The observation of two lipid diffusion coefficients means that the domains formed in this system are large so that the lipid exchange between domains is slow. The lines are merely guides for the eye and they indicate that the apparent activation energy is larger for the L_o phase. From **ref. 25**.

    result in a thin film that completely covers the glass plates when the solvent evaporates, but sometimes the lipid–solvent mixture forms a small drop during the evaporation, leading to a deposition of all lipid material in a small area on the glass plate. If this happens a change in the solvent composition and/or the evaporation temperature might improve the result.

2. Hydration by humid atmosphere is generally preferred over the use of liquid water because addition of liquid water disrupts the bilayers and results in the formation of vesicular structures. This occurs especially when more than one glass plate is needed for obtaining a sufficient S/N ratio. If the glass plates are stacked before hydration, addition of liquid water will disrupt the bilayers as water is sucked in between the plates by capillary forces. Attempts to stack prehydrated plates often results in mechanical disruption of the lipid bilayers. The oriented samples are fragile and should be handled carefully in order to keep the orientation. They can be stored for weeks at 5°C but should not be frozen, as this often disrupts the orientation.

3. The CORE package is written by Peter Stilbs and can be downloaded from his homepage (http://gamma.physchem.kth.se/~peter/). All parameters for running CORE are contained in the infile.txt. The CORE program can use several data formats as input (*see* the NTYPE variable in the manual). For our purpose a data file is exported from the spectrometer in ascii format (NTYPE = 42) and contains intensity data for all experiments consecutively in a column. The program then chops it up into individual

spectra using the no. of frequency channels spectrum (NFREQ) and no. of spectra (NK) variables. A minimal sample infile for analyzing a diffusion experiment is shown next. In this case the experiment consists of 20 spectra, each containing 512 data points, corresponding to different values of the gradient amplitude, whereas keeping δ = 3 ms and Δ = 15 ms. Data are fitted to two components with the initial guessed values of 3 and 300 $\mu m^2/s$. Note that all lines begin with a blank space or an exclamation mark, depending on whether the line should be treated as a comment or not. For more information see the manual provided with the program.

```
! STANDARD NAMELIST FOR DIFFUSION FITS
! LINES BEGINNING WITH ! ARE TREATED AS COMMENTS
$DATA
! EXPERIMENTAL PARAMETER INPUT
TPA=11.9,
AMP=.04,.08,.12,.16,.2,.24,.28,.32,.36,.4,
.44,.48,.52,.56,.6,.64,.68,.72,.76,.8,
SDLMS=20*3,
BDLMS=20*15,
! NO OF POINTS/SPECTRUM AND NO OF SPECTRA
NFREQ=512,NK=20,
! DEBUG, DATA FORMAT AND OUTPUT TYPE
NTRAC=-1,NTRAC1=-1,NTYPE=42,noutsp=2,
! NO OF COMPONENTS AND STARTING VALUES
NV=2,NV1=2,MODE=2,NOCOMP=2,
CUTOFF=-1,
X=3.E-12,3.E-10,1e-12,MASK=0,0,0,
LAST=1,
! NTIMES=20,
$end
```

Acknowledgments

This work was supported by The Swedish Research Council and The Knut and Alice Wallenberg Foundation.

References

1. Callaghan, P. T. (1991) Principles of Nuclear Magnetic Resonance Microscopy. Clarendon Press, Oxford.
2. Lindblom, G. and Orädd, G. (1994) NMR studies of translational diffusion in lyotropic liquid crystals and lipid membranes. *Progr. Nucl. Magn. Reson. Spectrosc.* **26,** 483–515.
3. Söderman, O. and Stilbs, P. (1994) NMR studies of complex surfactant systems. *Progr. Nucl. Magn. Reson. Spectrosc.* **26,** 445–482.
4. Stilbs, P. (1987) Fourier transform pulsed-gradient spin-echo studies of molecular diffusion. *Progr. Nucl. Magn. Reson. Spectrosc.* **19,** 1–45.

5. Lindblom, G. and Wennerström, H. (1977) Amphiphile diffusion in model membrane systems studied by pulsed NMR. *Biophys. Chem.* **6,** 167–171.
6. Orädd, G. and Lindblom, G. (2003) in *NMR on Macroscopically Oriented Lyotropic Systems*, (Burnell, E. E. and de Lange, C. A., eds.), Kluwer Academic Publishers, Amsterdam, The Netherlands, pp. 399–418.
7. Dvinskikh, S. V. and Furo, I. (2000) Combining PGSE NMR with homonuclear dipolar decoupling. *J. Magn. Reson.* **144,** 142–149.
8. Gaede, H. C. and Gawrisch, K. (2003) Lateral diffusion rates of lipid, water, and a hydrophobic drug in a multilamellar liposome. *Biophys. J.* **85,** 1734–1740.
9. Lindblom, G. (1996) in Nuclear magnetic resonance spectroscopy and lipid phase behavior and lipid diffusion. *Advances in Lipid Methodology, vol. 3*, (Christie, W. W., ed.), The Oily press Ltd., Dundee, Scotland, pp. 133–209.
10. Rajendran, L. and Simons, K. (2005) Lipid rafts and membrane dynamics. *J. Cell Sci.* **118,** 1099–1102.
11. Simons, K. and Toomre, D. (2000) Lipid rafts and signal transduction. *Nat. Rev. Mol. Cell Biol.* **1,** 31–39.
12. Simons, K. and Van Meer, G. (1988) Lipid sorting in epithelial cells. *Biochemistry* **27,** 6197–6202.
13. Ipsen, J. H., Karlström, G., Mouritsen, O. G., Wennerström, H., and Zuckermann, M. J. (1987) Phase equilibria in the phosphatidylcholine-cholesterol system. *Biochim. Biophys. Acta* **905,** 162–172.
14. Vist, M. R. and Davis, J. H. (1990) Phase equilibria of cholesterol/dipalmitoylphosphatidylcholine mixtures: ^{2}H nuclear magnetic resonance and differential scanning calorimetry. *Biochemistry* **29,** 451–464.
15. Orädd, G. and Lindblom, G. (2004) Lateral diffusion studied by pulsed field gradient NMR on oriented lipid membranes. *Magn. Reson. Chem.* **42,** 123–131.
16. Stejskal, E. O. and Tanner, J. E. (1965) Spin diffusion measurements: spin echoes in the presence of a time-dependent field gradient. *J. Chem. Phys.* **42,** 288–292.
17. Tanner, J. E. (1970) Use of the stimulated echo in NMR diffusion studies. *J. Chem. Phys.* **52,** 2523–2526.
18. Wästerby, P., Orädd, G., and Lindblom, G. (2002) Anisotropic water diffusion in macroscopically oriented lipid bilayers studied by pulsed magnetic field gradient NMR. *J. Magn. Reson.* **157,** 156–159.
19. Orädd, G., Westerman, P. W., and Lindblom, G. (2005) Lateral diffusion coefficients of separate lipid species in a ternary raft-forming bilayer: a pfg-NMR multinuclear study. *Biophys. J.* **89,** 315–320.
20. Stilbs, P., Paulsen, K., and Griffiths, P. C. (1996) Global least-squares analysis of large, correlated spectral data sets: application to component-resolved FT-PGSE NMR spectroscopy. *J. Phys. Chem.* **100,** 8180–8189.
21. Kärger, J., Pfeifer, H., and Heink, W. (1988) in Principles and applications of self-diffusion measurements by nuclear magnetic resonance, *Advances in Magnetic and Optical Resonance, vol. 13*, (Warren, W. S., ed.), Academic Press, Inc., San Diego, CA, pp. 1–89.

22. Powers, L. and Pershan, P. S. (1977) Monodomain samples of dipalmitoyl phosphatidylcholine with varying concentrations of water and other ingredients. *Biophys. J.* **20,** 137–152.
23. Moll, F., III. and Cross, T. A. (1990) Optimizing and characterizing alignment of oriented lipid bilayers containing gramicidin D. *Biophys. J.* **57,** 351–362.
24. Filippov, A., Orädd, G., and Lindblom, G. (2003) The effect of cholesterol on the lateral diffusion of phospholipids in oriented bilayers. *Biophys. J.* **84,** 3079–3086.
25. Filippov, A., Orädd, G., and Lindblom, G. (2004) Lipid lateral diffusion in ordered and disordered phases in raft mixtures. *Biophys. J.* **86,** 891–896.

11

Saturation-Recovery Electron Paramagnetic Resonance Discrimination by Oxygen Transport (DOT) Method for Characterizing Membrane Domains

Witold K. Subczynski, Justyna Widomska, Anna Wisniewska, and Akihiro Kusumi

Summary

The discrimination by oxygen transport (DOT) method is a dual-probe saturation-recovery electron paramagnetic resonance approach in which the observable parameter is the spin-lattice relaxation time (T_1) of lipid spin labels, and the measured value is the bimolecular collision rate between molecular oxygen and the nitroxide moiety of spin labels. This method has proven to be extremely sensitive to changes in the local oxygen diffusion-concentration product (around the nitroxide moiety) because of the long T_1 of lipid spin labels (1–10 µs) and also because molecular oxygen is a unique probe molecule. Molecular oxygen is paramagnetic, small, and has the appropriate level of hydrophobicity that allows it to partition into various supramolecular structures such as different membrane domains. When located in two different membrane domains, the spin label alone most often cannot differentiate between these domains, giving very similar (indistinguishable) conventional electron paramagnetic resonance spectra and similar T_1 values. However, even small differences in lipid packing in these domains will affect oxygen partitioning and oxygen diffusion, which can be easily detected by observing the different T_1s from spin labels in these two locations in the presence of molecular oxygen. The DOT method allows one not only to distinguish between the different domains, but also to obtain the value of the oxygen diffusion-concentration product in these domains, which is a useful physical characteristic of the organization of lipids in domains. Profiles of the oxygen diffusion-concentration product (the oxygen transport parameter) in coexisting domains can be obtained *in situ* without the need for the physical separation of the two domains. Furthermore, under optimal conditions, the exchange rate of spin-labeled molecules between the two domains could be measured *(10)*.

Key Words: Cholesterol; discrimination by oxygen transport (DOT); lipid raft; membrane domain; liquid-ordered phase; oxygen collision rate; saturation-recovery EPR; spin labeling.

1. Introduction

Because membranes, and thus membrane domains, are not really two-dimensional structures, knowledge of the molecular events in the depth dimension is important. This information is practically missing, from the research in the raft field. The discrimination by oxygen transport (DOT) method allows one not only to see the lateral organization of lipid membranes or differentiate between membrane domains, but also to obtain information about molecular dynamics and structure in the third dimension, namely, in the direction of the membrane depth. This is possible by using a very small probe, i.e., molecular oxygen and a variety of lipid spin labels (**Fig. 1**) incorporated into the membrane, at specific domains and specific depths. Thus, the DOT method provides information about the three-dimensional dynamic structure of membranes in the presence of coexisting membrane domains.

In the DOT approach, the rate of collision between molecular oxygen and the nitroxide moiety attached to a specific location on the lipid molecule is measured using the saturation-recovery electron paramagnetic resonance (EPR) technique. Molecular oxygen is paramagnetic and collisions of oxygen with the nitroxide spin label alter the resonance characteristics of the spin label, including the spin-lattice relaxation time. The oxygen collision rate, which is the product of the local concentration and the local diffusion coefficient of molecular oxygen within the membrane, is a very sensitive measure of the presence and movement of free volumes in the lipid bilayer, which may be very small, just sufficient to contain a single molecule of oxygen. The small size and appropriate level of hydrophobicity of molecular oxygen allows it to enter the small vacant pockets that are transiently formed in the lipid bilayer membrane. Therefore, collision rates between molecular oxygen and nitroxide spin labels at specific locations in the membrane are sensitive to the dynamics of *gauche-trans* isomerization of lipid hydrocarbon chains and to the structural nonconformability of neighboring lipids *(1–4)*. This unique approach was used to obtain profiles of oxygen collision rates across model and biological membranes, which provide useful information on the three-dimensional dynamic structure of membranes and membrane domains *(1,5–8)*. It was shown that the DOT method can be successfully applied to discriminate the slow oxygen transport (SLOT) domain from the bulk domain in reconstituted membranes crowded with integral membrane proteins *(9)* as well as the SLOT domain in the influenza virus envelope membrane, which is a cholesterol-rich

Fig. 1. *(Opposite page)* Chemical structures of selected spin labels, phospholipid-type: n-PC, T-PC, and n-SASL and cholesterol analog: CSL and ASL. Chemical structures of DMPC and cholesterol molecules are indicated to illustrate approximate localizations of nitroxide moieties across the membrane.

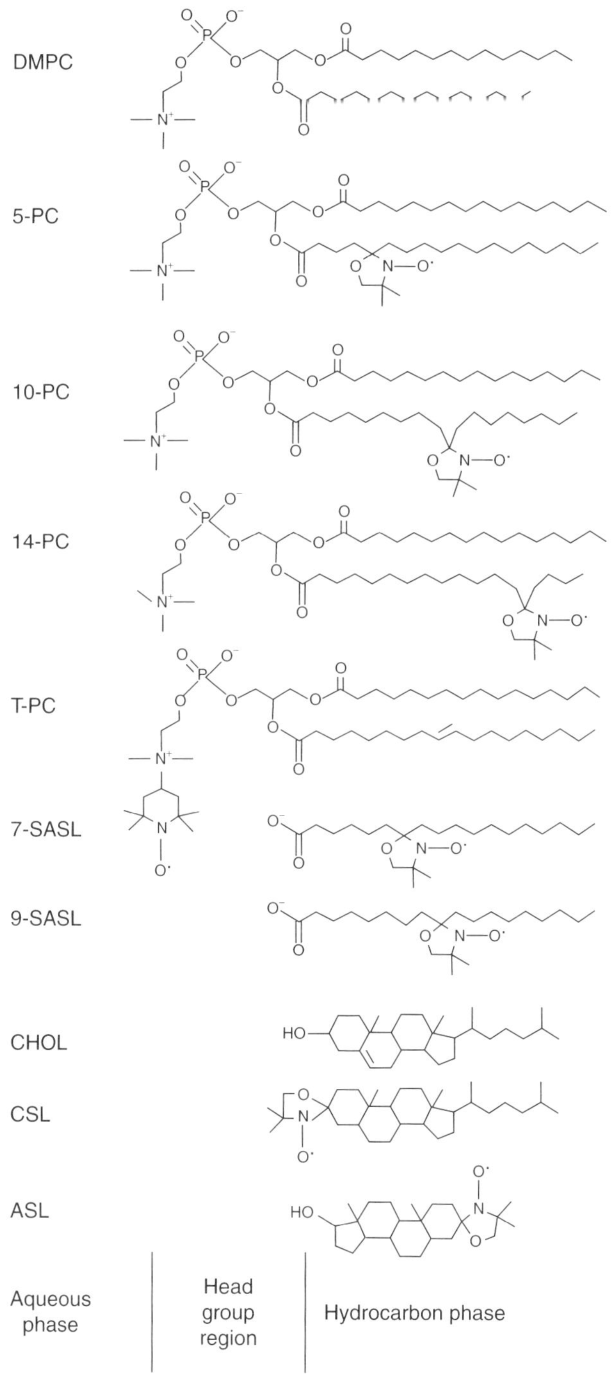

Lipid Domains: EPR Discrimination by Oxygen Transport 145

DMPC

5-PC

10-PC

14-PC

T-PC

7-SASL

9-SASL

CHOL

CSL

ASL

Aqueous phase | Head group region | Hydrocarbon phase

and protein-rich raft domain *(10)*. In model membranes made from binary mixtures of phosphatidylcholine and cholesterol or sphingomyelin and cholesterol, liquid-ordered, liquid-disordered, and solid-ordered domains are distinguished and characterized in different regions of a phase diagram when they form a single-phase or when two phases coexist *(11–13)*. In membranes made from the ternary raft-forming mixture, the SLOT raft domain was also distinguished from bulk lipids using the DOT method *(11,14)*.

1.1. Outline of Theory

The oxygen transport parameter, $W(x)$, was introduced as a convenient quantitative measure of the rate of the collision between the spin probe and molecular oxygen by Kusumi et al. *(15)* as

$$W(x) = T_1^{-1}(\text{Air}, x) - T_1^{-1}(N_2, x) \tag{1}$$

where the T_1s are the spin-lattice relaxation times of the nitroxides in samples equilibrated with atmospheric air and nitrogen, respectively. The collision rate is also proportional to the local oxygen concentration $C(x)$ and the local oxygen-diffusion coefficient $D(x)$ (thus, it is called the "transport" parameter and has nothing to do with active transport across the membrane) at a "depth" x in the membrane that is in equilibrium with atmospheric air:

$$W(x) = AC(x)D(x), A = 8\pi p r_0 \tag{2}$$

where r_o is the interaction distance between oxygen and the nitroxide moiety of the spin label (4.5 Å) *(16)* and p is the probability that an observable event occurs when a collision occurs (A is remarkably independent of the hydrophobicity and viscosity of the solvent and of spin label species *[17–19]*). Kusumi et al. *(15)* concluded that the oxygen transport parameter is a useful monitor of membrane fluidity that reports on translational diffusion of small molecules.

When located in two different membrane domains, the spin label alone most often cannot differentiate between these domains, giving very similar conventional EPR spectra and similar T_1 values. However, even small differences in lipid packing in these domains will affect oxygen partitioning and oxygen diffusion, which can be easily detected by observing the different T_1s from spin labels in these two locations in the presence of oxygen. In membranes equilibrated with air and consisting of two lipid environments with different oxygen transport rates, fast oxygen transport (FOT) domain and SLOT domain, the saturation-recovery signal is a simple double-exponential curve with time-constants of $T_1^{-1}(\text{Air}, \text{FOT})$ and $T_1^{-1}(\text{Air}, \text{SLOT})$ *(9,10, 13)*.

$$W(\text{FOT}) = T_1^{-1}(\text{Air}, \text{FOT}) - T_1^{-1}(N_2, \text{FOT}) \tag{3}$$

$$W(\text{SLOT}) = T_1^{-1}(\text{Air}, \text{SLOT}) - T_1^{-1}(N_2, \text{SLOT}) \tag{4}$$

Here "x" from **Eq. 1** is changed to the two-membrane domains, FOT and SLOT, and the depth is fixed (the same spin label is distributed between the FOT and SLOT domains). W(FOT) and W(SLOT) are oxygen transport parameters in each domain and represent the collision rate in samples equilibrated with air. Using lipid spin labels with the free radical nitroxide moiety attached at different positions on the lipid molecule (*see* **Fig. 1**), profiles of the oxygen transport parameter in coexisting domains can be obtained.

2. Materials

1. Phospholipid spin labels (1-palmitoyl-2-(n-doxylstearoyl)phosphatidylcholine (n-PC), where n = 5, 7, 10, 12, 14, or 16, or tempocholine-1-palmitoyl-2-oleoylphosphatidic acid ester [T-PC]) can be purchased from Avanti Polar Lipids, Inc. (Alabaster, AL). Doxylstearic acid spin labels (n-SASL, where n = 5, 7, 9, 10, 12, or 16), cholestane spin label (CSL), and androstane spin label (ASL) can be purchased from Sigma (St. Louis, MO). Spin labels are dissolved in chloroform at 1 mM and stored in a freezer at –70°C.
2. Stock solutions of lipids (phosphatidylcholine, sphingomyelin, and cholesterol) from Avanti Polar Lipids Inc. in chloroform (usually 20–50 mg/mL) are kept in a freezer at –70°C.
3. Buffers: typically, 10 mM PIPES and 150 mM NaCl (pH 7.0) is used as a buffer. For samples with n-SASL, 0.1 M borate at pH 9.5 is used as a buffer. In this case a rather high pH is chosen to ensure that all SASL probe carboxyl groups are ionized in lipid bilayer membranes *(20,21)*.

3. Methods

3.1. Spin Labeling

In these types of studies, two classes of spin labels are usually used (*see* **Fig. 1** for their structures and approximate localization in the lipid bilayer):

1. Spin labels that allow profiles of the oxygen transport parameter across the lipid bilayer (phospholipid spin labels, n-PC, T-PC and stearic acid spin labels, n-SASL) to be obtained. In these spin labels the nitroxide free radical fragment is located at different depths in the lipid bilayer from the polar head group (as in T-PC) to the membrane center (as in 16-PC or 16-SASL).
2. Spin-labeled cholesterol analogs (CSL and ASL). CSL is especially significant in the studies of the lateral organization of membranes containing raft domains because it is likely to locate itself like cholesterol and behave similar to cholesterol *(22,23)*. More caution is required for the interpretation of data using ASL *(22,23)*.
3. Before preparation of liposomes, chloroform solutions of lipids and spin labels are mixed to attain the final concentration of 0.5 or 1 mol% of spin labels in the lipid bilayer.

3.2. Preparation of Samples for EPR Measurements

The membranes used for EPR model measurements are multilamellar dispersions of lipids (multilamellar liposomes) prepared in the following way:

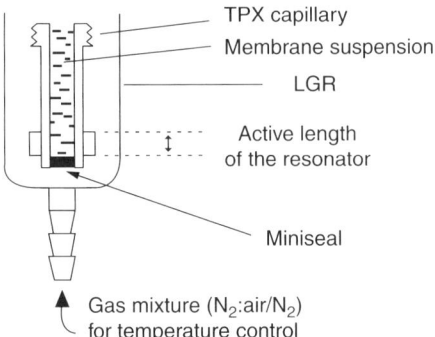

Fig. 2. Schematic drawing indicating position of the sample inside the LGR. The sample is equilibrated with the gas that is also used for temperature control.

1. Chloroform solutions of lipids and spin labels (containing 5–10 μmol of total lipid) are mixed to attain the desired compound concentrations.
2. Chloroform is evaporated with a stream of nitrogen, and the lipid film on the bottom of the test tube is thoroughly dried under reduced pressure (about 0.1 mmHg) for about 12 h.
3. A buffer solution (usually 0.5–1.0 mL) is added to the dried film at a temperature above the phase transition temperature of investigated membranes and vortexed vigorously.
4. The lipid dispersion is centrifuged briefly (15 min at 4°C with an Eppendorf bench centrifuge at 16,000g) and the loose pellet (about 20% lipid, [w/w]) is used for the EPR measurements.

3.3. Equilibration With Molecular Oxygen

1. The loose pellet is transferred to a capillary made of gas-permeable methylpentene polymer known as TPX (*see* **Note 1**) and the end of the capillary is sealed with Baxter Miniseal wax B4425.1 (*see* **Note 2**). This plastic is permeable to oxygen, nitrogen, and other gases and is substantially impermeable to water.
2. The TPX capillary is fixed inside the loop-gap resonator (LGR) of the X-band saturation-recovery EPR spectrometer with a special Teflon holder (*see* **Notes 3** and **4**).
3. Flow of dry nitrogen over the TPX capillary allows easy deoxygenating of samples to measure T_1.
4. Switching the gas to the air/nitrogen mixture allows one to easily equilibrate the sample with the required partial pressure of oxygen for oximetry measurements and for obtaining T_1s in the presence of molecular oxygen (*see* **Note 5**).
5. The mixture of air and nitrogen is adjusted with flowmeters (Matheson Gas Products, Montgomeryville, PA, Montgomaryville, PA, model 7631 H-604).
6. The schematic drawing in **Fig. 2** for a TPX capillary and the X-band LGR indicates the position of the sample relative to the active length of the resonator, as well as gas flow around the sample. The equilibration time for this geometry is typically 10 min, but can change with capillary wall thickness and temperature.

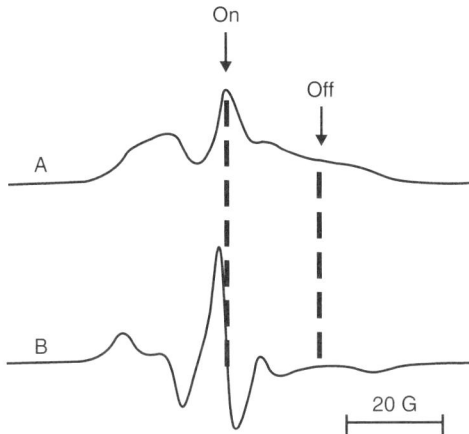

Fig. 3. EPR spectra of 5-PC in DMPC bilayer containing 20 mol% cholesterol recorded at 30°C. **(A)** absorption spectrum and **(B)** first-derivative spectrum (routinely called EPR spectrum) are displayed to better indicate the on-resonance position at which the saturation-recovery signals are recorded. In saturation-recovery measurements, a superimposed low-frequency (25 Hz) field modulation, stepping on- and off-resonance in absorption mode, is used to improve baseline stability. Both on- and off-resonance positions are indicated. Because the amplitude of this modulation is smaller than approx 20 G, the off-resonance position is set between the central and the high-field line, and the amplitude of the field modulation is changed to maximize the saturation-recovery signal.

3.4. Conventional EPR

1. Conventional EPR spectra are recorded before saturation-recovery measurements to ensure the correct magnetic field positions (on- and off-resonance) at which saturation-recovery signals are measured.
2. For the on-resonance position, the magnetic field should be set at the maximum of the EPR absorption (maximum of the central line of the EPR absorption spectrum for the nitroxide spin label, *see* **Fig. 3**).
3. For the off-resonance position, the magnetic field should be set outside the absorption spectrum (*see* **Fig. 3** for more details).

3.5. Saturation-Recovery EPR

The saturation-recovery EPR method of measuring electron spin-lattice relaxation (T_1) is a technique in which recovery of the EPR signal is measured at a low-level microwave field (weak observing microwave power) after the end of the saturating microwave pulse. The time-scale of this recovery is characterized by the spin-lattice relaxation time, T_1 (*see* **Notes 6** and **7**).

1. For saturation-recovery measurements, the sample in the TPX capillary is positioned in the LGR as shown in **Fig. 2** and values of the magnetic field are set to on- and off-resonance as shown in **Fig. 3**.

2. The sample is deoxygenated by blowing the nitrogen gas around the TPX capillary **(Fig. 2)** and the saturation-recovery signal is recorded at the required temperature for the spin label at a fixed depth in the membrane (only one type of spin label molecule is present in the lipid bilayer).
3. Next, the sample is equilibrated with certain partial pressure of oxygen at the required temperature by switching the gas flow to the air/nitrogen mixture **(Fig. 2)** and the saturation-recovery signal is then recorded (*see* **Note 8**).
4. The same procedure is repeated for other spin labels with the free radical nitroxide moiety at different depths in the membrane.
5. The T_1s of spin labels in the absence and presence of molecular oxygen are determined by analyzing the saturation-recovery signal of the central line obtained by short-pulse saturation-recovery EPR (*see* **Note 9**).
6. The pulse length for short-pulse experiments is in the range of 0.1–0.5 μs. Pump power is selected to maximize the amplitude of the saturation-recovery signal and is typically in the range of 2–3.5 G. Observing power is selected to be as high as possible without affecting the time-constant of the recovery. The minimum time between the end of the pulse and the beginning of observation of the recovery is determined by the ring-down time of the resonator and the switching transients and is usually longer than 0.1 μs. Typically, 10^5–10^6 decays are acquired with 512, 1024, or 2048 data points on each decay. Sample intervals are 2, 4, 8, 16, or 32 ns depending on sample, temperature, and oxygen tension. The total accumulation time is typically 2–5 min. The shortest time constant that can be measured is about 0.1 μs.
7. Saturation-recovery signals are fitted by single and double exponentials and compared. If no substantial improvement in the fitting is observed when the number of exponentials is increased from one, the recovery curves can be analyzed as is single exponential (*see* **Note 10**). This is often the case for samples equilibrated with nitrogen.
8. For samples equilibrated with different partial pressure of oxygen, the saturation-recovery signal often can be fitted successfully only with the double-exponential curve (as shown in **Fig. 4**) indicating the presence of two coexisting domains (*see* **Note 11**).

3.6. Calculation of the Oxygen Transport Parameter

The oxygen transport parameter (oxygen diffusion-concentration product) is the main experimental value that is measured in the DOT method.

1. T_1^{-1} values measured at fixed depth in membrane domains are plotted as a function of oxygen concentration (in air [%]) in the equilibrating gas mixture (**Fig. 5**) (*see* **Note 12**).
2. The oxygen transport parameter in each domain can be obtained by extrapolating the linear plots to the sample equilibrated with the atmospheric air (100% air) (*see* **Eq. 1**) as shown in **Fig. 5** (*see* **Note 13**).

3.7. Profiles of the Oxygen Transport Parameter in Different Membrane Domains

The final goal of the DOT measurements is not only to discriminate coexisting domains and characterize them by a single (at one depth) oxygen transport

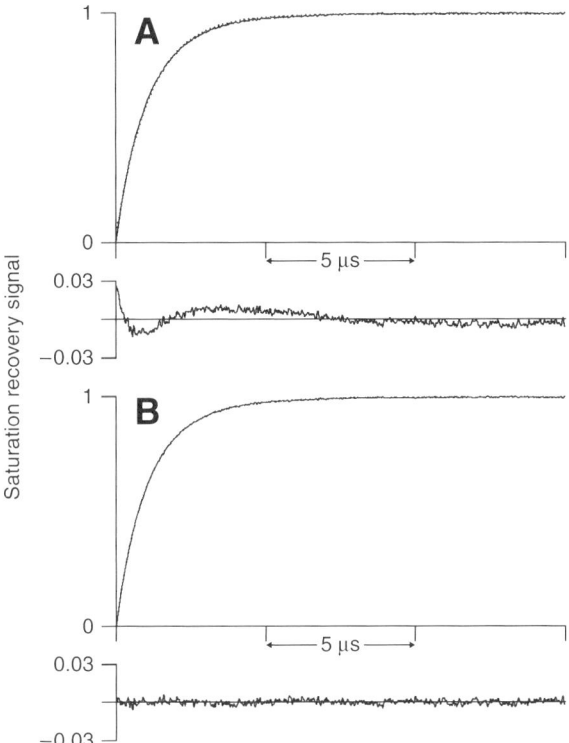

Fig. 4. Typical saturation-recovery signal from 5-PC in a DMPC membrane containing 20 mol% cholesterol at 30°C for a sample equilibrated with 50% air. In the absence of oxygen a single exponential is observed with a time-constant of 5.10 µs (data not shown). In the presence of oxygen, fitting the search to a single exponential mode (**A**) is unsatisfactory as shown by the residual. The fit (**B**) using the double-exponential mode (time-constants of 1.73 and 0.84 µs) is excellent. Typically, the time-constants are determined within an accuracy of ±3%. The double-exponential fit is consistent with two immiscible domains (phases) with different oxygen transport rates that are present at these conditions (liquid-ordered and liquid-disordered phases as shown in the phase diagram presented by Almeida et al. *[29]*).

parameter, but also to obtain detailed profiles of the oxygen transport parameter in both coexisting domains. These profiles contain unique information about the dynamic structure of membrane domains. Additionally, because these profiles can be obtained in coexisting domains without the need for their separation, the DOT method provides unique opportunities in studies of physical properties of raft domains *in situ*.

1. Profiles of the oxygen transport parameter in coexisting domains are constructed as shown in **Fig. 6**, in which, values of the oxygen transport parameter are plotted as a function of the depth in the membrane or the approximate position of

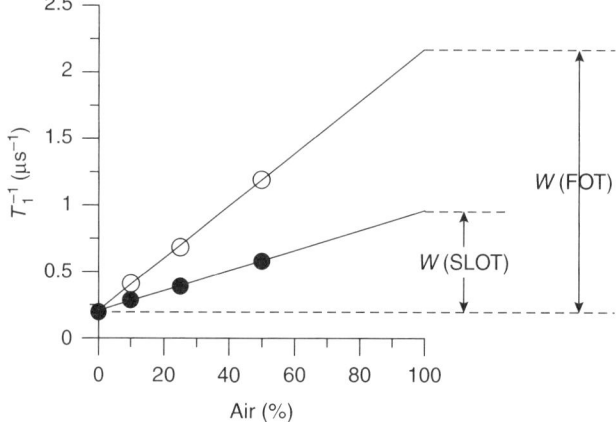

Fig. 5. Plot of T_1^{-1} for 5-PC in SLOT and FOT domains of DMPC membrane containing 20 mol% cholesterol at 30°C as a function of fraction of air in the equilibrating gas mixture. Experimental points show a linear dependence of up to 50% air, and extrapolation of up to 100% air is performed to indicate a way of calculating oxygen transport parameters. *W*(SLOT) and *W*(FOT) are oxygen transport parameters in SLOT (liquid-ordered) and FOT (liquid-disordered) domains.

     the nitroxide moiety of lipid spin labels for which oxygen transport parameters were measured.
2. With the use of the DOT method the main features of the oxygen transport parameter profiles can be obtained with nearly atomic resolution (*see* **Note 14**).
3. The oxygen transport parameter in the aqueous phase is also indicated for comparison (*see* **Note 15**).
4. Knowledge of profiles of the oxygen transport parameter (oxygen diffusion-concentration product) in membrane domains makes it possible to calculate a significant membrane characteristic—namely, the membrane permeability coefficient for oxygen in each domain (*see* **Note 16**).

4. Notes

1. For measurements at X-band these sample tubes are machined from TPX with dimensions of 0.6 mm internal diameter, 0.1 mm wall thickness, and 25 mm length. TPX rods from which the capillaries are machined can be purchased from Midland Plastic (Madison, WI).
2. This sealant is no longer commercially available, but can be found in many laboratories. Other tube sealant can be used including Critoseal (Fisher Scientific), (Hanover Park, IL) and X-Sealant (Bruker Biospin), (Billerica, MA).
3. TPX capillaries together with the Teflon holder can be obtained from Molecular Specialties (Milwaukee, WI).
4. For saturation-recovery measurements at Q-band, wherein the sample volume can be as small as 30 nL, a Teflon tube from Zeus (Orangeburg, SC) with an internal

Lipid Domains: EPR Discrimination by Oxygen Transport 153

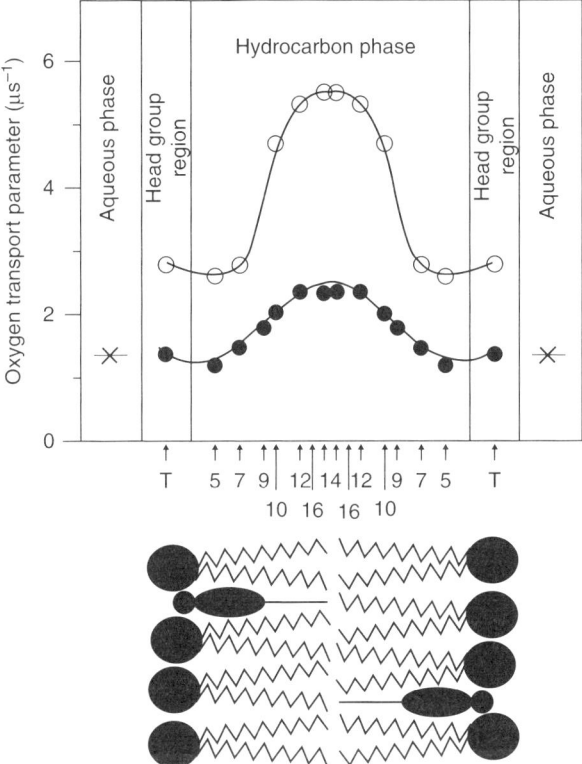

Fig. 6. Profiles of the oxygen transport parameter (oxygen diffusion-concentration product) across a DMPC membrane containing 15 mol% cholesterol obtained at 25°C. Symbols are as follow: ● indicates liquid-ordered (SLOT) domain; ○ indicates liquid-disordered (FOT) domain; and × indicates the oxygen transport parameter in the aqueous phase. Approximate localizations of nitroxide moieties of spin labels are indicated by arrows. T indicates T-PC. The nitroxide attached to C16 may pass through the center of the bilayer and stay in the other leaflet of the DMPC membrane. Schematic drawing indicates relative positions of regions and lipid molecules in the lipid bilayer.

diameter of 0.25 mm and wall thickness of 0.075 mm can be used *(24)*. Machining of a TPX capillary of this size is difficult. Handling samples with a volume of only a few microliters or less for oximetry measurements is described in *(25)*.
5. Because the same gas mixture is used for temperature control, samples are equilibrated with oxygen at the required temperature.
6. The home-built state-of-the-art X-band and Q-band saturation-recovery EPR spectrometers are available at the National Biomedical EPR Center, Medical College of Wisconsin, Milwaukee, WI. The mission of the Center is to make advanced EPR research resources available to investigators nationally, regionally, and locally (see the link for Center use: http://www.mcw.edu/display/router.asp?docid=3211). Another home-built X-band saturation-recovery spectrometer is located at the

Department of Biophysics, Faculty of Biochemistry, Biophysics, and Biotechnology, Jagiellonian University, Krakow, Poland. Presently Bruker produces EPR spectrometers capable for saturation-recovery measurements at X-band. Doing pulse saturation recovery is possible on an E-580 FT/EPR system equipped with the DC-AFC and LCW (low power CW arm) options combined with the AmpX CW microwave power amplifier. Saturation recovery is treated as an accessory to the E-580 and is not usually a stand-alone configuration.

7. By taking the difference between on- and off-resonance responses, the correction for the "background" signal and other artifacts is performed.
8. Routinely, measurements are made for 0, 10, 25, and 50% air in air/nitrogen mixture. However, when the plots of T_1^{-1} do not exhibit a linear dependence on percentage of air, a more detailed dependence on oxygen concentration should be obtained for evaluation of lipid exchange rates between domains. A theory developed by Kawasaki et al. *(10)* indicates that for certain conditions, lipid exchange between the raft domain and bulk lipids as well as the raft domain size could be evaluated. This theory is valid only for small domains, with 10–100 of lipid molecules, and for lipid exchange rates lying in the area of 10^4–10^7 s^{-1}.
9. The short-pulse method is favorable for multiexponential decays in oximetry measurements *(26,27)*. For a short pulse, only populations of the irradiated transition are affected; for a long pulse, all populations are altered because of transverse relaxations. Yin and Hyde *(28)* address the long- and short-pulse saturation-recovery methods in more detail.
10. Additional criteria for the goodness of a single exponential fit are the negligible pre-exponential coefficient for the second component, the large standard deviation of T_1 for the second component, and the repetition of the fit for different recording conditions such as number of points and time increment.
11. Although the saturation-recovery signals in the absence of molecular oxygen cannot differentiate between these two domains, in the presence of oxygen the recovery curves are very different in each domain, indicating that the collision rate of molecular oxygen (the oxygen transport parameter), is quite different in these two domains. This is a good illustration of why the method is named DOT, because different domains can be clearly discriminated and characterized only in the presence of molecular oxygen.
12. The linear dependence of T_1^{-1} on oxygen concentration in these plots indicates that the exchange rate of spin-labeled lipids between the two coexisting domains is slow (<10^4 s^{-1}—the upper time window limit of the DOT method—or the lifetime of each domain is long [>100 μs] *[10]*).
13. This process is required because accurate observation of saturation recovery becomes increasingly difficult as the oxygen partial pressure is increased, owing to faster recoveries.
14. This is possible not only because practically all atoms in the hydrocarbon chain of phospholipid could be labeled, but also because of the sensitivity of the DOT method. For example, at 50 mol% cholesterol in the dimyristoylphosphatidylcholine (DMPC)/cholesterol mixture (only the liquid-ordered phase present *[29]*)

the profile becomes practically rectangular with three to four times abrupt increase of oxygen transport parameter between the C9 and C10 positions *(13)*.
15. The temperature dependence of the oxygen transport parameter in water is almost absent owing to the opposite effect of temperature on oxygen solubility and translational diffusion.
16. The membrane permeability coefficient connects the oxygen flux across the lipid bilayer with the difference in oxygen concentration in water on each side of the bilayer. The method of calculation of the membrane permeability coefficient for oxygen was developed by Subczynski et al. *(5)* and does not require formation of an oxygen gradient.

Acknowledgments

This work was supported by grants EY015526, EB002052, and EB001980 of the NIH, and also by grant-in-aid from Scientific Research on Priority Areas from the Ministry of Education, Culture, Sport, Science and Technology of Japan and Strategic International Cooperative Program of Japan Science and Technology Agency (A.K.).

References

1. Subczynski, W. K., Hyde, J. S., and Kusumi, A. (1991) Effect of alkyl chain unsaturation and cholesterol intercalation on oxygen transport in membranes: a pulse ESR spin labeling study. *Biochemistry* **30,** 8578–8590.
2. Träuble, H. (1971) The movement of molecules across lipid membranes: A molecular theory. *J. Membr. Biol.* **4,** 193–208.
3. Pace, R. J. and Chan, S. I. (1982) Molecular motions in lipid bilayers. III. Lateral and transversal diffusion in bilayers. *J. Chem. Phys.* **76,** 4241–4247.
4. Altenbach, C., Greenhalgh, D. A., Khorana, H. G., and Hubbell, W. L. (1994) A collision gradient method to determine the immersion depth of nitroxides in lipid bilayers: application to spin-labeled mutants of bacteriorhodopsin. *Proc. Natl. Acad. Sci. USA* **91,** 1667–1671.
5. Subczynski, W. K., Hyde, J. S., and Kusumi, A. (1989) Oxygen permeability of phosphatidylcholine-cholesterol membranes. *Proc. Natl. Acad. Sci. USA* **86,** 4474–4478.
6. Subczynski, W. K., Hopwood, L. E., and Hyde, J. S. (1992) Is the mammalian cell plasma membrane a barrier to oxygen transport? *J. Gen. Physiol.* **100,** 69–87.
7. Subczynski, W. K., Lewis, R. N. A. H., McElhaney, R. N., Hodges, R. S., Hyde, J. S., and Kusumi, A. (1998) Molecular organization and dynamics of 1-palmitoyl-2-oleoylphosphatidylcholine bilayers containing a transmembrane α-helical peptide. *Biochemistry* **37,** 3156–3164.
8. Subczynski, W. K., Pasenkiewicz-Gierula, M., McElhaney, R. N., Hyde, J. S., and Kusumi, A. (2003) Molecular dynamics of 1-palmitoyl-2-oleoylphosphatidylcholine membranes containing transmembrane α-helical peptides with alternating leucine and alanine residues. *Biochemistry* **42,** 3939–3948.

9. Ashikawa, I., Yin, J.-J., Subczynski, W. K., Kouyama, T., Hyde, J. S., and Kusumi, A. (1994) Molecular organization and dynamics in bacteriorhodopsin-rich reconstituted membranes: discrimination of lipid environments by the oxygen transport parameter using a pulse ESR spin-labeling technique. *Biochemistry* **33,** 4947–4952.
10. Kawasaki, K., Yin, J.-J., Subczynski, W. K., Hyde, J. S., and Kusumi, A. (2001) Pulse EPR detection of lipid exchange between protein-rich raft and bulk domains in the membrane: methodology development and its application to studies of influenza viral membrane. *Biophys. J.* **80,** 738–748.
11. Subczynski, W. K., Wisniewska, A., Hyde, J. S., and Kusumi, A. (2004) Membrane microdomains as detected by "discrimination by oxygen transport" based on pulse EPR spin labeling. *Biophysical Society Discussions: Probing Membrane Microdomains*, Asilomar, California, Abstract P84-A. Online study book at, http://www.biophysics.org/discussions/2004/studybook.htm
12. Wisniewska, A, Subczynski, W. K. (2004) Lipid domains: EPR discrimination by oxygen transport. *Curr. Top. Biophys.* **28,** 89–94.
13. Subczynski, W. K, Wisniewska, A., Hyde, J. S., and Kusumi, A. (2007) Three-dimensional dynamic structure of the liquid-ordered domain in lipid membranes as examined by pulse-EPR oxygen probing. *Biophys. J.* **92,** 1573–1584.
14. Wisniewska, A. and Subczynski, W. K. (2006) Accumulation of macular xanthophylls in unsaturated membrane domains. *Free Radic. Biol. Med.* **40,** 1820–1826.
15. Kusumi, A., Subczynski, W. K., and Hyde, J. S. (1982) Oxygen transport parameter in membranes as deduced by saturation recovery measurements of spin-lattice relaxation times of spin labels. *Proc. Natl. Acad. Sci. USA* **79,** 1854–1858.
16. Windrem, D. A. and Plachy, W. Z. (1980) The diffusion-solubility of oxygen in lipid bilayers. *Biochim. Biophys. Acta* **600,** 655–665.
17. Hyde, J. S. and Subczynski, W. K. (1984) Simulation of ESR spectra of the oxygen-sensitive spin-label probe CTPO. *J. Magn. Reson.* **56,** 125–130.
18. Hyde, J. S. and Subczynski, W. K. (1989) Spin-label oximetry, in *Biological Magnetic Resonance*, vol. 8 (Berliner, L. J. and Reuben, J., eds.), Plenum Press, New York, pp. 399–425.
19. Subczynski, W. K. and Hyde, J. S. (1984) Diffusion of oxygen in water and hydrocarbons using an electron spin resonance spin-label technique. *Biophys. J.* **45,** 743–748.
20. Egreet-Charlier, M., Sanson, A., Ptak, M., and Bouloussa, O. (1978) Ionization of fatty acids at lipid-water interface. *FEBS Lett.* **89,** 313–316.
21. Kusumi, A., Subczynski, W. K., and Hyde. J. S. (1982) Effects of pH on ESR spectra of stearic acid spin labels in membranes: probing the membrane surface. *Fed. Proc.* **41,** 1394.
22. Kusumi, A. and Pasenkiewicz-Gierula, M. (1988) Rotational diffusion of a steroid molecule in phosphatidylcholine membranes: effects of alkyl chain length, unsaturation and cholesterol as studied by a spin label method. *Biochemistry* **27,** 4407–4418.

23. Pasenkiewicz-Gierula, M., Subczynski, W. K., and Kusumi, A. (1990) Rotational diffusion of a steroid molecule in phosphatidylcholine-cholesterol membranes: fluid phase microimmiscibility in unsaturated-phosphatidylcholine-cholesterol membranes. *Biochemistry* **29,** 4059–4069.
24. Hyde, J. S., Subczynski, W. K., Camenisch, T. G., Ratke, J. J., and Froncisz, W. (2004) Spin label EPR T_1 values using saturation recovery from 2 to 35 GHz. *J. Phys. Chem. B* **27,** 9524–9529.
25. Subczynski, W. K., Felix, C. C., Klug, C. S., and Hyde, J. S. (2005) Concentration by centrifugation for gas exchange EPR oximetry measurements with loop-gap resonators. *J. Magn. Reson.* **176,** 244–248.
26. Yin, J.-J. and Hyde, J. S. (1987) Spin-label saturation-recovery electron spin resonance measurements of oxygen transport in membranes. *Z. Phys. Chem.* (Munich) **153,** 57–65.
27. Hyde, J. S., Yin, J.-J., Feix, J. B., and Hubbell, W. L. (1990) Advances in spin label oximetry. *Pure Appl. Chem.* **62,** 255–260.
28. Yin, J.-J. and Hyde, J. S. (1989) Use of high observing power in electron spin resonance saturation-recovery experiments in spin-labeled membranes. *J. Chem. Phys.* **91,** 6029–6035.
29. Almeida, P. F. F., Vaz, W. L. C., and Thompson, T. E. (1992) Lateral diffusion in the liquid phases of dimyristoylphosphatidylcholine/cholesterol bilayers: a free volume analysis. *Biochemistry* **31,** 6739–6747.

12

Plasmon-Waveguide Resonance Spectroscopy Studies of Lateral Segregation in Solid-Supported Proteolipid Bilayers

Zdzislaw Salamon, Savitha Devanathan, and Gordon Tollin

Summary

Plasmon-waveguide resonance (PWR) spectroscopy is a high-sensitivity optical method for characterizing thin films immobilized onto the outer surface of a glass prism coated with thin films of a metal (e.g., silver) and a dielectric (e.g., silica). Resonance excitation by a polarized continuous wave (CW) laser above the critical angle for total internal reflection generates plasmon and waveguide modes, whose evanescent electromagnetic fields are localized on the outer surface and interact with the immobilized sample (in the present case a proteolipid bilayer). Plots of reflected light intensity vs the incident angle of the exciting light constitute a PWR spectrum, whose properties are determined by the refractive index (n), the thickness (t), and the optical extinction at the exciting wavelength (k) of the sample. Plasmon excitation can occur using light polarized both perpendicular (p) and parallel (s) to the plane of the resonator surface, allowing characterization of the structural properties of uniaxially oriented proteolipid films deposited on the surface. As will be demonstrated in what follows, PWR spectroscopy provides a powerful tool for directly observing in real-time microdomain formation (rafts) in such bilayers owing to lateral segregation of both lipids and proteins. In favorable cases, protein trafficking can also be monitored. Spectral simulation using Maxwell's equations allows these raft domains to be characterized in terms of their mass densities and thicknesses.

Key Words: Membrane thickness; microdomains; molecular sorting; protein trafficking; rafts; surface mass density.

1. Introduction

A number of biochemical and biophysical methodologies have been used to investigate microdomain formation in both whole cell and model membranes, e.g., low-temperature extraction with nonionic detergents, effects of cholesterol depletion, and various forms of visualization, including electron, atomic force, and fluorescence microscopies. All of these methods have potential problems *(1)*, and thus, despite a great deal of effort, there is no consensus yet on the size, lifetime, composition, or even existence of rafts in the plasma membrane of living cells *(cf. 2,3)*. As might be expected, the most definitive studies have been done with model membranes *(cf. 2)*, with atomic force microscopy (AFM) *(4,5)*, and fluorescence correlation spectroscopy *(6)* being particularly effective in such systems. These methodologies have clearly shown the spontaneous formation of laterally segregated domains within lipid bilayers and selective protein incorporation into such domains.

In this chapter, it will be demonstrated that plasmon-waveguide resonance (PWR) spectroscopy *(7–10)*, applied to solid-supported self-assembled lipid bilayers *(11–13)*, provides a uniquely useful method for observing raft formation in real-time, for obtaining information on the molecular composition of microdomains, and for monitoring protein incorporation into raft and nonraft regions. PWR allows the most important structural parameters of a lipid membrane, such as the thickness, the average surface area occupied per lipid molecule (i.e., molecular packing density), and the degree of long-range molecular order, to be characterized for a single lipid bilayer in both steady-state and kinetic modes *(14,15)*. Unlike other techniques for domain visualization, PWR can also provide insights into microenvironmental effects on protein functional properties *(16)*. The method involves resonance excitation by light, polarized perpendicular and parallel to a sensor surface, in plasmon and waveguide modes in a metal/dielectric film deposited on a glass prism. This generates a surface-localized evanescent electromagnetic field that interacts with molecules immobilized at the film surface *(7,14,15,17)*.

2. Materials

The only special requirement for buffer solutions used in PWR sample cells is that they should not interact with either the silica surface of the resonator or the deposited lipid bilayer. Sodium ions should be avoided where possible because these do bind to silica and alter its characteristics; potassium ions are preferable as the buffer cation (*see* **Note 1**). Organic solvents and detergents that incorporate into lipid films should be tested for their effects on PWR spectra, and kept at low concentrations when feasible. Octylglucoside has been found to be an effective detergent owing to its high critical micelle concentration and low tendency to disrupt bilayers. Other detergents can be used, but control experiments need to be performed to quantify their spectral effects (*see* **Note 2**). Control experiments also need to be done when using hydrophobic molecules as ligands for proteins incorporated into bilayers. Lipids should be as pure as possible and stored at low temperatures under an inert atmosphere to avoid oxidative degradation.

3. Methods
3.1. PWR Spectroscopy

Both the experimental arrangement and principles of PWR spectroscopy have been thoroughly described in previous publications *(7,14,15,17,18)* (*see* **Note 3**).

PWR Spectroscopy Studies in Solid-Supported Proteolipid Bilayers

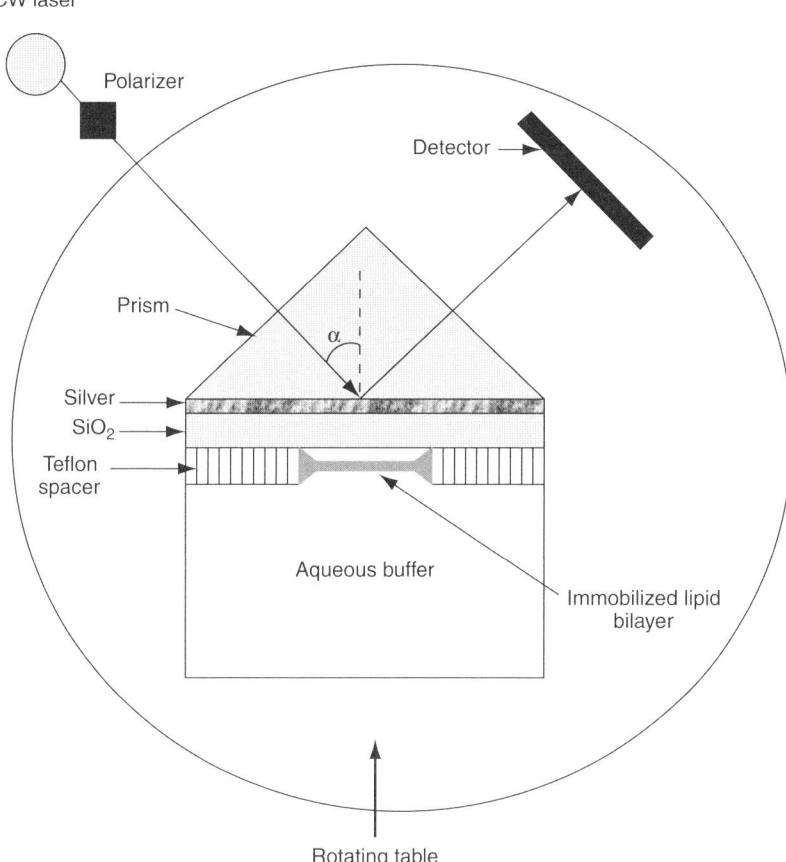

Fig. 1. Schematic illustration of a PWR apparatus, showing a lipid bilayer immobilized on the resonator surface in contact with an aqueous medium.

Those aspects that are especially relevant to the studies of lipid bilayer microdomains will be briefly reviewed. **Figure 1** shows a schematic view of a PWR spectrometer. As implemented by the authors, the method is based on the resonant excitation by polarized light from CW He–Ne laser (λ = 632.8 or 543.5 nm), passing through a glass prism under total internal reflection conditions, of plasmons in a thin metal film (Ag) deposited on the external surface of the prism, which is overcoated with a dielectric layer (SiO_2; *see* **Note 1**). The resonance is achieved by varying the light incident angle slightly above the critical angle for total internal reflection. The reflected light intensity as a function of the incident angle, monitored by a solid-state detector, constitutes the PWR spectrum. PWR spectra can be obtained with light polarized either perpendicular to the resonator surface (*p*-polarization) or parallel to this surface (*s*-polarization).

PWR spectra can be described by three parameters, the spectral position, the spectral width, and the resonance depth. These experimental features depend on the optical properties of the bilayer membrane, which are determined by the surface mass density (i.e., the amount of mass per unit surface area), the spatial mass distribution (i.e., the internal structure of the membrane, including molecular anisotropy and the long range molecular order of the bilayer), the membrane thickness, and the absorption or light scattering properties of the membrane at the plasmon excitation wavelength. These properties are described by three optical parameters: refractive index (n), extinction (or scattering) coefficient (k), and thickness (t) of the membrane, which can be evaluated by thin-film electromagnetic theory based on Maxwell's equations *(14,15,17,18–20)* (*see* **Note 4**). Inasmuch as both of the excitation wavelengths used are far removed from the absorption bands of the lipids and proteins being detected, a k value different than zero reflects a decrease of reflected light intensity owing only to scattering, resulting from imperfections in the membrane film (*see* **Subheading 3.3.**). It is important to recognize that for anisotropic thin films, such as the lipid bilayer membranes in the present work, scattering will be different for different exciting light polarizations.

The refractive index is a macroscopic quantity and is related to the properties of individual molecules through the molecular polarizability tensor, as well as to the environment in which these molecules are located (e.g., packing density and internal organization) *(12)*. Environmental properties are especially important when molecules are located in a matrix that has a nonrandom organization (such as a biomembrane) and thus possesses long-range spatial molecular order. Such molecular ordering creates an anisotropic system with a uniaxial optical axis having two (different) principal refractive indices, n_e (also denoted as n_p) and n_o (also referred to as n_s) *(12)*. The first of these indices is associated with a linearly polarized light wave in which the electric vector is polarized parallel to the optical axis. The second one is observed for light in which the electric vector is perpendicular to the optical axis. This is the fundamental basis on which measurement of refractive indices with polarized light can lead to the evaluation of the structural parameters of anisotropic systems such as a lipid bilayer membrane *(12)* (*see* **Note 4**). In the simplified case in which the molecular shape is rod-like (e.g., phospholipid molecules), and the molecules are ordered such that their long axes are parallel, one has an anisotropic system whose optical axis is perpendicular to the plane of the bilayer *(12)*. The values of the refractive indices measured with two polarizations of light (i.e., parallel, n_p, and perpendicular, n_s, to the optical axis) will describe this optical anisotropy (A_n) as follows *(17–20)*:

$$A_n = \left[(n_p)^2 - (n_s)^2\right] / \left[(n_{av})^2 + 2\right] \qquad (1)$$

where n_{av} is the average value of refractive index, and for a uniaxial system in which the optical axis is parallel to the membrane normal is given by:

$$n^2_{av} = 1/3\left[(n_p)^2 + 2(n_s)^2\right] \quad (2)$$

Thus, A_n reflects the spatial mass distribution created by both the anisotropy in the molecular polarizability and the degree of long-range order of molecules within the system *(30)*. For rod-shaped molecules, n_p is always larger than n_s. As can be seen from the Lorentz–Lorenz relation, n_{av} is also directly related to the surface mass density:

$$m = 0.1M/A\left\{t\left[(n_{av})^2 - 1/(n_{av})^2 + 2\right]\right\} \quad (3)$$

where M is molecular weight, A is molar refractivity, and t is thickness of the membrane. For the lipid molecules used in this work a reasonable approximation of M/A is 3.6 *(21)*. Thus, from the thickness of the membrane (t) and n_{av} one can calculate the surface mass density (or molecular packing density), i.e., mass per unit surface area (or number of moles/unit surface area *[12]*), which reflects the surface area occupied by a single molecule.

3.2. Formation of a Single Planar Lipid Membrane on the Resonator Surface

In the PWR studies, planar self-assembled solid-supported lipid membranes have been used *(11–13)*. The details of sample compartment design and the protocols for membrane preparation and protein incorporation have been described in earlier publications *(7,17–20)*. A short summary of these descriptions is presented herein. The method for membrane formation involves spreading 2–3 µL of lipid solution (typically 8–10 mg/mL lipid dissolved in 9:1 butanol:squalene) across an orifice (2 mm in diameter) in a Teflon block separating the silica surface of the PWR resonator from the aqueous phase (**Figs. 1** and **2**). The lipid composition can be varied over a wide range. The hydrated silica surface attracts the polar groups of the lipid molecules to form a monolayer with the hydrocarbon chains oriented toward the excess lipid solution. Spontaneous bilayer formation is initiated when the sample compartment of the resonator is filled with an aqueous solution, resulting in a thinning process to form the second monolayer of the lipid, and a plateau-Gibbs border consisting of lipid solution that anchors the bilayer to the Teflon block. This border allows excess lipid to flow into or out of the orifice in response to protein insertion and/or conformation changes.

When the appropriate lipid compositions were used (e.g., 1:1 mixtures of phosphatidylcholine and sphingomyelin [SM], both in the presence and in the absence of cholesterol), the bilayers produced PWR spectra that displayed two

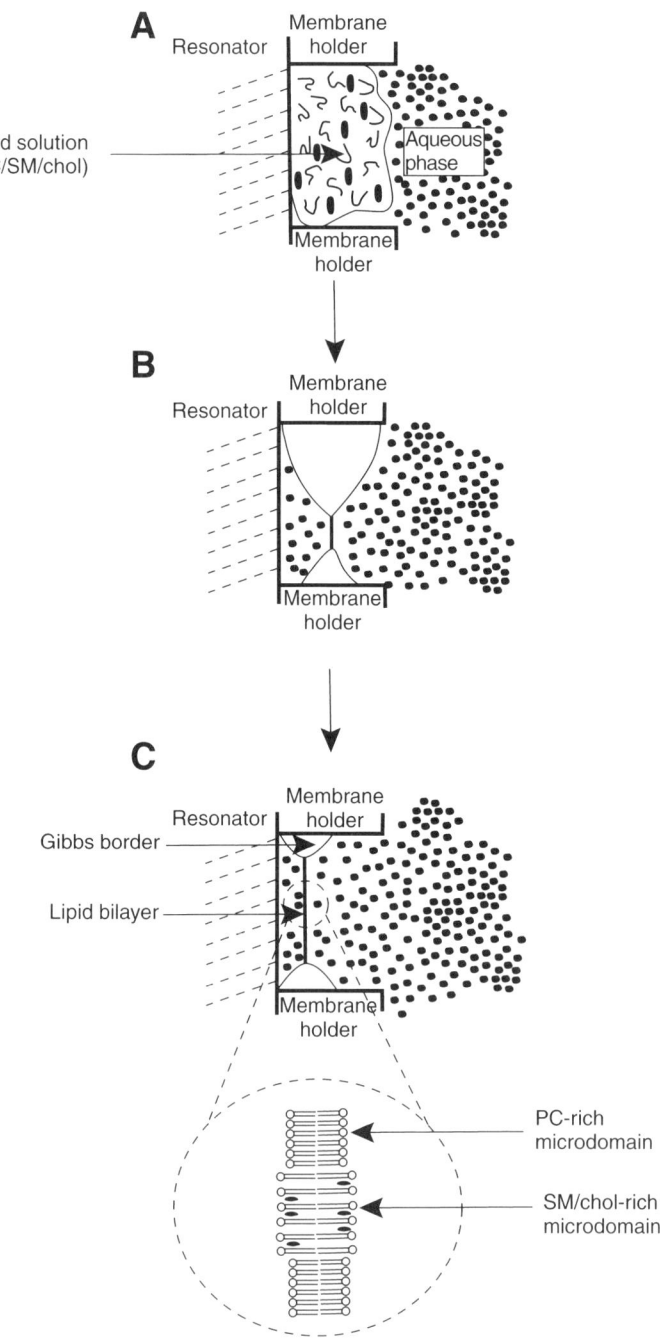

Fig. 2. Schematic illustration of the formation process of a lipid bilayer made up of a mixture of DOPC, SM, and cholesterol, resulting in microdomains.

resonances *(22–24)*. As it will be demonstrated later, these can be ascribed to the spontaneous formation of microdomains within the bilayer owing to lateral segregation of lipid molecules.

3.3. Distinguishing Between Irregularities (Imperfections) in the Lipid Bilayer and Microdomains

Application of PWR technology to studies of lipid rafts is based on the premise that one can observe overlapping PWR spectra resulting from optically dissimilar parts (i.e., raft and nonraft areas) of a single lipid membrane. This is owing to differences in mass density and in thickness between the microdomains. In this article all of the instrumental and technical requirements will not be dealt with that are necessary to observe such spectra, but instead will focus on issues that relate to a proper evaluation and interpretation of such complex PWR spectra, especially in the context of the self-assembling process that leads to the creation of a planar, solid-supported lipid bilayer membrane *(11–13)*.

Working with a single lipid membrane is always a challenging task, and is especially so in the context of lipid microdomains. In general, one may define microdomains as nonuniformities in the regular structure of a lipid bilayer resulting from lateral phase separation of different kind of lipids, leading to differences in both content and structure, and therefore in physical properties. Those who have practical experience with creation of self-assembled lipid membranes know how difficult it is to get a uniform bilayer, even when using a single type of lipid. This is especially problematic in the case of lipid mixtures. The self-assembly process (*see* **Fig. 2**) generally starts with a very thick film of lipid (usually tens of micrometers thick), spread within the aperture of a membrane holder, and by a thinning process ends up with two molecular layers approx 5.0 nm thick, perfectly arranged to form a regular bilayer attached to a fluid plateau-Gibbs border on the edge of the membrane holder *(11–13)*. The process itself is a rather complicated event occurring simultaneously along a number of pathways. Typically, it takes about 10–40 min to be completed. The annulus of lipid solution surrounding the bilayer is essential for its existence. It connects the bilayer to the aperture, so its mechanical properties have an effect on the ability to form and manipulate the membrane. The shape of the annulus is determined by the contact angle between the bilayer and the torus, as well as by the volume of solution placed within the aperture. In the case of solid-supported membranes there is an additional important parameter influencing the shape of the annulus and therefore the bilayer itself, namely, the contact angle between the torus and the solid support. This additional condition, as compared with a freely suspended planar bilayer between two aqueous phases, creates an asymmetry within the system. This provides an additional difficulty in designing the proper membrane holder and the solid support, in order to get acceptable bilayer membrane formation. It is important to note that the annulus has an average thickness much greater than that of the bilayer. These two phases not only depend on each other (i.e., if one does not form properly, the second will not either), but they also depend on ambient conditions (i.e., the membrane holder, the lipid mixture, and lipid solvent, as well as the properties of the aqueous phase). Furthermore, the Gibbs border and the bilayer tend toward chemical equilibrium with one another.

The process of bilayer creation involves not only removing a majority of the original lipid mass from the aperture (in most cases >99%), but also reassembling it from an isotropic fluid into a two-dimensional, ordered, highly anisotropic system. As a consequence of the complexity of this process, the self-assembling process may easily be influenced by number of ambient circumstances leading to disruption of membrane formation. This can lead either to a complete destruction of a lipid film or to the creation of a membrane with a number of structural and mass defects and imperfections. A problem then arises because such defects and imperfections may also lead to a complex resonance spectrum, which in some cases may be visually similar to that obtained with real microdomains. Thus, the question arises: how does one distinguish between real microdomain formation and the generation of randomly distributed structural defects? In order to do this, one must carry out an analysis of the PWR spectra to characterize the different optical properties of real microdomains as compared with membrane imperfections. This process is described in **Subheading 3.4.**

One of the key properties of bilayer microdomains is their high content of either SM *(2)* or SM and cholesterol (in the latter case they are usually called rafts) in mixtures with phosphatidylcholine, which leads to their being organized into what are referred to as gel phases (with SM alone) and liquid-ordered (L_o) domains (with SM/cholesterol mixtures). These are characterized by being more highly ordered and somewhat thicker than the surrounding liquid-disordered (L_d) regions of the membrane (*see* insert in **Fig. 2**, panel C). This is a consequence of the ordering influence of cholesterol and the presence in SM of a larger proportion of long saturated fatty acyl chains. These lipid phases can coexist within a single bilayer, giving rise to a heterogeneous pattern of islands of differing composition and physical properties.

Model membrane studies have shown that lipid–lipid interactions are sufficient to induce the formation of raft-like domains *(4)*. Thus, it is well established that phase separation can occur in binary lipid mixtures consisting of lipids that have different phase transition temperatures. Typically, an ordered gel phase, which is characterized by tightly packed lipids that have limited mobility, coexists with a fluid or liquid-disordered phase in which the lipids are loosely packed and have a high degree of lateral mobility. Addition of cholesterol has been reported to modify the gel phase component of such systems resulting in the formation of a liquid-ordered (L_o) phase in which the lipids are still tightly packed but acquire a relatively high degree of lateral movement. However, currently the requirement for cholesterol in raft formation in biological membranes is unclear *(4,25)*.

As is apparent from the earlier description and from **Fig. 2C**, there are three very distinct and important properties of real microdomains. First, they are characterized by an increased anisotropy as compared with a regular bilayer, especially in the horizontal plane. Second, they have a much increased surface mass density (or packing density), and third, they have specific dimensions (especially thickness) that are related to the molecular structure of the lipids that include them (e.g., fatty acyl chain length and degree of unsaturation). In contrast to this, because most imperfections occur as a result of disruption of the membrane creation process, they will be trapped in unfinished structures, and will be less anisotropic, will have lower packing density, and will have thicknesses that are not in the right range. In order to distinguish between these two possibilities, one has to carefully analyze the PWR spectra to quantitatively describe the properties of the membrane system. There are three approaches available to perform such analyses, which can be used depending on the specific experimental circumstances: (1) a graphical analysis of the resonance spectra *(26)*, (2) spectral simulation that compares a theoretical spectrum with an experimental spectrum *(22)*, and (3) nonlinear least-squares fitting of a theoretical spectrum to the experimental one *(7,18–20)*. As the simulation approach has been used most often in work dealing with membrane microdomains, this technique will be focused on in **Subheading 3.4.**

3.4. Spectral Simulation

This methodology involves simulating experimental spectra by theoretical resonance curves in order to evaluate the optical parameters of the bilayer membrane, (i.e., n_p, n_s, t, and k), and then using these values (*see* **Eqs. 1–3**) to calculate the surface mass density (or packing density) and the refractive index anisotropy. This provides a detailed description of the membrane structure. Such simulation is based on two facts. First, the PWR spectrum can be described by the classical electromagnetic theory of thin films and the equations describing such resonance curves can be used in the simulation to obtain the component curves. Second, the number of measured parameters (i.e., position, width, and depth of the curve) equals the number of unknown optical parameters. Hence, one can uniquely determine the latter from the simulated spectra. This approach has been used in the authors previous publications *(7,18–20)* for nonlinear least-squares fitting of a theoretical resonance curve to the experimental spectra. This is relatively easy to do when the spectrum corresponds to a single resonance. However, this approach is more difficult when the experimental spectrum is a complex one, i.e., it consists of more than one (usually two) single resonance curves, corresponding to the microdomain composition. Therefore, the simulation has to be done in two steps. First, one must calculate single resonance curves for the components, and second, one must sum such single resonance curves using appropriate ratios to fit the complex spectrum.

This first step results in an evaluation of the optical parameters (n, k, and t) that describe the physical properties of those parts of the lipid membrane that contribute to this particular single resonance curve *(7,12,18–20)*. These can be obtained with a high degree of accuracy because they are well separated in their effects on the plasmon resonance spectra. Theoretical analysis of the effect of each optical parameter on the resonance spectra allows the evaluation of error limits for each of these parameters *(18)*. The values of such errors have been included in the analysis presented below. The second step provides information about the ratio of the bilayer surfaces covered by the two different kinds of membrane that are exposed to the excitation laser beam. This is based on the fact that the area under a resonance curve is constant, and therefore a single resonance will result in a narrow and deep resonance curve, whereas two resonances occurring simultaneously will broaden the spectrum and make it shallower.

There is another important consequence of the simulation process in the case of PWR measurements with a single lipid membrane. Although repetition of the measurements in separate experiments with the same type of membrane will result in similar values of the optical properties of the components, the ratio of the surface areas covered by the different membrane components that are exposed to the excitation laser beam may vary from membrane to membrane. Therefore, the final spectrum may change its visual characteristics. This implies that one can average the parameter values obtained from different measurements and use them to calculate physical quantities describing the properties of such bilayers.

3.5. Specific Examples of Segregation of Lipids and Proteins Into Microdomains

PWR spectroscopy has been used together with simulation methods to analyze results obtained with solid-supported lipid bilayers consisting of binary mixtures of dioleoylphosphatidylcholine (DOPC) and SM, and of palmitoyloleoylphosphatidylcholine (POPC) and SM *(22,23)*. Experiments have also been carried out with DOPC/SM/cholesterol ternary mixtures, although in this case simulation analyses were not carried out owing to insufficient results with binary mixtures having varying cholesterol contents. As will be seen later, mixtures of these lipids spontaneously generate segregated microdomains that can be directly observed and characterized by PWR spectroscopy. Such lipid bilayers were then used to study the partitioning of two proteins within the membrane: a glycosylphosphatidylinositol (GPI)-linked protein (placental alkaline

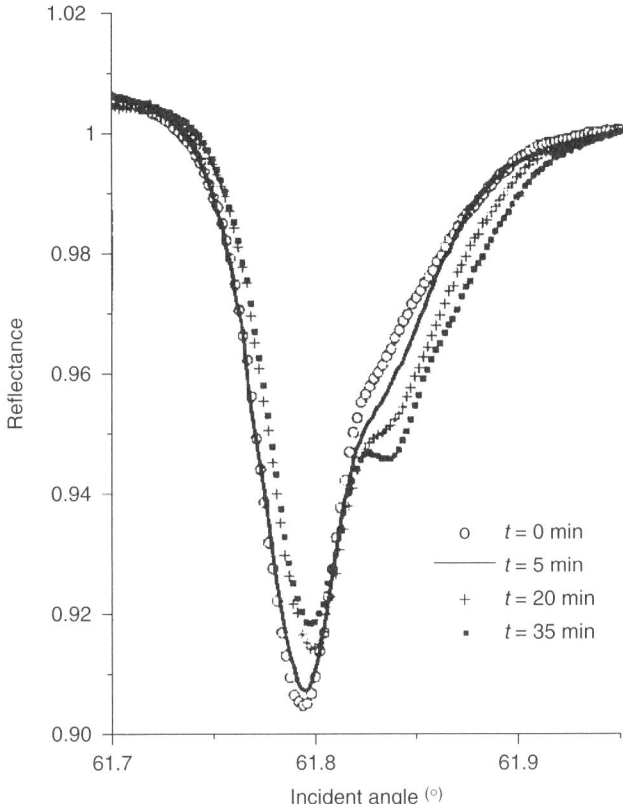

Fig. 3. Time-resolved PWR spectra of a lipid bilayer made up of a 1:1 mixture of DOPC and SM (from brain) showing the spontaneous generation of two resonances through lateral segregation. The resonance at smaller incident angles corresponds to a PC-rich microdomain and the resonance at larger angles to an SM-rich microdomain (*see* text and **Fig. 4**).

phosphatase [PLAP]) *(22)*, and human δ-opioid receptor (hDOR) *(23)*. All experimental details of this work can be found in the earlier publications. Some of the key results will be summarized herein, thereby demonstrating how the PWR technique can be applied to studies of lateral phase segregation leading to lipid microdomains and protein trafficking between them.

Figure 3 shows time-resolved spectra illustrating the dynamics of domain formation using a DOPC/SM mixture at a molar ratio of 1.5:1. These spectra clearly indicate superposed resonances, indicating lateral phase separation of lipids occurring over a period of minutes *(22)*. Although not shown herein, the kinetics of the formation of laterally segregated phases was dependent on the molar ratio of the two lipids. Thus, the kinetics was significantly faster with a

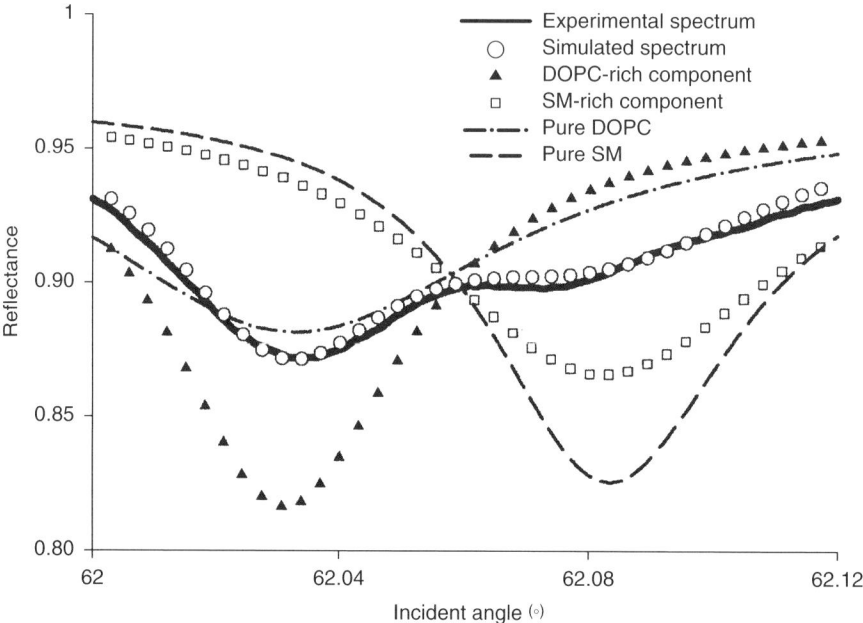

Fig. 4. Simulation of the PWR spectrum shown in **Fig. 3** (35 min) by a superposition of DOPC-rich and SM-rich component curves. Shown for comparison are the PWR spectra obtained from bilayers formed from pure DOPC and pure SM. The optical parameters obtained from these simulated spectra are given in Table 1.

larger excess of DOPC than when a larger amount of SM was present, perhaps as a consequence of a lower microviscosity in the former than in the latter. However, the thermodynamic equilibrium of the DOPC-enriched system was shifted away from that of segregated microdomains, probably as a result of coalescence, i.e., after transient lateral separation, the system equilibrated into what appeared to be a single phase consisting of a mixture of these two lipids. In contrast, the system with a larger amount of SM remained able to form stable segregated microdomains.

As indicated earlier, in order to confirm that what appears to be a microdomain spectrum is not because of random imperfections, a simulation analysis was carried out. **Figure 4** shows an experimental spectrum for this mixture (solid line) simulated by two single spectra describing DOPC-rich domains (triangles) and SM-rich domains (squares), which when superposed result in a complex resonance curve (circles). The spectra of the two-component microdomains can be compared with PWR spectra obtained from pure single-lipid bilayer spectra: pure DOPC (dash-dot curve), and pure SM (dash curve). As can be seen from this figure, the experimental resonance curve for the binary mixture is very well simulated by the

Table 1
Optical Parameters Obtained From Simulation Analysis of Lipid Membranes Made up Either of Single Lipids or of Binary Mixtures of Lipids

Lipid bilayer composition	Thickness t_{av} (nm ± 0.1)	Surface area/molecule (nm² ± 0.01)	Optical anisotropy A_n (±0.003)
DOPC	5.0	0.53	0.01
POPC	5.4	0.48	0.02
SM	6.1	0.39	0.04
DOPC/SM (1:1)			
DOPC-rich domain	5.2	0.51	0.015
SM-rich domain	5.9	0.40	0.025
POPC/SM (1:1)			
POPC-rich domain	5.4	0.46	0.025
SM-rich domain	6.1	0.39	0.025

From **ref. 22**.

simulated spectrum (circles). However, the component spectra describing the two microdomains (triangles and crosses) differ significantly in width and depth from the single lipid component spectra, although the angular positions are not different. Thus, the component DOPC spectrum (triangles) is narrower and has greater depth, whereas the component SM spectrum (crosses) is broader and shallower, as compared with the curves obtained from simulation of the single lipid spectra (dash-dot and dash).

A comparison of the optical parameters resulting from the simulation of the PWR spectra of lipid microdomains with those obtained with membranes consisting of the single-lipid components (**Table 1**), reveals a clear pattern *(22)*. Thus, the DOPC region of the bilayer has increased values of both the thickness and packing density, whereas the SM values for the microdomains decrease as compared with the single lipid. Also important to note are the opposite changes in the refractive index anisotropies, which are consistent with the packing density and thickness alterations. Similar results have been obtained with POPC/SM mixtures (*see* **Fig. 5**). These simulations lead to the conclusion that the experimental spectra for mixtures can be interpreted as a sum of two resonance curves that are produced by small modifications of the single component parameters, suggesting that the lipid compositions in the microdomains are modified by the inclusion of very small amounts of the other component. However, it must be pointed out that these small amounts are significant enough to considerably alter the resonance spectra of the pure single components and to allow formation of two stable membrane phases. This would suggest that microdomains made up of the pure lipids are not thermodynamically or kinetically stable. It is possible that mixing in small amounts of

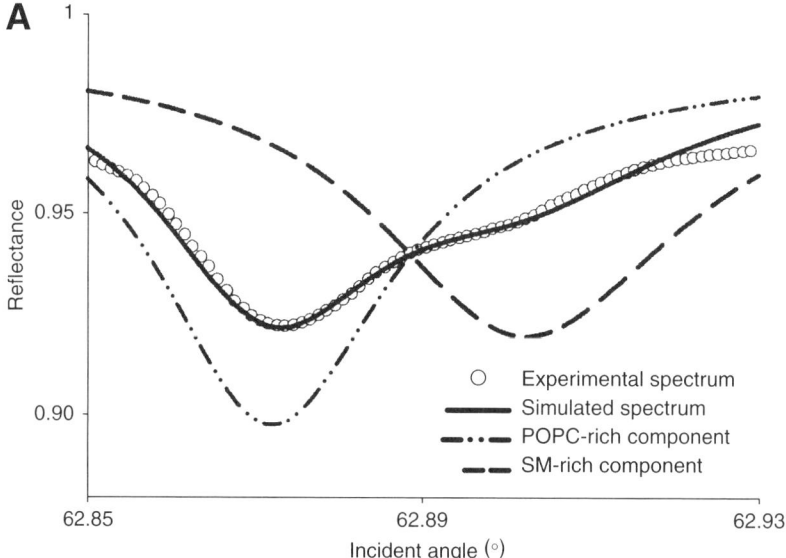

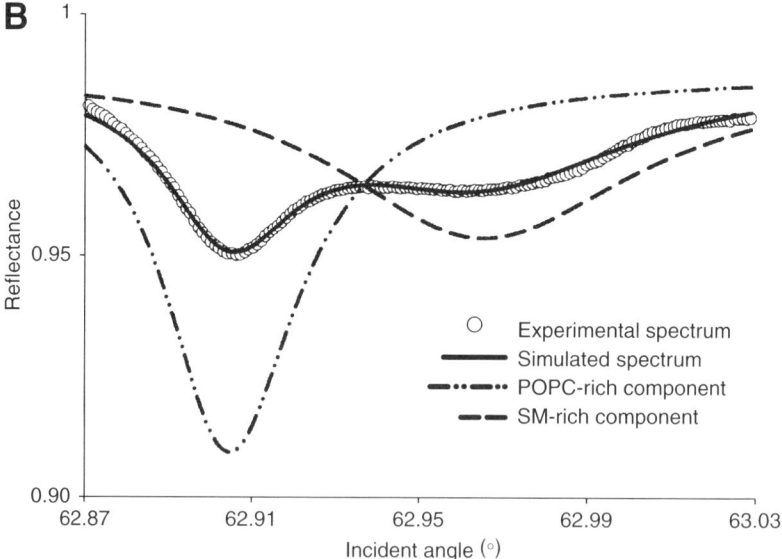

Fig. 5. Simulations of PWR spectra obtained from a bilayer made up of a 1:1 mixture of POPC and brain SM in before **(panel A)** and after **(panel B)** incorporation of an agonist-bound (DPDPE) hDOR. The mass density values obtained from the simulations are as follows: 0.85 and 1.85 pmoles/cm^2 for POPC-rich and SM-rich domains, respectively.

the other component can act to stabilize the system by filling in the gaps in the bilayer structure, somewhat like the mixture of amino acid side chains in the interior of a globular protein molecule allows a more compact structure to be formed.

These results are consistent with earlier observations obtained by fluorescence and AFM indicating that microdomains can be formed in supported lipid monolayers as well as in both supported and unsupported bilayers. Both these techniques have provided information about the thickness and overall dimensions of the membrane domains, which are also consistent with the PWR results. However, the PWR technique provides another important capability, i.e., it allows one to quantify important structural parameters such as packing density and internal organization, which permit a fuller description of the properties of such membrane domains. In addition, PWR used in a kinetic mode (*see* **Fig. 3**), allows one to quantify the formation process of lateral segregation.

3.6. Protein Segregation Within Lipid Microdomains

In this section, the ability of PWR spectroscopy to characterize the insertion of protein molecules into lipid microdomains, and to follow protein trafficking between such microdomains will be described. **Figure 5B** shows a PWR spectrum *(23)* obtained on insertion of an agonist-bound (DPDPE) hDOR (a G protein coupled receptor mediating pain responses in the brain) into a POPC/SM bilayer characterized by the PWR spectrum shown in **Fig. 5A** (open circles). The latter spectrum has been simulated very well (**Fig. 5A**, solid line) by two single-component spectra corresponding to the two lipid microdomains that exist within the bilayer (i.e., POPC-rich; dash-dot and SM-rich; dash). The parameters obtained from this simulation are given in **Table 1**. As can be seen by comparing the spectrum in **Fig. 5A** with that in **Fig. 5B**, incorporation of the protein influenced both the main maximum and the shoulder, thereby changing the overall shape of the spectrum. Although not shown, similar alterations occurred in both *p*- and *s*-polarizations *(23)*. Such spectral modifications are owing to both mass density changes and structural alterations of the proteolipid membrane. Note that larger shifts were obtained for the SM-rich than for the POPC-rich domain on incorporation of the agonist-bound receptor into the bilayer. Quantification of these changes is shown by the simulations in **Fig. 5A,B** dash-dot for the POPC-rich and dash for the SM-rich domains. From these component spectra, it is possible to calculate the mass and anisotropy changes that occur as a consequence of incorporating the receptor into the two microdomains. The values for mass density in these two domains are given in the legend to **Fig. 5** *(23)*. These show that the agonist-bound receptor prefers the SM-rich environment compared with the POPC-rich microdomain. It should also be pointed out that experiments performed with another δ-opioid agonist *(23)*, gave similar results, with preferential partitioning of the receptor into the SM-rich phase.

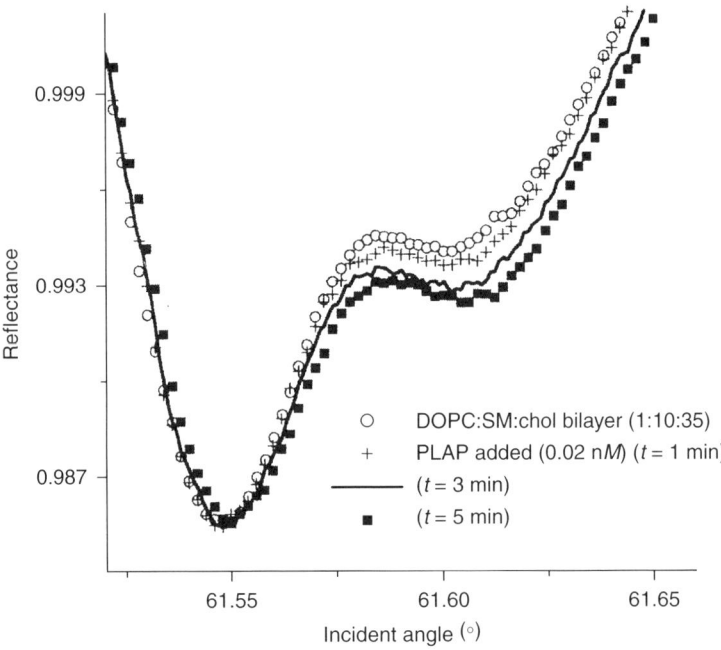

Fig. 6. Time-resolved PWR spectra obtained from a bilayer containing a mixture of DOPC, brain SM, and cholesterol (1:1:0.35) before and after addition of PLAP.

In the case of unliganded receptor incorporation, the opposite is true; i.e., more receptor is inserted into the POPC-rich phase than into the SM-rich domain (23). Subsequent addition of DPDPE (or other agonists) to the proteolipid system containing unliganded receptor leads to enrichment of receptor in the SM-rich domain. The distribution of protein was approximately the same regardless of whether the ligand was bound before or after incorporation of the receptor into the bilayer. Furthermore, as it has been demonstrated by PWR experiments described elsewhere (27), the G protein has a much higher affinity for agonist-bound receptor inserted into the SM-rich domain (0.6 nM) than one present in the POPC-rich phase (19 nM). Thus, the activity of the receptor is strongly modulated by the lipid microenvironment.

It should be noted that these experiments were performed without the inclusion of cholesterol in the bilayers. Although this lipid clearly plays a role in microdomain formation in bilayers (both in vivo and in vitro), the precise structural basis for this is at present unclear (28). In some preliminary experiments, PWR experiments with ternary lipid mixtures containing cholesterol (DOPC:SM:cholesterol, 1:1:0.35) have also been carried out. **Figure 6** shows an example of the kinetics of the insertion of the GPI-linked protein PLAP into such a bilayer. Although at present these spectra are unable to be simulated, owing to insufficient information on the spectral properties of bilayers formed from binary mixtures containing varying amounts of POPC, SM, and cholesterol, it is clear that the major changes with time occur in the higher-angle shoulder, indicating the preferential insertion of this protein into the domain corresponding to this resonance (i.e., most likely the SM/cholesterol-rich microdomain).

The aforementioned results demonstrate the ability of PWR spectroscopy to monitor protein insertion into lipid microdomains, to measure the influence of ligand binding on protein distribution, and to characterize the influence of lipid microenvironment on functional properties such as G protein binding. These observations have been rationalized based on the following observations. First, as previous studies performed in the laboratory using PWR have demonstrated, the opioid receptor adopts different conformations on binding to different classes of ligands *(20)*. Second, as demonstrated both by the PWR results *(22,23)* and by AFM studies *(28,29)*, the SM-rich domain is thicker than the POPC-rich domain, as a consequence of the presence of longer, largely saturated acyl chains in SM. As the agonist-activated receptor is more elongated than the unliganded or antagonist-bound receptor, implying a larger hydrophobic thickness, it is hypothesized that hydrophobic matching between the receptor and the bilayer provides the basis for the different receptor partitioning in each case. Furthermore, as the SM-rich domain favors the elongated structure produced by agonist activation, it is logical that the binding of G protein is stronger in this domain than in the POPC-rich domain.

In summary, it has been shown that PWR spectroscopy has distinct advantages over other methods of visualizing lipid microdomains in that it allows information to be obtained in real-time regarding the domain structure and composition, including the ability to distinguish between domains formed from various lipid species. The uniqueness of this spectroscopic technique is that it allows the most important structural parameters of a lipid membrane, such as the thickness, the average surface area occupied by one lipid molecule (or molecular packing density), and the degree of long-range molecular order, to be characterized for a single lipid bilayer in both steady-state and kinetic modes. Unlike other techniques for domain visualization, PWR can also provide insights into microenvironmental effects on protein functional properties.

4. Notes

1. The sensor surface is a thin layer of a dielectric material exposed to aqueous buffer, and thus, may be sensitive to ions present in solution. For instance, in the case of silica, one has to be aware that sodium ions are particularly damaging to the sensor surface causing its deterioration. Such problems can be avoided by either replacing one type of ion with another (e.g., sodium with potassium ions),

or replacing one kind of dielectric with another type of material. The choice of the latter will be dependent on the sensor optical properties required for a particular experiment, because changes of the dielectric material will always result in alterations of the spectral characteristics of the sensor *(7–10)*.
2. As mentioned, the insertion of proteins into a membrane is performed by diluting a detergent in which the membrane protein is dissolved to below its critical micelle concentration. Using this protocol, one has to be aware that any detergent may interact with a lipid membrane. Such interaction can interfere with measurements of protein insertion. Therefore, one has to always perform a control experiment with the detergent used.
3. PWR is a relatively new optical technique that may be regarded as analogous to surface plasmon resonance (SPR) *(30,31)*. The latter method has been commercially available for about 15 yr, and thus, there is a broader awareness in the scientific community of this technology as compared with PWR. In fact, the analogy between these two methods is quite narrow because PWR is able to study both optically isotropic as well as anisotropic thin films and interfaces deposited on the sensor surface, whereas SPR can only be applied to isotropic samples. Therefore, both the principles on which these technologies are based, and the experimental setups and data analysis procedures are quite different. Furthermore, the amount of information that can be obtained about systems using PWR is much larger than with SPR.
4. The analysis of PWR spectra using Maxwell's equations allows one to obtain a set of structural parameters of a lipid membrane deposited on the resonator surface *(7,17–20)*. In general, there is no direct comparison between the values of such parameters obtained with different techniques. The reason is twofold: first, they are usually obtained with different kinds of lipid membranes (e.g., in PWR experiments one has a single "black-lipid membrane," which in most of the cases will also contain some of the solvent in which the lipids were dispersed). Second, the definition of the parameters is directly related to the technology with which they are obtained. For example, thickness in the case of PWR represents an averaged value for an interfacial region characterized by a refractive index different from that of the bulk solution. For a lipid membrane deposited on the resonator surface this will include both hydrocarbon and head-group regions, together with water molecules located at both the silica surface of the sensor (*see* **Fig. 1**), and bound to the lipid membrane.

References

1. Pike, L. J. (2004) Lipid rafts: heterogeneity on the high seas. *Biochem. J.* **378**, 281–292.
2. Simons, K. and Vaz, W. L. C. (2004) Model systems, lipid rafts, and cell membranes. *Annu. Rev. Biophys. Biomol. Struct.* **33**, 269–295.
3. Munro, S. (2003) Lipid rafts: elusive or illusive? *Cell* **115**, 377–388.
4. Saslowsky, D. E., Lawrence, J., Ren, X., Brown, D. A., Henderson, R. M., and Edwards, J. M. (2002) Placental alkaline phosphatase is efficiently targeted to rafts in supported lipid bilayers. *J. Biol. Chem.* **277**, 26,966–26,970.

5. Milhiet, P., Vie, V., Giocondi, M., and Le Grimellec, C. (2001) AFM characterization of model rafts in supported bilayers. *Single Mol.* **2,** 119–121.
6. Bacia, K., Scherfeld, D., Kahya, N., and Schwille, P. (2004) Fluorescence correlation spectroscopy relates rafts in model and native membranes. *Biophys. J.* **87,** 1034–1043.
7. Salamon, Z., Macleod, A. H., and Tollin, G. (1997) Coupled plasmon-waveguide resonators: A new spectroscopic tool for probing proteolipid film structure and properties. *Biophys. J.* **73,** 2791–2797.
8. Salamon, Z., Macleod, A. H., and Tollin, G. (1999) Coupled plasmon-waveguide resonance spectroscopic device and method for measuring film properties. US Patent No. 5,991,488.
9. Salamon, Z. and Tollin, G. (2001) Coupled plasmon spectroscopic device and method for measuring film properties in the ultraviolet and infrared spectral ranges. US Patent No. 6,330,387 B1.
10. Salamon, Z. and Tollin, G. (2002) Coupled plasmon spectroscopic device and method for measuring film properties in the ultraviolet and infrared spectral ranges. US Patent No. 6,421,128 B1.
11. Salamon, Z., Wang, Y., Tollin, G., and Macleod, H. A. (1994) Assembly and molecular organization of self-assembled lipid bilayers on solid substrates monitored by surface plasmon resonance spectroscopy. *Biochim. Biophys. Acta* **1195,** 267–275.
12. Salamon, Z. and Tollin, G. (2001) Optical anisotropy in lipid bilayer membranes: Coupled plasmon-waveguide resonance measurements of molecular orientation, polarizability, and shape. *Biophys. J.* **80,** 1557–1567.
13. Salamon, Z., Schmidt, R. A., Tollin, G., and Macleod, H. A. (1996) Reusable biocompatible interface for immobilization of materials on a solid support. US Patent No. 5,521,702.
14. Salamon, Z. and Tollin, G. (1999) Surface plasmon resonance, theory, in *Encyclopedia of Spectroscopy and Spectrometry, vol. 3,* (Lindon, J. C., Tranter, G. E., and Holmes, J. L., eds.), Academic Press, NY, pp. 2311–2319.
15. Salamon, Z. and Tollin, G. (1999) Surface plasmon resonance, applications, in *Encyclopedia of Spectroscopy and Spectrometry, vol. 3,* (Lindon, J. C., Tranter, G. E., and Holmes, J. L., eds.), Academic Press, NY, pp. 2294–2302.
16. Alves, I. D., Ciano, K. A., Boguslavski, V., et al. (2004) Selectivity, cooperativity and reciprocity in the interactions between δ-opioid receptor, its ligands, and G-proteins. *J. Biol. Chem.* **279,** 44,673–44,682.
17. Salamon, Z., Brown, M. F., and Tollin, G. (1999) Plasmon resonance spectroscopy: probing interactions within membranes. *Trends Biochem. Sci.* **24,** 213–219.
18. Salamon, Z. and Tollin, G. (2001) Plasmon resonance spectroscopy: probing molecular interactions at surfaces and interfaces. *Spectroscopy* **15,** 161–175.
19. Salamon, Z., Huang, D., Cramer, W. A., and Tollin, G. (1998) Coupled plasmon-waveguide resonance spectroscopy studies of the cytochrome b_6f/plastocyanin system in supported lipid bilayer membranes. *Biophys. J.* **75,** 1874–1885.
20. Salamon, Z., Cowell, S., Varga, E., Yamamura, H. I., Hruby, V. J., and Tollin, G. (2000) Plasmon resonance studies of agonist/antagonist binding to the human

δ-opioid receptor: new structural insights into receptor-ligand interactions. *Biophys. J.* **79**, 2463–2474.
21. Cuypers, P. A., Corsel, J. W., Janssen, M. P., Kop, J. M. M., Hermens, W. T., and Hemker, H. C. (1983) The adsorption of prothrombin to phosphatidylserine multilayers quantitated by ellipsometry. *J. Biol. Chem.* **258**, 2426–2431.
22. Salamon, Z., Devanathan, S., Alves, I. D., and Tollin, G. (2005) Plasmon-waveguide resonance studies of lateral segregation of lipids and proteins into microdomains (rafts) in solid-supported bilayers. *J. Biol. Chem.* **280**, 11,175–11,184.
23. Alves, I. D., Salamon, Z., Hruby, V. J., and Tollin, G. (2005) Ligand modulation of lateral segregation of a G-protein-coupled receptor into lipid microdomains in sphingomyelin/phosphatidylcholine solid supported bilayers. *Biochemistry* **44**, 9168–9178.
24. Devanathan, S., Salamon, Z., Lindblom, G., Gröbner, G., and Tollin, G. (2006) Effects of sphingomyelin, cholesterol and zinc ions on the binding, insertion and aggregation of the amyloid $A\beta_{1-40}$ peptide in solid-supported lipid bilayers. *FEBS J.* **273**, 1389–1402.
25. Lawrence, J. C., Saslowsky, D. E., Edwardson, J. M., and Henderson, R. M. (2003) Real-time analysis of the effects of cholesterol on lipid raft behavior using atomic force microscopy. *Biophys. J.* **84**, 1827–1832.
26. Salamon, Z. and Tollin, G. (2004) Graphical analysis of mass and anisotropy changes observed by plasmon-waveguide resonance spectroscopy can provide useful insights into membrane protein function. *Biophys. J.* **86**, 2508–2516.
27. Alves, I. D., Varga, E., Salamon, Z., Yamamura, H. I., Tollin, G., and Hruby, V. J. (2003) Direct observation of G-protein binding to the human δ-opioid receptor using plasmon-waveguide resonance spectroscopy. *J. Biol. Chem.* **278**, 48,890–48,897.
28. Milhiet, P. E., Giocondi, M. -C., and Le Grimellec, C. (2002) Cholesterol is not crucial for existence of microdomains in kidney brush-border membrane models. *J. Biol. Chem.* **277**, 875–878.
29. Muresan, A. S., Diamant, H., and Lee, K. Y. C. (2001) Effect of temperature and composition on the formation of nanoscale components in phospholipid membranes. *J. Am. Chem. Soc.* **123**, 6951–6952.
30. Salamon, Z., Macleod, H. A., and Tollin, G. (1997) Surface plasmon resonance spectroscopy as a tool for investigating the biochemical and biophysical properties of membrane protein systems. I: Theoretical principles. *Biochim. Biophys. Acta* **1331**, 117–129.
31. Salamon, Z., Macleod, H. A., and Tollin, G. (1997) Surface plasmon resonance spectroscopy as a tool for investigating the biochemical and biophysical properties of membrane protein systems. II: Applications to biological systems. *Biochim. Biophys. Acta* **1331**, 131–152.

13

Fluorescence Recovery After Photobleaching Studies of Lipid Rafts

Anne K. Kenworthy

Summary

Fluorescence recovery after photobleaching (FRAP) is a microscopy-based technique that can be used to ask how lipid rafts impact protein and lipid diffusion in cells. This chapter, describes how to perform FRAP measurements of putative raft and nonraft proteins and lipids using a confocal microscope. Methods have been outlined for (1) transfecting cells with plasmids encoding for the expression of green fluorescent protein-tagged proteins, (2) labeling cells with fluorescent lipid analogs or with the lipid-binding toxin cholera toxin B-subunit, (3) depleting and loading cholesterol into cell membranes using methyl-β-cyclodextrin, and (4) performing and analyzing confocal FRAP measurements.

Key Words: Cholera toxin B-subunit; cholesterol; confocal microscopy; DiI; fluorescence recovery after photobleaching; GFP; lateral diffusion; lipid rafts; methyl-β-cyclodextrin.

1. Introduction

Fluorescence recovery after photobleaching (FRAP, also referred to as fluorescence photobleaching recovery) is a technique that can be used to monitor the diffusional mobility of proteins and lipids in cell membranes or artificial bilayers. In FRAP experiments, proteins or lipids on a cell membrane are fluorescently labeled, and molecules in a small region of the membrane are subjected to a pulse of high intensity laser excitation, which renders them irreversibly photobleached. Using low laser intensity, the exchange of bleached molecules with unbleached molecules in the surrounding area of membrane is then monitored over time (**Fig. 1**). Two parameters are typically gleaned from a FRAP recovery curve: a diffusion coefficient and a mobile fraction (M_f). The diffusion coefficient provides a measure of the mean squared displacement per unit time, whereas the M_f reports on what fraction of fluorescent molecules are able to diffuse over the time-course of the experiment.

From: *Methods in Molecular Biology, vol. 398: Lipid Rafts*
Edited by: T. J. McIntosh © Humana Press Inc., Totowa, NJ

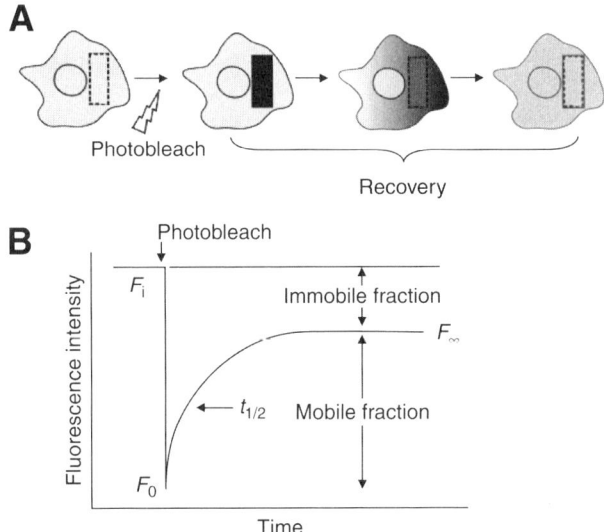

Fig. 1. Principles of fluorescence recovery after photobleaching experiments. (**A**) Schematic depiction of a confocal FRAP experiment. A population of fluorescent molecules is irreversibly bleached by repetitively scanning a region of interest (rectangle). Recovery of fluorescence in the bleach region occurs by diffusional exchange of fluorescent molecules outside of the bleach region with the bleached molecules. (**B**) Characteristics of fluorescence photobleaching and recovery curves. The recovery of fluorescence in the bleach region is characterized by two parameters, the characteristic recovery time ($t_{1/2}$), which is related to the diffusion coefficient, and the mobile fraction, a measure of the fraction of molecules that are able to diffuse over the time-course of the experiment. Figure adapted from **ref. 18** Kenworak 2006 with © permission of Elsevier.

Over the last 30 yr, FRAP studies have been instrumental in showing that in cell membranes, the diffusion of proteins and lipids is slow compared with diffusion in artificial membranes, and is also characterized by the presence of immobile fractions of molecules. A major implication of such studies is that cell membranes are heterogeneous and compartmentalized in ways that prevent the free diffusion of proteins *(1,2)*. Recent studies have now turned to the question of how lipid rafts fit into this picture of membrane structure (reviewed in **refs. 3–5**). In particular, FRAP is now being used to ask if the diffusional mobility of raft and nonraft proteins can be distinguished, as well as how protein and lipid diffusion is regulated by cholesterol *(6–15)*.

Until recently, FRAP measurements could only be performed on dedicated FRAP microscopes using fluorescently labeled Fab fragments as probes. Thus, the technique was not readily available to most researchers, and FRAP measurements were limited to plasma membrane proteins for which specific antibodies were available. Over the past few years, FRAP has begun to enjoy more widespread use. Two technological developments have contributed to the resurgence of interest in this technique. First, many commercial confocal microscopes are designed to perform photobleaching measurements. Second, because of the advent of green fluorescent protein (GFP), it is now possible to fluorescently tag a wide range of proteins, including those localized inside cells. Thus, FRAP can now be used to address a wide range of questions related to protein and lipid dynamics *(16–18)*.

This chapter describes how to perform confocal FRAP measurements of lipid rafts has been described. It begins by outlining procedures for transfecting cells with plasmids encoding for the expression of GFP-tagged proteins. Next, methods are discussed for labeling cells with fluorescent lipid analogs or with the lipid-binding toxin cholera toxin B-subunit (CTXB), a commonly used marker for lipid rafts. Methods for cholesterol depletion and loading are outlined next. Finally, a detailed protocol is presented for performing FRAP measurements and subsequent data analysis.

2. Materials
2.1. Cell Culture and Transfection

1. Cell culture media: Dulbecco's modified Eagle's medium (DMEM) supplemented with 10% fetal bovine serum.
2. Media for transfection: DMEM (serum-free).
3. FuGene 6™ (Roche Diagnostics, Indianapolis, IN).
4. Microscope cover slips, no. 1.5, chambered cover glasses (e.g., LabTek, Nagle Nune International, Rochester, NY, or glass bottom dishes, MatTek).

2.2. Labeling With Fluorescent Protein or Lipid Analogs

1. Fluorescent CTXB: purchase commercially (Sigma-Aldrich) or prepare from unlabeled toxin (Sigma-Aldrich) using a kit designed for fluorescent labeling of proteins (e.g., Cy3 monoreactive dye packs, GE HealthCare Biosciences, Amersham Biosciences Corporation, Piscataway, NJ).
2. DiI cell-labeling solution (DiIC16) (Invitrogen Molecular Probes. Carlsbad, CA): prepare a 100 µg/mL stock solution in ethanol, store at −20°C. Prepare a working solution by diluting to 0.3–1.5 µg/mL in serum-free imaging buffer containing 0.1% fatty acid-free bovine serum albumin (BSA) (Sigma-Aldrich, St. Louis, MO).
3. Vybrant™ DiIC18 (Invitrogen Molecular Probes, Carlsbad, CA) (supplied as a stock solution by the manufacturer).
4. Serum-free imaging buffer: phenol red-free DMEM supplemented with 25 mM HEPES and 0.1% w/v BSA.

2.3. Cholesterol Depletion and Cholesterol Loading

1. Cholesterol depletion using methyl-β-cyclodextrin (MβCD) (Sigma-Aldrich): prepare a 10 mM working solution in serum-free imaging buffer (*see* **2.3.3**).
2. Cholesterol loading using water-soluble cholesterol (MβCD–cholesterol complexes) (Sigma-Aldrich); prepare a working stock solution 300 mM in MβCD (*see* **Note 1**).
3. Serum-free imaging buffer: phenol red-free DMEM supplemented with 25 mM HEPES and 0.1% w/v BSA (Sigma-Aldrich).

2.4. Fluorescence Recovery After Photobleaching

1. Live-cell imaging buffer: phenol-red free DMEM supplemented with 10% fetal cafl serum (FCS) and 25 mM HEPES.
2. Live-cell mounting chamber (if cover slips are used) (*see* **Note 2**).

3. Methods
3.1. Transfection

A detailed protocol is outlined for transfecting COS-7 cells with plasmids encoding for expression of GFP fusion proteins using Fugene 6. Other cell types may be more efficiently transfected using other transfection reagents or protocols. Cells are plated on cover slips on day one, transfected on day two, and imaged on day three.

1. Day one.
   a. Plate cells onto cover slips (in a six-well plate) or Lab-tek chambers the day before they are to be transfected so that they will be at 30–70% confluence the following day.
2. Day two.
   a. Vortex Fugene 6. Allow the Fugene to come to room temperature (RT) before use, about 15 min.
   b. Aliquot the DNA into sterile microfuge tubes.
      i. A typical amount of DNA to use is 1 μg per well of a six-well plate, at a Fugene:DNA ratio of 3 μL/1 μg.
      ii. The DNA stock should be at a concentration of 0.1–2 mg/mL.
   c. Prepare a working solution of Fugene:
      i. The amount will depend on the surface area of each dish and the number of dishes. For each well of a six-well dish, plan on using 100 μL total volume of serum-free media containing 3 μL Fugene6 (97 μL media + 3 μL Fugene). *See* manufacturer's instructions for other size dishes.
      ii. Add the media to a sterile tube, then after vortexing the Fugene again add the Fugene, without touching the sides of the tube, then incubate for 5 min at RT.
      iii. One will need 100 μL Fugene dilution for each well of plate. If multiple wells are to be transfected, prepare enough for all the wells at one time.
   d. Add the diluted Fugene solution (100 μL) onto each DNA sample in a sterile microfuge tube dropwise and mix by tapping the tube. Incubate for 15 min at RT.

e. Add the Fugene:DNA mixture to the cells; swirl to mix.
   f. If multiple cover slips have been plated in the same well, make sure that handling of the dish has not caused them to overlap one another. If they are overlapping, gently separate them using flame-sterilized forceps.
   g. Replace cells in incubator and allow them to express the protein overnight.
3. Day three.
   a. Confirm GFP-protein expression by fluorescence microscopy before FRAP studies (*see* **Note 3**).

3.2. Labeling Cells With Fluorescent Lipid Analogs or Lipid-Binding Proteins

Plate cells onto cover slips (in a six-well plate) or Lab-tek chambers 1 or 2 d before they are to be labeled.

3.2.1. Fluorescent CTXB Labeling

1. Prepare a working dilution of fluorescent CTXB: 0.1–1 µg/mL in imaging buffer.
2. Remove cells from growth medium and place cover slip in new dish. Rinse cover slip two times in cold imaging buffer.
3. Label cells with fluorescent CTXB for 15 min on ice (*see* **Note 4**).
4. Wash 3 × 5 min in cold imaging buffer before imaging.
5. Chose appropriate excitation and emission settings depending on the fluorophores used to label the CTXB.

3.2.2. DiIC$_{16}$ Labeling

1. Prepare a working dilution of DiIC16: 0.3–1.5 µg/mL in serum-free imaging buffer with 0.1% fatty acid-free BSA.
2. Remove cells from growth medium and place cover slip in new dish. Rinse cover slip two times with serum-free imaging buffer.
3. Incubate cells in DiIC$_{16}$ 5 min at RT (*see* **Note 5**).
4. Wash two times in serum-free imaging buffer without BSA before imaging.
5. Image using Cy3/rhodamine settings.

3.2.3. DiIC$_{18}$ Labeling

1. Prepare a working dilution of Vybrant DiI at a ratio of 5 µL of supplied dye solution to 1 mL serum-free imaging buffer with 0.1% fatty acid-free BSA (*see* **Note 6**).
2. Remove cells from growth medium and place cover slip in new dish. Rinse cover slip two times with serum-free imaging buffer.
3. Incubate cells in prewarmed Vybrant DiIC$_{18}$ for 10 min at 37°C (*see* **Note 5**).
4. Quickly rinse once in warmed serum-free imaging buffer fatty acid-free BSA.
5. Rinse two times for 1 min with warmed serum-free imaging buffer plus fatty acid-free BSA before imaging.
6. Image using Cy3/rhodamine settings.

3.3. Preparation of Labeled Cholesterol-Depleted, -Loaded, and -Repleted Cells

Cells are cholesterol depleted or loaded just before FRAP experiments (*see* **Note 7**).

3.3.1. Mock-Treated Cells

1. Remove cells from growth medium and place cover slip in new dish. Rinse cover slip two times with serum-free imaging buffer.
2. Remove serum-free imaging buffer, add 1 mL of fresh serum-free imaging buffer, and incubate for 30 min at 37°C.
3. If desired, label cells with fluorescent CTXB or DiI as described in **Subheading 3.2**.
4. Transfer cells to live cell imaging chamber and image in serum-free imaging buffer.

3.3.2. Cholesterol-Depleted Cells

1. Prepare a 10 m*M* MβCD solution in serum-free imaging buffer.
2. Remove cells from growth medium and place cover slip in new dish. Rinse cover slip two times with serum-free imaging buffer.
3. Remove serum-free imaging buffer, add 1 mL of 10 m*M* MβCD, and incubate for 30 min at 37°C.
4. If desired, label cells with fluorescent CTXB or DiI as described in **Subheading 3.2**.
5. Transfer cells to live cell imaging chamber and image in serum-free imaging buffer.

3.3.3. Cholesterol-Loaded Cells

1. Prepare stock-loading solution: 300 m*M* MβCD–cholesterol complex in serum-free imaging buffer.
2. Remove cells from growth medium and place cover slip in new dish. Rinse cover slip two times with serum-free imaging buffer.
3. Remove serum-free imaging buffer, add 1 mL of fresh serum-free imaging buffer. Directly pipet in cholesterol loading solution to yield a final concentration of 10 m*M*. Incubate for 30 min at 37°C.
4. If desired, label cells with fluorescent CTXB or DiI as described in **Subheading 3.2**.
5. Transfer cells to live cell imaging chamber and image in serum-free imaging buffer.

3.3.4. Cholesterol-Repleted Cells

1. Prepare cholesterol depletion and cholesterol loading solutions as described earlier.
2. Remove cells from growth medium and place cover slip in new dish. Rinse cover slip two times with serum-free imaging buffer.
3. Remove serum-free imaging buffer, add 1 mL of 10 m*M* MβCD in serum-free imaging buffer, and incubate 30 min at 37°C.
4. Remove MβCD solution and rinse two times with serum-free imaging buffer.
5. Remove serum-free imaging buffer, add 1 mL of fresh serum-free imaging buffer

with cholesterol-loading solution diluted to a final concentration of 10 m*M* directly in well, and incubate for 30 min at 37°C.
6. If desired, label cells with fluorescent CTXB or DiI as described in **Subheading 3.2.**
7. Transfer cells to live cell imaging chamber and image in serum-free imaging buffer.

3.4. Basic Confocal FRAP Protocol

This section assumes that the experimenter has a working knowledge of confocal microscopy and focuses on experimental details that are required to obtain reproducible, high-quality FRAP data.

3.4.1. Determine the Appropriate Settings for the Fluorophore and Cell Type

Choice of the laser excitation, dichroic mirrors, and emission filters are determined by the fluorophore. Settings for widely used fluorophores such as GFP and Cy3 are often available as preprogrammed configurations on many commercial confocal microscopes. The choice of the microscope objective, pinhole settings, and zoom will be dictated by the cell type and the size of the structures being studied. Once the basic imaging settings have been determined, they should be held constant in all subsequent experiments. One exception is the detector gain, which can be changed in order to image cells with varying protein expression levels/labeling intensities. The detector gain should be set to maximize the fluorescence signal, whereas at the same time minimizing the number of saturated pixels.

3.4.2. Choose an Appropriate Bleach Region

For measurements of plasma membrane protein diffusion, either a spot or a strip are useful bleach region geometries (*see* **Subheading 3.5.4.**). Recovery in the bleach region can be monitored either by collecting images of the surrounding area of the cell or by imaging only the bleach region (*see* **Note 8**). The size of the bleach region determines the rate of fluorescence recovery: the larger the bleach region, the longer it will take for fluorescence recovery to occur (*see* **Subheading 3.5.4.**). Strips of width 4 µm have been sucessfully used to monitor diffusion of plasma membrane proteins and lipid analogs with diffusion coefficients ranging from 0.1 to more than 1 µm^2/s (*6,7*).

3.4.3. Determine Conditions Required for Bleaching

For bleaching, the bleach region is repetitively scanned at high laser power. A fixed sample should be used to determine how many scans are required to bleach the fluorescence to near background values. This can vary substantially depending on the laser power, microscope objective, and optics. Typically, it has been found that with a Zeiss LSM 510 (Carl Zeiss MicroImaging, Thornwood, NY), approx 10 scans at 100% transmission at 488 nm line of a 40-mW argon laser are required to fully bleach enhanced GFP (EGFP) for a ×40 1.4 numerical aperture objective at ×4 zoom. The 488-nm laser line is among the most powerful of most confocal microscopes and is excellent not only for bleaching EGFP but also Cy3 and DiI (which are normally excited at 543 nm or similar laser lines). Note that the minimum number of scans necessary to bleach should be used. This is important in order to minimize artifacts that can occur as the result of depletion of fluorescent molecules in the region surrounding the bleach region of interest ROI (*see* **Note 9**).

3.4.4. Determine Conditions for Imaging Fluorescence Recovery

Once the bleach region and bleaching conditions have been established, the next step is to determine how long to collect data in the recovery phase of the experiment. To do this, the bleach is incorporated into a time series. Several images are collected before bleaching the ROI, and then multiple images are collected after the bleach (**Fig. 1**). It is important to visualize as much of the early part of the fluorescence recovery as possible and to continue to collect data until recovery is complete. In order to ensure that the early part of the recovery can be readily resolved, it may be necessary to increase the size of the bleach region (*see* **Subheading 3.4.2.**). Photobleaching of the sample during this phase of the experiment needs to be kept to a minimum. Ideally, in control experiments in which the sample is imaged without bleaching the ROI, the fluorescence intensity should remain constant (*see* **Note 10**).

3.4.5. Perform the Experiment

Once appropriate conditions for bleaching and monitoring the recovery phase have been established, it is finally time to do the actual experiment. Typically, data is collected for 10 cells per treatment for each experiment and experiments are repeated on at least three separate days. For ease of data analysis it is recommended to keep the position of the bleach region constant from cell to cell, moving the cells as needed rather than moving the bleach region. In order to yield reproducible data, the bleaching and recovery conditions should not be changed between experiments (except for adjusting the detector gain for different cells if needed).

3.5. FRAP Data Analysis

3.5.1. Quantitate Fluorescence Intensities

The first step in data analysis is to extract the fluorescence intensities from the time series of images. Collect the mean fluorescence intensity for the bleach region as well as the surrounding region of the cell (**Fig. 2**). Fluorescence intensities should also be measured for a background region positioned outside of the cell. The fluorescence intensity vs time datasets can then be imported into a

Fluorescence Recovery After Photobleaching Studies of Lipid Rafts

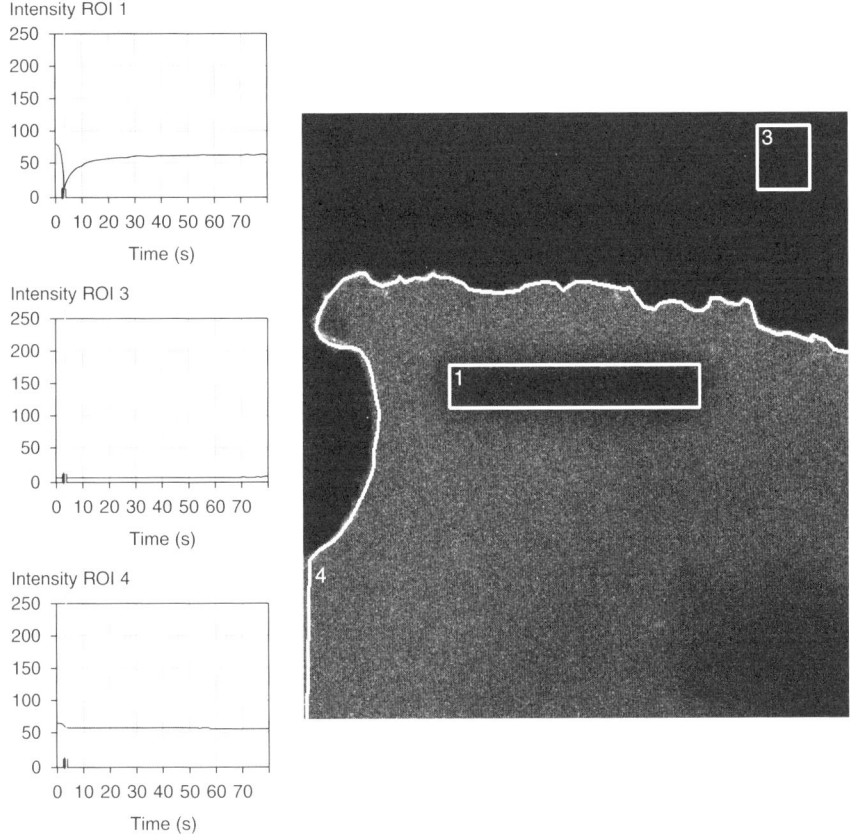

Fig. 2. Example of how to quantitate fluorescence intensities from a confocal FRAP experiment. Examples of ROIs for bleach region, surrounding cell, and background and corresponding plots of mean fluorescence intensity vs time for each are shown.

graphing and data analysis program, such as Excel (Microsoft, Redmond,WA) or KaleidaGraph (Synergy Software, Reading, PA), for further analysis.

3.5.2. Normalize the Data

In order to directly compare recovery curves for different treatments, it is convenient to normalize the data to correct for background fluorescence F_{bkgd}, loss of fluorescence during the bleach, and variations in protein expression levels between cells:

$$F(t)_{norm} = 100 \times \frac{F(t)_{ROI} - F_{bkgd}}{F(t)_{cell} - F_{bkgd}} \times \frac{F_{i\_cell} - F_{bkgd}}{F_{i\_ROI} - F_{bkgd}} \qquad (1)$$

The first part of the normalization accounts for irreversible loss of molecules owing to the bleach event (*see* **Note 11**). To correct this, the bleached ROI intensity $F(t)_{ROI}$ is divided by the whole cell intensity $F(t)_{cell}$ for each time-point $F(t)$. The second part of the normalization rescales the data in terms of percentage of initial fluorescence by dividing by the prebleach intensity (F_i) and multiplying by 100. The resulting normalized data then can be averaged for different cells and the associated standard error or standard deviation can be calculated. The first time-point after the bleach should be set to $t = 0$. This is an approximation because the time required to bleach may be significant depending on the number of bleach iterations. In such a case, further corrections can be made (e.g., *see* **ref. 19**).

3.5.3. Calculate the M_f

M_f is defined as the fraction of molecules that recover during the time-course of the experiment (*see* **Note 12**). M_f can be calculated from data obtained from the normalized recovery curves **(Eq. 1)** as

$$M_f = (F_\infty - F_0)/(F_i - F_0) \qquad (2)$$

where F_∞, F_0, and F_i are the normalized fluorescence intensities at the asymptote, immediately following the bleach, and before the bleach, respectively (**Fig. 1**).

3.5.4. Calculate the Diffusion Coefficient

The choice of method for calculation of diffusion coefficients depends on the bleach region geometry and type of data collected. Several examples are described next.

1. *Halftime of recovery:* as a first approach to data quantitation, FRAP recovery curves can be simply described in terms of the half time of recovery ($t_{1/2}$). This can be approximated by fitting to the equation

$$F(t) = 100\left[F_0 + F_\infty \left(t/t_{1/2}\right)\right]/\left[1 + \left(t/t_{1/2}\right)\right] \qquad (3)$$

Here, F_0 is the fluorescence intensity immediately after the photobleach and F_∞ is the intensity at the asymptote of the fluorescence recovery after bleaching *(20)*.

2. *Uniform spot:* a Gaussian intensity profile is typically assumed for so-called spot photobleach measurements *(21)*. However, when bleaching is performed using a confocal microscope, the radius of the bleach region can be chosen to be significantly larger than a diffusion-limited spot. As such, one can analyze the recovery curves making the simplifying assumption that the bleach spot corresponds to a uniform circular disk. For this case, D is given by *(21,22)*

$$D = 0.224 \ r^2/t_d \qquad (4)$$

where r is the radius of the bleached region and t_d is the characteristic diffusion time. Examples of where this equation have been applied to confocal FRAP measurements can be found in **refs. 23** and **24**.

3. *Strip:* it can sometimes be useful to monitor diffusion into a strip of width w (with length much greater than w) in order to approximate one-dimensional recoveries. By making the simplifying assumptions that the bleach is complete and that there is no immobile fraction as well as certain geometric constraints *(25,26)*, fluorescence recovery for this bleach geometry can be described by

$$F(t) = F_\infty \left\{ 1 - [w^2(w^2 + 4\pi Dt)^{-1}]^{1/2} \right\}. \tag{5}$$

However, note that this equation is only an approximation *(25)*.

4. *Computer simulations:* confocal FRAP allows one to collect images of parts of the cell surrounding the bleached region. It is possible to make use of this information to calculate diffusion coefficients using computer simulations. One such program, called Diffuse, simulates diffusion in inhomogeneous media by using images collected from the actual FRAP experiment to simulate diffusive recovery of fluorescent molecules into the bleach region *(25)*. It takes into account cell geometry, allows for arbitrary choice of bleach region, and does not require a complete bleach *(25)*. Although this program allows for choice of arbitrary bleach region geometries, for ease of data analysis it is convenient to maintain a constant geometry (strip) across experiments *(6)*. Further details on how to obtain and use this simulation can be found in **ref. 19**.

4. Notes

1. Water-soluble cholesterol exhibits biphasic solubility. Therefore, it is prepared as a concentrated stock and then diluted as needed. The solution may become slightly cloudy on dilution owing to the release of cholesterol from the MβCD.
2. To make a simple, reusable chamber to image cells grown on cover slips, punch a hole out of a square piece of silicon sheet trimmed to fit on a slide. The silicon can then be attached to the slide using petroleum jelly. To use, fill the hole with a drop of imaging buffer. Pick up the cover slip with a pair of forceps and use a kimwipe to remove excess liquid. Invert the coverlip onto the buffer, then press gently with a folded kimwipe to remove the excess buffer. Surface tension will cause the cover slip to remain attached to the silicon.
3. Three factors have routinely been observed that can give rise to low transfection efficiencies using Fugene 6: (1) the cell density is too high at the time of transfection, (2) the Fugene 6 solution is old, or (3) the DNA has deteriorated. These problems can be addressed by (1) plating cells at lower densities, (2) using a tube of new Fugene 6 for transfection, and (3) thawing a new aliquot of or preparing a fresh prep of the plasmid DNA.
4. Fluorescent CTXB labeling is performed at low temperature to prevent internalization during the labeling process. On warming up the cells, internalization of CTXB will be apparent first as a series of punctate structures present throughout

the cell, and at later times by the accumulation of fluorescent toxin in perinuclear recycling compartments and the Golgi complex.
5. Note that, like CTXB, both DiIC16 and Vybrant DiI are internalized over time.
6. Vybrant DiI can also be diluted in normal growth medium. The use of serum-free media is recommended in experiments in which cells have previously been subjected to cholesterol depletion or loading.
7. It is important to note that the physiological effects of cholesterol loading and depletion are not limited to their presumed effect on lipid raft integrity. Cholesterol depletion inhibits both raft and nonraft mediated endocytosis *(27)*, and cholesterol loading accelerates endocytosis *(6,7,28)*. In addition, it is also becoming increasingly clear that MβCD can influence protein diffusion in ways that are not necessarily related to cholesterol depletion *(6,7,14)*. These two factors should be carefully considered when analyzing the effects of these treatments on protein or lipid diffusion.
8. There are several advantages to imaging the region surrounding the bleach ROI. It offers a useful quality control check to verify that the cell did not move and the focal plane was maintained during data aquisition. In addition, by collecting images of the region of the cell surrounding the bleach region, it is possible to directly measure how much fluorescent material was lost as the result of the bleach (*see* **Subheading 3.5.2.**). In contrast, for the case of small bleach regions and/or for rapidly diffusing molecules, higher time resolution is needed to monitor the initial phase of recovery, and it is more appropriate to image only the bleach ROI. Alternatively, some confocals now include a separate bleaching laser, which enables continued visualization of the sample during the bleach.
9. Depletion of fluorescent molecules in the region surrounding the bleach ROI is one type of artifact in confocal FRAP experiments that can arise as the result of the finite time required to bleach and image the sample *(29)*. Also note that a recent report suggests that GFP is reversibly bleached by the process of imaging *(30)*. This effect has been observed by the author as well, but the magnitude of reversible photobleaching is small under most conditions compared with the amount of irreversible bleaching that occurs in a FRAP experiment. Nevertheless, it is important to be aware of this possibility.
10. If photobleaching is observed, try increasing the time interval between images and/or decreasing the laser power and simultaneously increasing the detector gain. Drift of the focal plane during the recovery phase can also occur when sample is not in thermal equilibrium with the microscope.
11. Unlike in spot photobleaching, wherein the size of the bleach spot is so small that it is essentially negligible compared with the size of the pool of unbleached molecules, in confocal FRAP the bleach region can be made arbitrarily large. When a significant fraction of fluorescent molecules is eliminated by the bleach, recovery in the bleach region will obviously never be able to reach 100% of the prebleach value. If this is not appropriately corrected, one could erroneously conclude that an immobile fraction is present.
12. Note that the presence of endocytic vesicles can give rise to an apparent immobile fraction. This effect has been routinely observed in measurements of CTXB diffusion *(6)*.

Acknowledgments

Thanks to past and present members of the Kenworthy laboratory, who helped to refine the techniques described herein. Supported by ROI No. GM073846.

References

1. Jacobson, K., Sheets, E. D., and Simson, R. (1995) Revisiting the fluid mosaic model of membranes. *Science* **268,** 1441–1442.
2. Edidin, M. (1992) Patches, posts and fences: proteins and plasma membrane domains. *Trends Cell Biol.* **2,** 376–380.
3. Kenworthy, A. K. (2005) Fleeting glimpses of lipid rafts: how biophysics is being used to track them. *J. Invest. Med.* **53,** 312–317.
4. Lommerse, P. H., Spaink, H. P., and Schmidt, T. (2004) In vivo plasma membrane organization: results of biophysical approaches. *Biochim. Biophys. Acta* **1664,** 119–131.
5. Lagerholm, B. C., Weinreb, G. E., Jacobson, K., and Thompson, N. L. (2005) Detecting Microdomains In Intact Cell Membranes. *Annu. Rev. Phys. Chem.* **56,** 309–336.
6. Kenworthy, A. K., Nichols, B. J., Remmert, C. L., et al. (2004) Dynamics of putative raft-associated proteins at the cell surface. *J. Cell Biol.* **165,** 735–746.
7. Goodwin, J. S., Drake, K. R., Remmert, C. L., and Kenworthy, A. K. (2005) Ras diffusion is sensitive to plasma membrane viscosity. *Biophys. J.* **89,** 1398–1410.
8. Niv, H., Gutman, O., Henis, Y. I., and Kloog, Y. (1999) Membrane interactions of a constitutively active GFP-Ki-Ras 4B and their role in signaling. Evidence from lateral mobility studies. *J. Biol. Chem.* **274,** 1606–1613.
9. Rotblat, B., Prior, I. A., Muncke, C., et al. (2004) Three separable domains regulate GTP-dependent association of H-ras with the plasma membrane. *Mol. Cell Biol.* **24,** 6799–6810.
10. Niv, H., Gutman, O., Kloog, Y., and Henis, Y. I. (2002) Activated K-Ras and H-Ras display different interactions with saturable nonraft sites at the surface of live cells. *J. Cell Biol.* **157,** 865–872.
11. Roy, S., Plowman, S., Rotblat, B., et al. (2005) Individual palmitoyl residues serve distinct roles in H-ras trafficking, microlocalization, and signaling. *Mol. Cell Biol.* **25,** 6722–6733.
12. Hao, M., Mukherjee, S., and Maxfield, F. R. (2001) Cholesterol depletion induces large scale domain segregation in living cell membranes. *Proc. Natl. Acad. Sci. USA* **98,** 13,072–13,077.
13. Meder, D., Moreno, M. J., Verkade, P., Vaz, W. L., and Simons, K. (2006) Phase coexistence and connectivity in the apical membrane of polarized epithelial cells. *Proc. Natl. Acad. Sci. USA* **103,** 329–334.
14. Shvartsman, D. E., Gutman, O., Tietz, A., and Henis, Y. I. (2006) Cyclodextrins but not compactin inhibit the lateral diffusion of membrane proteins independent of cholesterol. *Traffic* **7,** 917–926.
15. Nicolau, D. V., Jr., Burrage, K., Parton, R. G., and Hancock, J. F. (2006) Identifying optimal lipid raft characteristics required to promote nanoscale

protein-protein interactions on the plasma membrane. *Mol. Cell Biol.* **26,** 313–323.
16. Lippincott-Schwartz, J., Snapp, E., and Kenworthy, A. K. (2001) Studying protein dynamics in living cells. *Nat. Rev. Mol. Cell Biol.* **2,** 444–456.
17. Lippincott-Schwartz, J., Altan-Bonnet, N., and Patterson, G. H. (2003) Photobleaching and photoactivation: following protein dynamics in living cells. *Nat. Cell Biol.* **(Suppl)** S7–S14.
18. Goodwin, J. S. and Kenworthy, A. K. (2005) Photobleaching approaches to investigate diffusional mobility and trafficking of Ras in living cells. *Methods* **37,** 154–164.
19. Snapp, E. L., Altan, N., and Lippincott-Schwartz, J. (2003) *Current Protocols in Cell Biology*, vol. 3, (Bonifacino, J. S., Dasso, M., Harford, J. B., Lippincott-Schwartz, J., and Yamada, K., eds.), pp. 21.21–21.24.
20. Feder, T. J., Brust-Mascher, I., Slattery, J. P., Baird, B., and Webb, W. W. (1996) Constrained diffusion or immobile fraction on cell surfaces: a new interpretation. *Biophys. J.* **70,** 2767–2773.
21. Axelrod, D., Koppel, D. E., Schlessinger, J., Elson, E., and Webb, W. W. (1976) Mobility measurement by analysis of fluorescence photobleaching recovery kinetics. *Biophys. J.* **16,** 1055–1069.
22. Soumpasis, D. M. (1983) Theoretical analysis of fluorescence photobleaching recovery experiments. *Biophys. J.* **41,** 95–97.
23. Adams, C. L., Chen, Y. -T., Smith, S. J., and Nelson, W. J. (1998) Mechanisms of epithelial cell-cell adhesion and cell compaction revealed by high-resolution tracking of E-cadherin-green fluorescent protein. *J. Cell Biol.* **142,** 1105–1119.
24. Schmidt, K. and Nichols, B. J. (2004) A barrier to lateral diffusion in the cleavage furrow of dividing mammalian cells. *Curr. Biol.* **14,** 1002–1006.
25. Siggia, E. D., Lippincott-Schwartz, J., and Bekiranov, S. (2000) Diffusion in inhomogeneous media: theory and simulations applied to whole cell photobleach recovery. *Biophys. J.* **79,** 1761–1770.
26. Ellenberg, J., Siggia, E. D., Moreira, J. E., et al. (1997) Nuclear membrane dynamics and reassembly in living cells: targeting of an inner nuclear membrane protein in interphase and mitosis. *J. Cell Biol.* **138,** 1193–1206.
27. Subtil, A., Gaidarov, I., Kobylarz, K., Lampson, M. A., Keen, J. H., and McGraw, T. E. (1999) Acute cholesterol depletion inhibits clathrin-coated pit budding. *Proc. Natl. Acad. Sci. USA* **96,** 6775–6780.
28. Sharma, D. K., Brown, J. C., Choudhury, A., et al. (2004) Selective stimulation of caveolar endocytosis by glycosphingolipids and cholesterol. *Mol. Biol. Cell* **15,** 3114–3122.
29. Weiss, M. (2004) Challenges and artifacts in quantitative photobleaching experiments. *Traffic* **5,** 662–671.
30. Sinnecker, D., Voigt, P., Hellwig, N., and Schaefer, M. (2005) Reversible photobleaching of enhanced green fluorescent proteins. *Biochemistry* **44,** 7085–7094.

14

Single-Molecule Tracking

Marija Vrljic*, Stefanie Y. Nishimura*, and W. E. Moerner

Summary

The current models of eukaryotic plasma membrane organization separate the plasma membrane into different environments created by lipids and interactions between membrane proteins and the cytoskeleton, but characterization of their physical properties, such as their sizes, lifetimes, and the partitioning of membrane components into each environment, has not been accomplished. Single-molecule (fluorophore) tracking (SMT) experiments are well suited to the noninvasive study of membrane properties. In SMT experiments, the position of a single fluorescently labeled protein or lipid probe is followed optically as it moves within the membrane. If the motion of the probe is unhindered, then the spatial trajectory of the molecule will follow two-dimensional Brownian motion. If the probe encounters a structure that in some way inhibits its movement, then the probe's trajectory will deviate from Brownian motion. It is likely that even if a certain type of lipid or protein partitions strongly into one environment, each individual lipid or protein will spend some fraction of its lifetime in the less favorable environment. Because SMT follows the motion of an individual probe over a large area (~10×10 μm^2), transitions between environments can be observed directly by monitoring the path of each protein or lipid. Additionally, heterogeneity owing to multiple populations of molecules permanently residing in different states may be distinguished from a single population of molecules transitioning between different states. By judicious choice of label, such that the motion of the labeled protein or lipid is unaffected by the label itself, and through the use of probes with different affinities for each membrane environment, SMT measurements in principle can reveal the structure of the plasma membrane.

Key Words: Fluorophore; membrane environment; single molecule tracking; Brownian motion; diffusion; lipid domain; confinement.

1. Introduction
1.1. Detection of Different Environments

In the current models for membrane organization, domains are described by several general cases: static, mobile, unstable/dynamic, or not present *(1–4)* (**Fig. 1**). The presence and characteristics of membrane organization can be

---
*Contributed equally

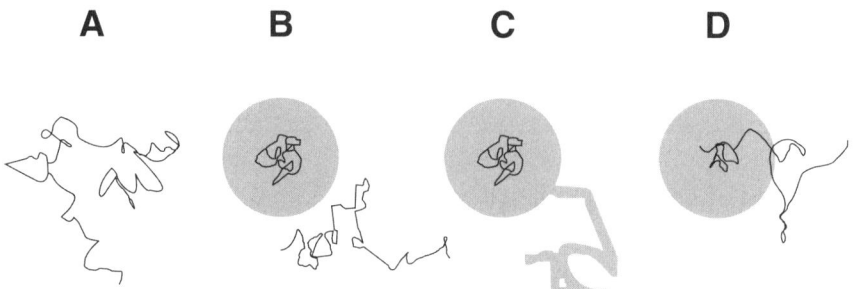

Fig. 1. Cartoons representing possible types of domains. The gray circle and white background represent two different environments. The positional trajectory of the membrane probe is depicted as a solid black line. **(A)** No domains are present, so the diffusion of the probe is Brownian. **(B)** Static domain with an impermeable boundary. The probes partition exclusively inside or outside of the domain. The diffusion of the probe may appear confined and Brownian, respectively. **(C)** Mobile domain with an impermeable boundary. A probe trapped within the domain will report on diffusion within the domain (black line) coupled with the diffusion of the domain itself (gray line). **(D)** Domains with permeable boundaries or short-lived domains. The diffusion of the probe will report on the environment inside and outside of the domain. The transitioning may be detected by looking for two populations of diffusion coefficients within a single trajectory or by locating non-Brownian segments of a trajectory.

determined by monitoring the diffusion of a probe localized within an environment. If only one environment exists, then the probe will undergo Brownian motion within the membrane **(Fig. 1A)**. In the presence of more than one environment, a probe could be trapped within a small, immobile domain **(Fig. 1B)**, trapped within a mobile domain **(Fig. 1C)**, or transition between environments **(Fig. 1D)**.

Herein the view is taken of the membrane appropriate for time-scales long compared with the time for individual collisions, whereas a membrane may be characterized by a macroscopic average viscosity. The motion of the membrane probe is analyzed in terms of a two-dimensional (2D) model of Brownian motion. This model was originally described by Saffman and Delbruck *(5)* and extended by Hughes et al. *(6)*. The model relates the theoretical diffusion coefficient to the radius of the diffuser, membrane thickness, and membrane viscosity. Thus, changes in the observed diffusion coefficient may be used to calculate the underlying change in membrane viscosity, membrane thickness, or diffuser size.

Single-molecule tracking (SMT) directly probes the motion of individual membrane molecules, albeit on a time-scale (ms) long compared with individual collisions. It is the only technique capable of monitoring the behavior of an individual membrane molecule during the course of its lifetime, as long as the molecule is labeled. The method of SMT can be applied to a variety of systems, which utilize different proteins, lipids, or other probe molecules of interest located in different cell types or different model-lipid membranes. Specific requirements will be presented for obtaining trajectories of individual membrane proteins in a fibroblast plasma membrane. The more general discussion of how to improve signal-to-noise and obtain longer trajectories through judicious choice of label, cell type, imaging set-up, and data analysis can be applied to the reader's unique system of interest. Then the basic analyses used to determine the diffusion coefficients in a population of diffusers, regions of confinement in a single trajectory, confinement in a mobile domain, and advice on how to correctly identify Brownian vs non-Brownian diffusion are presented.

Using single-molecule trajectories to identify deviations from Brownian motion is nontrivial because Brownian motion itself is highly stochastic. Experimental and computational controls are necessary to confirm that putative non-Brownian behavior is statistically improbable for a Brownian diffuser. Experimental controls that demonstrate abolishment of putative confinement regions would strengthen the argument that non-Brownian behavior was observed. Random walk simulations should be used as a control to show that the putative non-Brownian behavior is indeed non-Brownian. Simulated random walks may also be used to check the programs used to analyze experimental single-molecule trajectories and calibrate one's intuition concerning Brownian motion. Because Brownian motion is a statistical process, a large sample pool and statistically rigorous data analysis are required to differentiate between Brownian and non-Brownian behavior.

2. Materials

2.1. Cell Culture of Chinese Hamster Ovary Cells

1. Roswell Park Memorial Institute (RPMI) 1640 phenol-red free media (Gibco BRL, Long Island, NY) supplemented with 10% fetal bovine serum (HyClone, Logan, UT) (*see* **Notes 1** and **2**).
2. Eight-well chambered cover glass (Nalge Nunc International, Naperville, IL).
3. Fibronectin (CalBiochem, San Diego, CA).
4. Enzymatic oxygen scavengers (1% [v/v] glucose [Sigma, St. Louis, MO], 1% [v/v] glucose oxidase [Sigma], 1% [v/v] catalase [Sigma], and 0.5% [v/v] 2-mercaptoethanol [Sigma]).

2.2. Cell Labeling

1. *Protein of interest:* chinese hamster ovary (CHO) cells were transfected with either transmembrane or GPI-linked major histocompatibility (MHC) class II proteins (class II, I-E^k) (*see* **Note 3**).
2. *Protein probe design:* the extracellular domain of a MHC class II protein can bind only one peptide. The peptide sequence was chosen such that the half-time of the protein–peptide complex was much longer than the imaging time (>200 h). A peptide from moth cytochrome-*c* (residues 95–103), MCC 95–103 was synthesized using standard F-moc chemistry and labeled with Cy5 dye using standard NHS-chemistry.

High-resolution matrix-assisted laser desorption and ionization mass spectrometry was used to verify identity and dye:peptide ratio (1:1).
3. *Plasma membrane labeling:* fluorescent lipid analogs, 1,1′-didodecyl-3,3,3′3′-tetramethylindocarbocyanine perchlorate (DiIC$_{12}$), 1,1′-dioctadecyl-3,3,3′,3′-tetramethylindocarbocyanine perchlorate (DiIC$_{18}$), and N-(6-tetramethylrhodaminethiocarbamoyl)-1,2-dihexadecanoyl-sn-glycero-3-phosphoethanolamine (TRITC-DHPE) (Invitrogen, Carlsbad, CA) and egg-phosphatidylcholine (Avanti Polar Lipids Inc., Alabaster, AL) were dissolved in chloroform (1 and 20 mg/mL, respectively).
4. An extruder and 100 nm diameter pore filters (Avestin Inc., Ottawa, ON, Canada).
5. Dulbecco's Phosphate Buffer Solution (PBS) (Gibco BRL).

2.3. Imaging Setup

1. Inverted microscope (Nikon, Melville, NY).
2. ×100, 1.4 NA, oil-immersion objective (Nikon).
3. 645 nm or 545 nm dichroic beamsplitter and 640 alpha-epsilon longpass (AELP) or 545 longpass (LP) emission filters (Omega Optical Inc., Brattleboro, VT).
4. *Light source:* 633 nm HeNe laser (MellesGriot, Carlsbad, CA) and 532 nm diode-pumped solid-state laser (Intelite Inc., Genoa, NV) provided approx 2 kW/cm^2 at the sample plane.
5. *Camera:* I-Pentamax intensified Si charge-coupled device (CCD) (Princeton Instruments, Trenton, NJ), back-illuminated electron-multiplying Si electron multiplying charge-coupled device (EMCCD) camera, Cascade 512B (Roper Scientific, Tucson, AZ), and iXon (Andor, South Windsor, CT).

2.4. Data Analysis

Custom programs utilizing mathematical program for data analysis, MATLAB (The MathWorks, Natick, MA).

3. Methods
3.1. Preparation of Sample for Imaging

1. For imaging, 1×10^4 cells per well were plated on eight-well chambered cover glass coated with fibronectin to facilitate adhesion of CHO cells to the glass substrate. Cells were cultured for 12–18 h, 37°C in supplemented RPMI 1640 media. CHO cells adhere well to the fibronectin-treated glass surface, exhibiting the fibroblast morphology, and spreading out to dimensions of approx $30 \times 10 \times 5$ μm^3. The bottom and the top planes of the plasma membrane are parallel to the focal plane of the microscope and can be treated as 2D planes (*see* **Notes 1–3**).
2. For labeling of I-E^k proteins, CHO cells expressing I-E^k were incubated with 0.5–1.0 μg/mL Cy5 labeled moth cytochrome c peptide residues 95–103 (IAYLKQATK) (MCC95-103) peptide for 15 min at 37°C in supplemented RPMI 1640 media. Cells were thoroughly rinsed to remove unbound Cy5-peptide. The peptide concentration was adjusted such that a maximum of 0.3 labeled I-E^k molecules/μm^2 were observed.
3. For labeling the plasma membrane with fluorescent lipid analogs, the cells were incubated with lipid vesicles containing small amounts of fluorescent lipid analogs: DiIC$_{12}$, DiIC$_{18}$, or TRITC-DHPE. Lipid vesicles were prepared by lyophilizing

200–300 μL of egg phosphatidylcholine and a low concentration of fluorescent lipid analogs (1–10 mol%) for 10 h and then reconstituting the film in 300 μL of Dulbecco's PBS at pH 7.4. The mixture was passed 20–30 times through an extruder with a 100 nm pore filter. Cells were incubated with a 2% (v/v) solution of vesicles in PBS for 10 min at room temperature. Cells were then washed with PBS followed by supplemented RPMI 1640 media.

4. Enzymatic oxygen scavengers were added to RPMI media before imaging in order to extend fluorescence on time of Cy5. CHO cells can cycle between aerobic and anaerobic metabolism without effects on their viability for times approx 1–2 h (*[7]* and unpublished data). Oxygen scavengers were not used with fluorescent lipid analogs.

3.2. Imaging Setup

The ability to detect a domain depends on the interplay between the diffusion coefficient of a probe, the lifetime of the probe, the size of the domain, and the temporal (integration time) and spatial resolution of the camera. For instance, if the membrane probe with a diffusion coefficient of 1 μm^2/s is confined to a domain 100 nm in diameter, then the probe will diffuse the length of the domain in 2.5 ms on average. To observe the true diffusion coefficient of the probe within the domain, the temporal resolution of the detection system must be much faster than 2.5 ms per frame. The spatial resolution may be improved by fitting the fluorescence spot to a Gaussian profile *(8)*. In general, a spatial resolution of at least 50 nm per pixel is acceptable.

1. Wide-field epifluorescence imaging of cells was performed using an inverted Nikon Eclipse TE300 microscope **(Fig. 2A)**. Laser illumination at wavelengths of 633 nm or 532 nm provided an intensity of 1–2 kW/cm^2 at the sample plane. The illumination intensity was optimized to yield the lowest acceptable signal-to-noise ratio (*see* below) in order to reduce bleaching of the fluorophore (*see* **Note 4.**). The collimated excitation light was focused at the back focal plane of the objective using a approx 0.3 m focal length lens to obtain a 10 μm by 10 μm illumination area at the sample plane. An appropriate dichroic beamsplitter (645 nm or 545 nm) reflected the excitation light into a ×100, high numerical aperture, oil-immersion lens. Fluorescence was collected through the same objective, passed through the dichroic beamsplitter, and appropriate filters were chosen to reject fluorescence from Rayleigh scattering (640 AELP or 545 LP).
2. Transmission imaging using white light illumination allowed direct visualization of the cell edges **(Fig. 2B)**. The fluorescent photons were collected continuously on an intensified CCD camera (I-Pentamax) at a frame rate of 10 Hz or on a Cascade512B Si CCD with on-chip multiplication with a maximum frame rate of 100 Hz **(Fig. 2C)**. The temporal resolution is limited by the number of collected photons required to obtain a good signal-to-noise ratio for the particular dye (*see* **Notes 4** and **5**). The signal-to-noise ratio (SNR) for fluorescence detection of 1 mol is given by:

$$\mathrm{SNR} = \frac{D\Phi_F \left(\dfrac{\sigma}{A}\right)\left(\dfrac{P_0}{h\upsilon}\right)\Delta t}{\sqrt{\dfrac{D\Phi_F \sigma P_0 \Delta t}{Ah\upsilon} + c_b P_0 \Delta t + N_d \Delta t}} \quad (1)$$

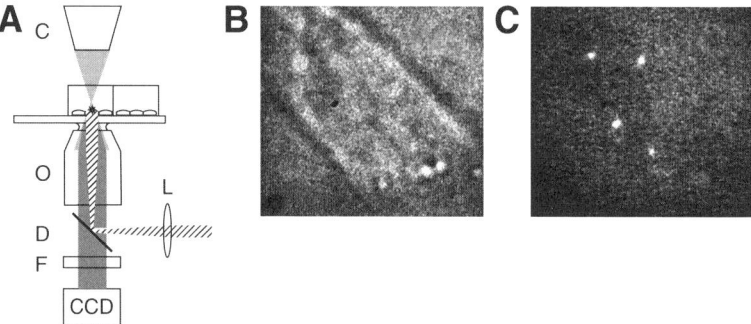

Fig. 2. Imaging set-up and an example of a cell image. (**A**) Representative wide-field epiillumination microscope with a defocusing lens (L), a white light condenser (C), a high numerical aperture, oil immersion objective (O), a dichroic mirror (D), emission filter (F), and CCD camera (CCD). (**B**) Transmission image (13×13 μm^2) of the top, central region of an oblong CHO cell. The image shows the outline of the cell running from the upper left to the lower right of the panel, with some intracellular structures. (**C**) Fluorescence image (13×13 μm^2) of the same CHO cell showing the corresponding fluorescence signal from the Cy5-labeled I-E^k proteins on the cell surface (white spots). The spatial positions of the fluorescent dots change with time indicating that the proteins are mobile (modified with permission from Vrljic et al. *[19]*).

where D is the collection factor, Φ_F is the fluorescence quantum yield, σ is the absorption cross-section, A is the beam area, $P_0/h\upsilon$ is the number of excitation photons per second, Δt is the counting interval (averaging time), C_b is the background count rate per Watt of laser power, and N_d is the dark count rate. Several comprehensive reviews of SNR considerations have been published *(9,10)*.

3. As an alternative to SNR, the signal-to-background ratio (SBR) can be used as a measure of a signal size. Herein, SBR is the signal of interest/background scattering. With the aforementioned imaging conditions, SBR of 1.6 was generally obtained. For single copies of the labeled peptide, Cy5-MCC, imaged on the I-Pentamax, the average signal in 100 ms without background was 751 ± 206 counts, and the average background was 477 ± 77 counts. The diffraction-limited spot size for immobile particles had a diameter of approx 300 nm, whereas mobile particles had a typical diameter of approx 500 nm during 100 ms.

3.3. Data Analysis

The diffusion parameters of single-molecule trajectories are extracted from the data and then compared with Brownian diffusion in order to characterize their diffusion as free, constrained, or a variety of other behaviors (*see* **Note 6**). All the analyses that will be described have originated from the characteristic 2D probability distribution of radial displacements r of a Brownian diffuser from some origin, $p(r,i\Delta t)$, at a time $i\Delta t$, (where i is the time step index, Δt is the time interval between observations, and $i\Delta t$ is the time lag), and have been discussed in detail *(11–20)*.

This distribution has the form r times a Gaussian centered at the origin, which broadens with time lag with an average mean-square displacement, $\langle r^2 \rangle$, equal to $4D(i\Delta t)$, where D is the diffusion coefficient.

$$p(r, i\Delta t) = \frac{1}{4\pi D i\Delta t} \exp(-r^2/4D\, i\Delta t)$$

The mean of this probability density function may be used to easily calculate the mean diffusion coefficient for a set of displacements. A cumulative probability distribution may also be computed. Information about the distribution of displacements contained in the cumulative probability density function is used to identify the presence of one or more populations of diffusers.

3.3.1. Random Walk Simulation

Because analytical solutions of non-Brownian motion that address the subtleties of specific diffusion models are currently not available, the influence of environmental structure on the motion of a random walker has been demonstrated numerically using Monte Carlo simulations *(11,13)*. A simple random-walk simulation is presented that may be used to approximate the expected behavior of a Brownian diffuser.

1. The tracer starts at the origin (0, 0) and takes steps on a square lattice with equal probability of moving either up, down, left, or right. The MATLAB statistical function rand uniformly generates a random number between 0 and 1. If the generated number is less than or equal to 0.25, then the tracer steps one lattice unit in the positive y-direction. If the generated number is greater than 0.25, but less than or equal to 0.5, then the tracer steps one lattice unit in the positive x-direction, and so on. The unit length of the lattice is given by $r = \sqrt{4D\Delta t / \text{steps}}$ where D is diffusion coefficient, Δt is experimental time lag, and steps is the number of steps before the coordinates are saved.
2. In order to simulate angular freedom, the diffuser's coordinates are recorded every 1000th step. The steps in the output trajectory file will vary in both distance and angle **(Fig. 3)**.

3.3.2. Record the Trajectories of the Probes

1. Single-molecule trajectories are mapped by determining the center of the brightest area of the fluorescent spots (x, y) in each frame as a function of time **(Fig. 1D)**.
    a. In these experiments, the center position was determined by the eye in each frame with an accuracy of approx 53 nm for the I-Pentamax, and approx 39 nm for the Cascade 512B (diameter of one pixel). This spatial resolution is sufficient

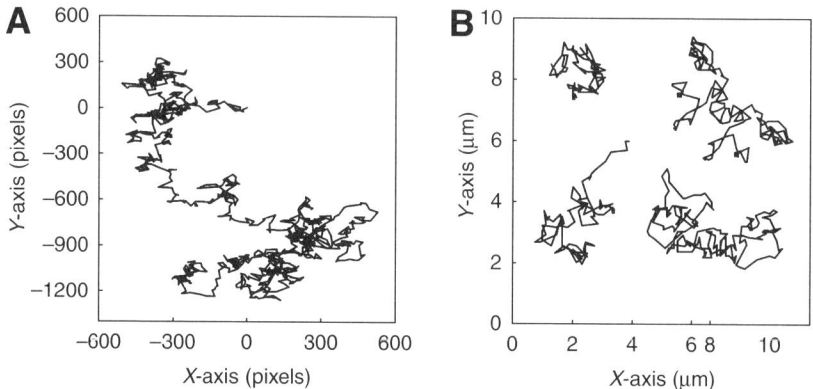

Fig. 3. Examples of 2D trajectories. (**A**) Representative simulated random walk. The average step size is 31.6 pixels or 1.25 μm. Time interval for each step is 1 s. Notice how some regions may appear confined. (**B**) Examples of characteristic trajectories of GPI-linked (top two tracks) and transmembrane I-E^k (bottom two tracks) labeled with MCC-Cy5 (coordinates are 100 ms apart). Each trajectory is made up of 20–30 steps. (modified with permission from Vrljic et al. *[19]*).

       in case of I-E^k proteins in CHO cells because the average displacement of I-E^k proteins between frames was 300 nm (~6 pixels).
   b. Higher spatial resolution can be achieved when the intensity profile of the fluorescence spot is fit to the point spread function for the microscope.
2. Only molecules that blink or exhibit digital photobleaching are tracked to insure that aggregates are not included.
3. As a general rule one should follow fluorescent spots only while they are in focus. Out of focus spots indicate movement away from a 2D plane and will have a lower apparent diffusion coefficient in data analysis. Thus, only molecules on the upper or lower membrane surfaces and away from the cell edges are tracked.
4. If two molecules merge (diffraction limited spots overlap) a conservative approach requires one to stop following the molecule unless one can spatially resolve each position.

3.3.3. Calculate Radial Displacements for a Particular Time Lag

For any time interval (termed a time lag, $i\Delta t$, where Δt is the time resolution and i is the time step index with $i = 1,2,3,\ldots, N_{total}$ and N_{total} = total number of frames-1) along a trajectory, displacements of a single spot from its arbitrary origin are calculated from its *x*- *y*-coordinates *(15,17,18)*. A radial displacement (r) is defined as the distance between some origin and a point $i\Delta t$ away. Any point along the trajectory may be treated as an origin, not only the first point in the trajectory. Origins are selected such that the resulting displacements are

Single-Molecule Tracking

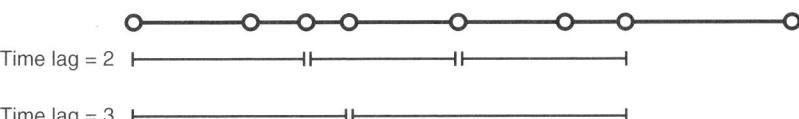

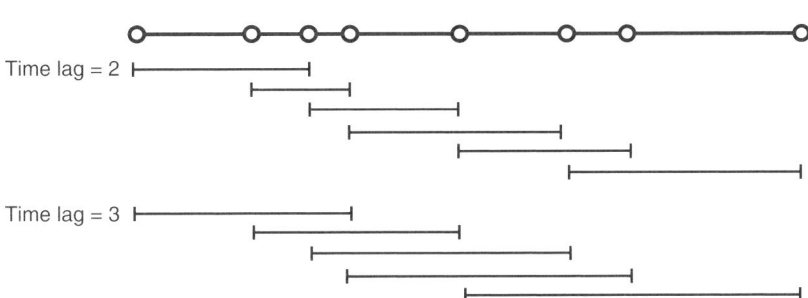

Fig. 4. A schematic representation of independent and all displacements ($N_{Total} = 7$). A trajectory is represented by a black line with open circles. Open circles represent the positions of a molecule in each frame. (**A**) Independent displacements are shown for time lags 2 and 3 (N_I [time lag of 2] = 3). Note that the points along the trajectory used as origins do not overlap. (**B**) All displacements are shown for time lags 2 and 3 using all points along the trajectory as origins (N_A [time lag of 2] = 6).

either independent or overlapping (**Fig. 4**) (*see* **Note 7a**). For independent displacements at time lag, $i\Delta t$:

$$r(j) = \left| \vec{r}(j\,i\Delta t) - \vec{r}\left[(j-1)i\Delta t\right] \right|, \text{ for } j = 1, 2, 3, \ldots, N_I \quad (2)$$

where the number of nonoverlapping displacements of length $i\Delta t$ is defined as:

$$N_I = \left[N_{total} / i\Delta t \right]$$

For all pairs of points (which includes overlapping displacements) at time lag, $i\Delta t$:

$$r(j) = \left| \vec{r}\left[(j+i)\Delta t\right] - \vec{r}(j\Delta t) \right|, \text{ for } j = 0, 1, 2, \ldots, (N_A - 1) \quad (3)$$

where the number of overlapping displacements of length $i\Delta t$ is defined as:

$$N_A = N_{total} - i\Delta t + 1$$

3.3.4. Mean Square Displacements to Calculate the Diffusion Coefficient

The mean square displacements may be used to calculate the mean apparent diffusion coefficient. This can be done for an individual trajectory or all trajectories pooled together. Mean square displacement vs time lag plots may be used to determine if the motion of the membrane molecule is Brownian (*see* **Notes 7b** and **7c**).

1. Use displacements calculated in **Subheading 3.3.3.**
2. Calculate the mean square displacements by averaging over independent pairs of points:

$$\langle r^2(i\Delta t) \rangle = \frac{1}{N_1} \sum_{j=1}^{N_1} \left[\vec{r}(j i\Delta t) - \vec{r}\left[(j-1)i\Delta t\right] \right]^2 \quad (4)$$

or by averaging over all pairs of points:

$$\langle r^2(i\Delta t) \rangle = \frac{1}{N_A} \sum_{j=0}^{N_s-1} \left[\vec{r}\left[(j+i)\Delta t\right] - \vec{r}(j \Delta t) \right]^2 \quad (5)$$

The difference between averaging over all pairs or independent pairs has been discussed *(15,17)*.

3. Plot the mean square displacements as a function of time lag $i\Delta t$ and extract the diffusion coefficient. For a Brownian diffuser in two dimensions, the mean squared displacement is related to the diffusion coefficient by: $\langle r^2(i\Delta t) \rangle = 4D i\Delta t$.

3.3.5. Histograms of Individual Diffusion Coefficients

A critical aspect of any single-molecule histogram is the need to explicitly determine the meaning of the widths of the observed distribution. Often, owing to the limitations of the measurement, a range of values would be observed arising solely from statistical effects. In the case of observed diffusion coefficients from single-molecules, measured values could differ from molecule to molecule owing to the limited length of the observed trajectories, so calculation of the expected distribution of observed values is important.

In this analysis, the distribution of individual diffusion coefficients is found by calculating the diffusion coefficient for each trajectory. The resulting histogram is compared with the expected distribution of diffusion coefficients for a homogeneous population of diffusers, and static heterogeneity is observed as deviation from the homogeneous distribution (*see* **Note 7d**). The method includes:

1. All trajectories used in this analysis must have the same length because each trajectory must contribute the same number of displacements, N.
2. Calculate the $\langle r^2 \rangle$ as a function of time lag (**Eq. 4**).
3. Use the $\langle r^2 \rangle$ to calculate the diffusion coefficient (D_e) for a particular time lag, for individual trajectories using: $D_e = \langle r^2 \rangle / 4i\Delta t$.

4. Use the calculated D_e's from individual trajectories to create histograms of diffusion coefficients for particular time lags (**Fig. 5**).
5. Compare the experimental histogram with the theoretical probability distribution of observed individual diffusion coefficients for the Brownian walk *(17,19)*.

$$p(D_e)dD_e = \frac{1}{(N-1)!} \cdot \left(\frac{N}{D_o}\right)^N \cdot (D_e)^{N-1} \cdot \exp\left(\frac{-ND_e}{D_o}\right) \cdot dD_e \quad (6)$$

where $N = N_{total}/i\Delta t$, N_{total} is length of trajectories, $i\Delta t$ is time lag, D_o is true mean diffusion coefficient, D_e is apparent or experimental diffusion coefficient for an individual trajectory. In order to plot **Eq. 6**, the arithmetic mean of D_e's from all trajectories, for a respective time lag, is used as an estimate for D_o.

3.3.6. Cumulative Radial Distribution Function to Identify One (or More) Populations

The cumulative radial distribution of displacements gives information about the heterogeneity of the experimental diffusers, but a reasonable number of displacements must be used to build the distribution (*see* **Note 7c**). The distribution of radial displacements can be created using displacements from a single trajectory if N_{total} is large. If displacements from one trajectory are used, then multiple populations in the distribution indicate that the molecule has transitioned between several environments during the course of the trajectory lifetime. If the trajectories are short, then displacements from all trajectories should be used to create the distribution. In this case, multiple populations in the distribution indicate that either the molecules transitioned between environments or multiple populations of diffusers are present. In addition, the ability to detect transitioning will depend on the fraction of time a molecule spends diffusing with each diffusion coefficient.

1. Use displacements, r, for a particular time lag, calculated in **Subheading 3.3.** using either independent or all pairs of points.
2. Construct the cumulative radial distribution function (CDF) $P(r, i\Delta t)$ for a particular time lag, $i\Delta t$, by counting the fraction of displacements with values less than or equal to r (**Fig. 6**).
3. Fit the CDF to the theoretical distribution and calculate the apparent diffusion coefficient for each time lag, $i\Delta t$.

$$P(r, i\Delta t) = 1 - \exp\left[(-r^2)/4D(i\Delta t)\right] \quad (7)$$

where D is the diffusion coefficient, and r is displacement. There is a minimal number of the displacements required for an accurate fit. Monte Carlo random walk simulations can be used to determine that number. For example, random walk simulations suggest that for diffusion coefficients in the range 0.02–0.2 µm²/s

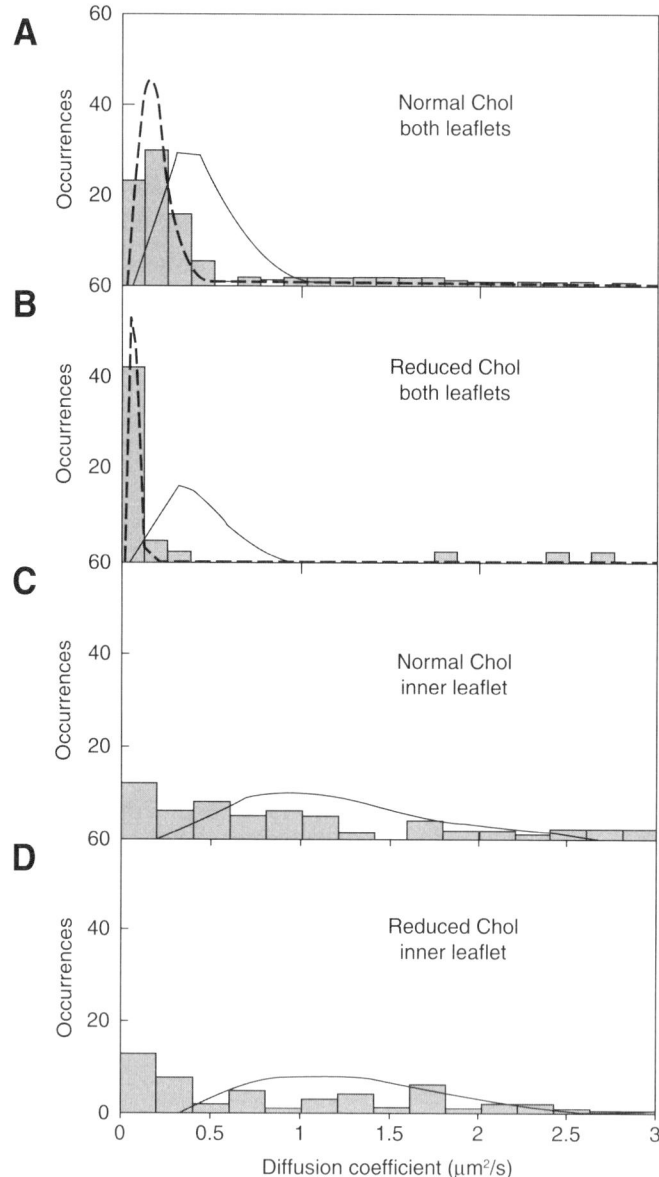

Fig. 5. Distributions of individual diffusion coefficients for fluorescence of a lipid analog, TRITC-DHPE. The observed distributions of individual diffusion coefficients for TRITC-DHPE under four experimental conditions are shown. The diffusion coefficients were calculated using five displacements so each individual TRITC-DHPE trajectory was clipped to be 10 steps long. The expected homogeneous distribution of Ds (from **Eq. 6**) given the average diffusion coefficient for the full population (solid lines) and the expected two population distribution of Ds (modified from **Eq. 6**) (dashed line) are shown.

a minimum of 10 displacements are necessary. If fewer than 10 displacements are used to fit the probability distribution function, then the value of the diffusion coefficient will be lower than the true value owing to a fitting artifact of the CDF *(19)*.

If the trajectories are of different lengths, then the calculated D will be skewed toward the longest trajectories, particularly at large $i\Delta t$ (they provide the most number of displacements). In order to minimize the higher weighting of the long trajectories use a minimum of 50 displacements per fit. Alternately, you can clip all the trajectories to be the same length. Saxton *(15)* suggests that the longest time lag, $i\Delta t$, used for a fit should be one quarter of the total length of a trajectory.

4. Plot the residuals of the fit, and evaluate the goodness of fit (**Fig. 6**). If deviation from the fit is present, two (or more) populations may be present.
5. If the presence of two or more populations is suspected, then fit the data to the CDF for two populations of diffusion coefficients *(21)*:

$$P(r, i\Delta t) = 1 - \left(\exp\left[(-r^2)/4D_1(i\Delta t) \right] + \exp\left[(-r^2)/4D_2(i\Delta t) \right] \right) \quad (8)$$

6. In order to confirm that the second (or multiple) population is real, create random walks using parameters based on experimental data (e.g., diffusion coefficients, number of trajectories, the length of trajectories), and fit them to both one and two-population fits to determine regions of validity of a two-population model. For example, random walks created using a single $D = 0.2$ μm^2/s, $t = 5$ s, show a second diffusion coefficient with the $D_2 = 0.13 \pm 0.11$ μm^2/s and $D_2(\%) = 15 \pm 15$ when fitted to **Eq. 8** *(19)*.

3.3.7. Determine Whether Diffusion is Brownian or Anomalous

If the diffusion is Brownian, then the mean square displacement is linear with time. If the diffusion is anomalous then the mean square displacement is

---

Fig. 5. *(Continued)* The observed distribution of Ds is broader than expected for all cases suggesting the presence of multiple populations. (**A**) Histogram of Ds for TRITC-DHPE at normal total cell cholesterol concentration. A large slow population and smaller fast population are observed. (**B**) After cholesterol reduction, the mean of the slower population in **panel A** decreases, indicating that the slower population of TRITC-DHPE is affected by cholesterol depletion. TRITC-DHPE partitions into both leaflets of the plasma membrane, and thus, the earlier distributions reflect diffusion in both plasma membrane leaflets. (**C**) A membrane impermeable fluorescence quencher, 2,4,6-trinitrobenzenesulfonic acid, was used to quench fluorophores in the outer leaflet at normal total cell cholesterol concentration, and (**D**) reduced total cell cholesterol. The number of molecules in the slower population decreased relative to before the addition of 2,4,6-trinitrobenzenesulfonic acid, indicating that the faster population resides in the inner leaflet. No change in the distributions of Ds in **panels C** and **D** was observed suggesting that the diffusion of TRITC-DHPE in the inner leaflet is not affected by cholesterol depletion (Reprinted in part with permission from Nishimura et al. *[34]*. © 2006 American Chemical Society).

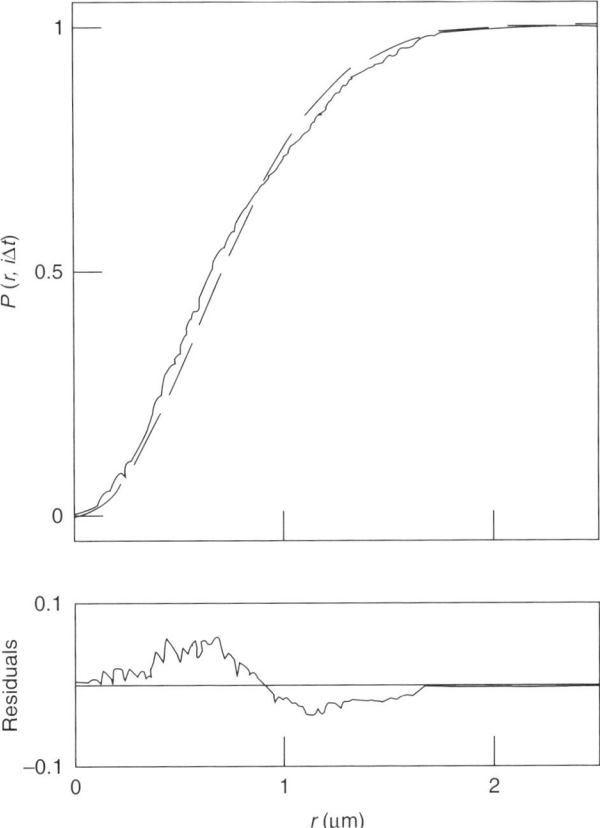

Fig. 6. Analysis of trajectories by cumulative radial probability distribution. One population radial distribution fit (**Eq. 7**) exemplified for transmembrane I-E^k (200 single trajectories, from 25 cells) at time lag = 1.0 s. The solid line represents the data points. Dashed line represents the least-squares fit to **Eq. 7**. The residuals of the fits to **Eq. 7** are shown in the inset (modified with permission from Vrljic et al. *[19]*).

nonlinear with time, and is usually evaluated by the parameter α is obtained using $\langle r^2 \rangle = 4D_0 t^\alpha$, $D = D_0 t^{\alpha - 1}$ where D_0 is the true diffusion coefficient and D is the mean apparent diffusion coefficient *(11,22,23)*. For 2D Brownian motion $0.9 < \alpha < 1.1$, for anomalous diffusion $0.1 < \alpha < 0.9$, for directed motion $\alpha < 0.9$ and for an immobile population $\alpha < 0.1$ *(23)*. Several detailed discussions on the α-parameter have been published *(11,24)*.

1. Using diffusion coefficients from CDF fits (**Subheading 3.6.**) plot log$D$ as a function of log $i\Delta t$ (**Fig. 7**). If you have more than a single diffusion coefficient, plot logD_1 vs log time lag and logD_2 vs log time lag.

Single-Molecule Tracking

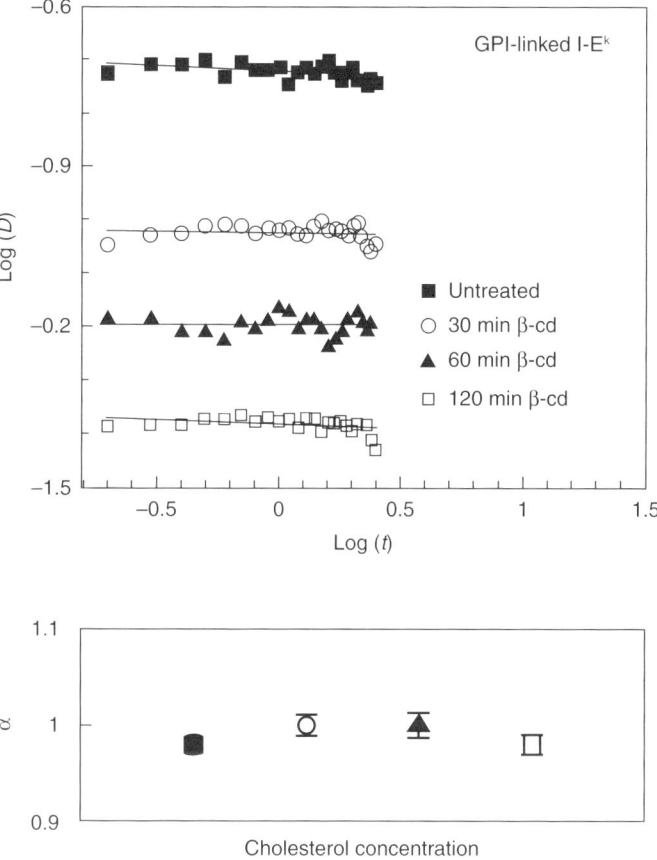

Fig. 7. Assessment of Brownian motion, and calculation of the α parameter. Log of diffusion coefficients for GPI-linked I-E^k at different cholesterol concentrations (expressed as incubation time with β-cyclodextrin) were plotted vs log of time lag. Diffusion coefficients were calculated from fits to the CDF (**Eq. 7**) and (**Fig. 6**). 50–200 trajectories contribute to each time lag. Time resolution is 100 ms per frame. α-parameters for different cholesterol concentrations are shown in the inset. α-parameters are approx 1 for all cholesterol concentrations, suggesting that although the diffusion coefficient of the whole population decreases as cholesterol decreases the motion remains Brownian. The additional information these plots reveal are the limits on the size of the putative domains. For example, at normal cholesterol concentration this data suggests that there are no static impermeable barriers characterized by areas in the range of 0.01–4.0 μm^2. The lower limit is based on the pixel size of 53 nm. The upper limit was calculated using $\langle r^2 \rangle = 4Dt$, where $t = 5$ s and $D = 0.2$ μm^2/s. If a conservative estimate is used based on the method of Saxton (*12*) then the upper limit would have an area of 0.3 μm^2. The results certainly do not rule out static barriers with areas larger than 0.3–4.0 μm^2 or domains with permeable boundaries (modified with permission from Vrljic et al. *[26]*).

2. Determine the value of the anomalous diffusion parameter α by fitting the data to:

$$\log D = \log D_0 + (\alpha - 1)\log(i\Delta t) \qquad (9)$$

3. If the diffusion is anomalous, analyze the shape of the D vs time lag plots or the individual trajectories. Consider the interplay between time lag, spatial resolution, lifetime of the probe, and other properties of the system such as the concentration and size of the immobile obstacles and the mobility, size, and permeability of the domains.
   a. Plot mean square displacements from **Subheading 3.4.** as function of time.
   b. Plot diffusion coefficients from **Subheading 3.6.** as a function of time.

3.3.8. Identifying Confinement Regions

The following analysis allows the reader to identify regions of an individual trajectory whereby the molecule is confined (Based on Simson, et. al. *[25]* and Saxton *[12]*). It does so by identifying those regions where the molecule resides for times longer than a Brownian diffuser is expected to remain, given a particular diffusion coefficient. The theoretical probability that a Brownian diffuser (diffusion coefficient is D) will explore a given distance over a specified amount of time is well established. Each experimental trajectory will be broken into segments of a specified number of frames (this can be thought of as a time lag), and then the probability is calculated that a Brownian diffuser would have explored a corresponding distance during the specified amount of time given the mean diffusion coefficient of the trajectory.

The analysis begins by generating simulated random walks with a range of diffusion coefficients that match those observed in the experimental data (for instance: 1000 random walks and 1000 steps long). Then each random walk is segmented, and the probability is calculated of observing the maximum displacement from the first point in each segment, given the length of time in the segment and the mean diffusion coefficient of the trajectory.

The researcher will use these random walks to set a threshold probability level (Ψ_c) and a maximum time segment (S_m), to use as criteria to identify confinement regions in the experimental data. The threshold probability level is set such that no instances of confinement are identified in a set of random walks. Shorter time segments will give more regions that have low probabilities and longer time segments will average over more displacements and give higher probabilities, so this value must be optimized.

Finally, the experimental trajectories are segmented (with the largest segment defined by S_m) and the probabilities of the segments are calculated. If a point in the experimental trajectory has a probability lower than the threshold value from the random walks, then that point is labeled as a confinement region.

3.3.8.1. DETERMINE THE CONFINEMENT CRITERIA USING RANDOM WALKS

1. Determine the mean diffusion coefficient for each experimental trajectory.
2. Create random walks with the same range of diffusion coefficients as those observed in the experimental data. Use a large number of trajectories (1000) with a large number of steps (1000).
3. Choose a value for the time segment (from 3 to maximum segment, S_m, length), and use that value to segment the trajectory ($t = i\Delta t$). Use each point in the trajectory as an origin for the segment. For each segment, determine the largest displacement from the origin of the segment (R).
4. Using segments originating at each point in the trajectory, calculate the probability that an unconstrained diffuser would remain within R for t, given D (Saxton *[12]*):

$$\log \Psi = 0.2048 - 2.5117\left(\frac{Dt}{R^2}\right) \quad (10)$$

5. Plot the probability vs point in the trajectory.
   a. Set the threshold probability, Ψ_c, such that no instances of confinement are observed in the random walk population.
   b. Simson et al. *(25)* found $\Psi_c = 0.007\%$ for 1×10^{-11} cm^2/s $< D < 2 \times 10^{-10}$ cm^2/s.
6. Repeat **steps 3–5**, and determine the optimal segment length. If the segment length is too small, then even points in the random walks will appear to be confined. If the segment is too large, then areas of actual confinement in the experimental data will no longer be judged as confinement regions.

3.3.8.2. JUDGING CONFINEMENT IN THE EXPERIMENTAL TRAJECTORIES

1. Begin with a particular value for the time segment (t) and divide an individual trajectory into segments of this length using each point in the trajectory as an origin. For each segment, determine the largest displacement from the origin (i.e., the first point in the segment), R. Using the segment that originates at each point in the trajectory, calculate the probability that an unconstrained diffuser would remain within R for t, given D using **Eq. 10**.
2. Plot the probability vs time (point in the trajectory), and identify regions where the probability falls below the Ψ_c.
3. The percent of trajectories with confinement regions can be plotted against probability level or segment length, and the effect of varying the other parameter can be investigated *(25)*. The percentage of trajectories with confinement regions will decrease as each of these parameters is increased. Look for regions of invariance to verify real confinement, and to determine the time of confinement and confinement radius.

3.3.9. Relative Diffusion Between Two Molecules—Detection of Two Molecules Trapped in the Mobile Domain

If a probe is localized in a mobile domain, then the measured diffusion coefficient will reflect the diffusion coefficient of the probe and the domain, and the analyses that have been discussed so far may not identify the molecule as confined. In order to verify the existence of a mobile domain, the *relative* motion of two probes is analyzed. In this case the motion of the two probes is re-expressed as the motion of one probe relative to an origin fixed at the location of the other probe. The intermolecular distance between two Brownian diffusers will grow with a well-defined probability *(19)*. Additionally, the intermolecular distance for pair of confined molecules will grow more slowly than the Brownian pair. In this analysis, a histogram is built of the number of times a pair of molecules is found to be closer than twice their initial separation distance, given a particular time lag and the relative diffusion coefficient of the pair. If the experimental pair is found to be within twice their initial separation distance more often than the unconstrained pair, then the molecules are confined to a domain.

1. Record the trajectories of two molecules, each diffusing with diffusion coefficient D. Calculate D from **Subheading 3.3.4.** or **3.3.6.** Time $t = 0$ is defined as the first time that two molecules are within a distance ρ_0 of each other. For example, in **Fig. 8**, time $t = 0$ is defined as the first time the two labeled proteins are within 0.3–1.0 μm of each other (where 300 nm is the diameter of a diffraction limited spot).
2. Re-express the motion of the two probes as the motion of one with respect to a fixed origin. For Brownian motion the moving particle has a relative diffusion coefficient of $E = 2D$.
3. Generate histograms of the dimensionless time parameter $\varepsilon = 4Et/\rho_0^2$. As a histogram is being built of the number of times a pair of molecules is found within twice their initial separation distance only those times when the interprotein distance is $\leq 2\rho_0$ are scored as positive hits. Any re-entry events are counted relative to the first frame of the pair. The proteins in the pair are monitored until one of the fluorophores bleaches (**Fig. 8**).
4. Compare the experimental histogram with the theoretical probability of finding the second protein within a distance R from the origin (first protein) at time t. The explicit form of this relationship for $R = 2\rho_0$ is given by (*see* **ref. 19** for a derivation):

$$P(\rho' \leq 2, t) = \int_0^2 \frac{2}{\varepsilon} \exp\left[\frac{-(1+\rho'^2)}{\varepsilon}\right] I_0\left[\frac{2\rho'}{\varepsilon}\right] \rho' d\rho' \qquad (11)$$

where, $\varepsilon = 4Et/\rho_0^2$, $\rho' = \rho/\rho_0$, $R' = R/\rho_0$.

5. The integration is performed numerically. If the histogram of $\varepsilon = 4Et/\rho_0^2$ follows **Eq. 11**, then the two particles are Brownian diffusers. If the histogram is above **Eq. 11**, then molecules are staying together longer than expected by the Brownian diffusion (e.g., are trapped in a domain), and if the histogram is below **Eq. 11** then the molecules are staying together shorter than expected by the Brownian diffusion (e.g., repulsing each other). Note short trajectories can cause the occurrences to fall below **Eq. 11** *(19)* **Fig. 8**. Use random walk simulations to show that the observed data is not an artifact of the length of the experimental trajectories.

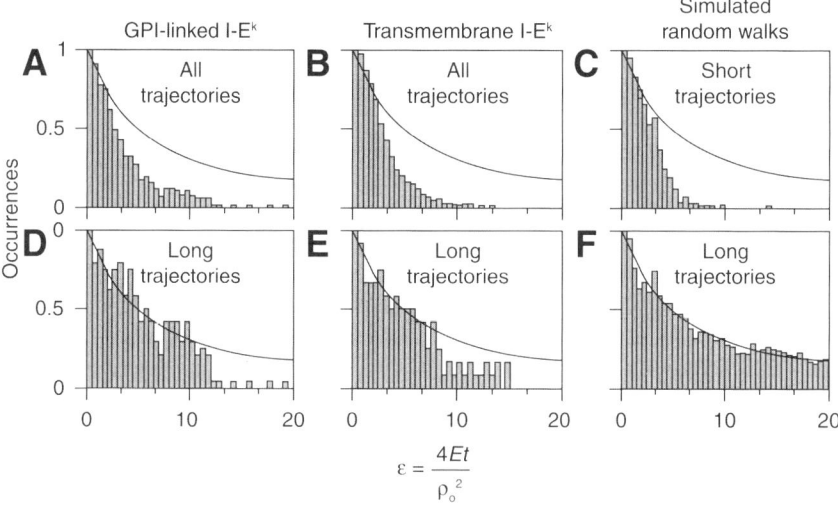

Fig. 8. Analysis of relative diffusion between pairs of membrane proteins at normal cholesterol concentration. **(A)** GPI-linked I-E^k. Histogram represents probability distribution of the data obtained from 74 pairs (from 11 cells). $D = 0.22$ μm^2/s was used to calculate ε. Pairs with separation distances (ρ_0) between 0.3–1 μm at ms were analyzed. Minimum length of analyzed trajectories was 600 ms. Average length of a pair was 1.8 s. Given the mean diffusion coefficients for both transmembrane I-E^k and GPI-linked I-E^k, 600 ms is a mean time required for the distance between two molecules to reach $2\rho_0$ if they started $\rho_0 = 1$ μm apart. Solid line is the plot of **Eq. 11**. **(B)** Transmembrane I-E^k. $D = 0.18$ μm^2/s was used to calculate ε. For the 96 pairs used, the average length of a pair was 1.5 s (from 7 cells). All other parameters as in A. **(C)** Simulated random walks. One point tracer, which begins distance ρ_0 from the origin walks on a square lattice with diffusion coefficient equal to E, with equal probability of moving in any direction on the lattice. Initial separation distances are given by experimental data. $D = 0.2$ μm^2/s, length 2.0 s, 65 pairs. Solid line is **Eq. 11**. **(D)** same as A, except that all pairs are at least 2 s long (from 8 cells). **(E)** same as **B**, except that all pairs are at least 4 s long (from 3 cells). **(F)** Length of the 65 pairs was 500 s. These results show that the data follow the theoretical distribution for small ε, but fall below the theoretical curve at larger ε. This effect could indicate that the distance between proteins grows faster than expected and could be the result of the presence of repulsive forces between the proteins. However, random walk simulations have shown that the effect is owing to the length of trajectories (compare **panels C** and **F**). These results suggest that the motion of pairs of proteins separated by distances in the range of 0.3–1.0 μm for times up to 3 s is Brownian. Thus, pairs of proteins cannot be restricted to small, freely diffusing domains with diameters in this range (modified with permission from Vrljic et al. *[26]*).

4. Notes

1. *Choice of cell type:* optimal samples for SMT are live cells containing fluorescently labeled probes in the plasma membrane. Some care is necessary in selecting the appropriate cell type for the experiment, as some cells adhere to surfaces, others do not, and some crawl. Whereas probe trajectories can be obtained from mobile cells, it is convenient to choose a cell type, such as fibroblasts, which can be easily immobilized on a transparent cover slip.

    Because probe diffusion in the membrane is most easily analyzed with a 2D model for Brownian motion, the shape of the cell's surface is important. If the cells have a high degree of curvature, then even when the top of the cell is observed, some of the probe's motion will be in the plane perpendicular to the microscope plane. Owing to the current technical limitations motion in this plane is not typically resolved, and the particle could appear to be anomalously diffusing. The trajectories should be obtained from flat areas of the plasma membrane parallel to the optical plane.

2. *Tissue culture:*
    a. Some of the common reagents used for tissue culture media, like phenol-red and fetal calf serum (FCS), fluoresce in the visible region and may contribute to the autofluorescent background. Cells should be passaged in phenol-red free media. Omitting the phenol-red from the media just for imaging is not ideal. Omitting FCS during imaging reduces the background fluorescence.
    b. The amount of various lipids in different lots of FCS is quite variable and may lead to modification in the cell membrane lipid composition. It has been observed that the mean diffusion coefficients of I-E^k proteins are slightly different between CHO cells cultured in different lots of serum *(26)*.
    c. Check whether any cell media supplements interfere with the assay. For example, FCS contains lipids and proteins that can bind lipids. Vrljic et al. *(26)* have observed that diffusion coefficients drop faster when cells are incubated with β-cyclodextrin in the absence of FCS than in the presence of FCS. Nystatin, used as antifungal, binds cholesterol in the plasma membrane.

3. *Label selection:*
    a. The fluorophore of choice needs to be a good single-molecule fluorophore, having a long lifetime, high quantum yield (which results in a better signal-to-noise ratio using less laser pumping power), large extinction coefficient, and low residency time in the triplet state (produces interruptions of fluorescence emission). In general, the red fluorophores, although they may be less photostable than green and blue dyes (e.g., Cy3B vs Cy5), are better for cell imaging owing to lower cellular autofluorescence in the red compared with the blue and green regions of the spectrum.
    b. Typically, fluorophores (autofluorescent protein fusions and dyes), quantum dots, or scattering beads (polystyrene or gold) are chosen as labels and attached to the membrane proteins or lipids of interest. Whereas beads and quantum dots can only be attached to a molecule of interest through antibodies or biotin–strepavidin conjugation, fluorophores can be attached through covalent fusion with fluorescent proteins, an antibody that binds to the protein or the lipid, a small ligand

that binds to the protein of interest, or by direct covalent attachment (to cysteines, free amines, and sugars on the surface of the protein and lipid tails, or free amines on the lipid headgroups). The choice of labeling scheme is of paramount importance because some labels can induce cross-linking or inhibit biological function of the molecule. For example, fluorescent analogs of cholesterol or tail-labeled lipid probes, in model lipid monolayers, do not exhibit the same phase behavior as their unlabeled counterparts (Radhakrishnan and Okonogi, personal communication). Antibodies and cholera toxin can cause crosslinking.

The choice of beads, quantum dots, or fluorophores as labels is governed by the type of information desired from the collected data. Beads, quantum dots, and fluorophores can be used to construct 2D positional trajectories; however, quantum dot and bead trajectories have superior length to fluorophores (several seconds to minutes for beads and quantum dots vs several hundred milliseconds to seconds for fluorophores) and beads have superior time resolution (microseconds for beads and milliseconds for fluorophores and quantum dots). Thus, the bead and quantum dots can be good probe choices. However, bead and quantum dot labels are large (the smallest used beads are 40 nm in diameter) and cross-linking is a common side effect *(27)*. Cross-linking creates probes with radii larger than single probes, and potentially changes the interactions of the probe with putative domains and other membrane components.

Experiments studying mobile small domains (several nanometers in diameter), colocalization of the probes, and properties of the cytoplasmic leaflet necessitate the use of fluorophores. Cross-linking is not an issue with judicious choice of fluorophore and the attachment of a single fluorophore to a protein or lipid of interest can create probes that interact with other membrane components in biologically relevant ways. For example, the fluorescently labeled peptide–MHC complex interacts with its ligand, the T-cell receptor, with the same kinetics as peptide–MHC complex in the absence of a label, and head-labeled lipids behave as unlabeled ones in the lipid monolayers (Rabinowitz and Radhakrishnan personal communication).

When a fluorophore is used as a label, additional information may be obtained from the intensity of the fluorescence emission as a function of time. If the fluorophore changes its properties based on, for example, the viscosity or pH of the environment, then the intensity can be used to sense certain types of membrane inhomogeneity. However, intensity analysis can be challenging because of photophysically induced intensity fluctuations of some fluorophores (e.g., isomerization of Cy5).

c. A new class of fluorophores that possesses an amphiphilic structure similar to lipid and is well suited for single-molecule studies has been reported *(28–30)*. These molecules, termed dicyanomethylenedihydrofuran (DCDHFs), consist of an amine donor and a dicyanomethylenedihydrofuran acceptor linked by a conjugated unit (benzene, naphthalene, styrene, and so on). The most notable feature of the DCDHFs fluorophores is that their emission wavelengths and fluorescence quantum yields are sensitive to solvent polarity and local rigidity *(28,30)*. For example, in polymer films, high photostabilities (photobleaching quantum yield from 7.5×10^{-7} to 14×10^{-7}) and high fluorescence quantum yields (0.39–0.95) have been reported, whereas the fluorescence quantum yields for the same DCDHFs in solution are significantly lower *(28)*. Recently, seven DCDHF lipid analogs were demonstrated to be reporters of mobility in the plasma membrane of CHO cells, and the diffusion coefficients of five of these were reported *(31)*. A representative image of single molecules of the DCDHF-N-12 derivative in the plasma membrane of a CHO cell is shown in **Fig. 9A**. Single molecules of the DCDHF-N-12 derivative ($\lambda_{abs\ max}$ 543/$\lambda_{em\ max}$ 657 in ethanol) provide total numbers of photons and signal-to-noise on a bar with the conventional lipid analog TRITC-DHPE *(31)* (**Fig. 9C,D**). The distribution of diffusion coefficients for single-molecules of DCDHF-N-12 in the CHO cell plasma membrane are shown in **Fig. 9B**.

4. *On-time of the probe:* in order to obtain information on membrane structure, the probe must be observed for long enough times such that it interacts with barriers to its diffusion. For instance, the on-time of a confined probe before blinking or photobleaching determines the maximum domain size that can be detected *(12)*, so the trajectory needs to be sufficiently long to detect the domain boundary. For probes diffusing in the presence of obstacles, the observed diffusion coefficient will vary with time *(11)*, so again long on-times are desirable.

5. *Temporal resolution:* the integration time of the detector needs to be sufficiently fast to allow resolution of the diffusional motion within small domains and transitions between environments. If the integration time is too large compared with the motion of the probe during the integration time, the diffraction-limited spot smears and estimation of the probe's position is challenging. The point spread function of the single-molecule spot should be an airy disk in the best case, well approximated by a Gaussian profile.

   If the integration time is too long compared with the time it takes the molecule to traverse the domain, then the observed diffusion coefficient will be lower than the actual diffusion coefficient. In order to obtain the true diffusion coefficient and the size of the domain, the integration time needs to be smaller than the time it takes the probe to traverse the domain ($\Delta t < \langle r^2 \rangle/4D$) *(13)*. If the probe is excluded from the domain, at short times a plot of its diffusion coefficient vs time lag will appear non-Brownian, but at longer times it will appear Brownian with a lower diffusion coefficient than its actual value *(11)*. To resolve such behavior, short integration times are necessary. Thus, data should be taken at the fastest possible integration time. If the probe is transitioning between different environments, then at short integration times displacements from both environments can be recorded, whereas at longer integration times the displacements are averaged and the diffusion appears Brownian with a lower than actual diffusion coefficient. This has been nicely illustrated *(27,32)*.

6. *Statistics:* the ability to detect subtle deviations from Brownian motion and to differentiate between small populations depends on having a large number of collected trajectories or total number of displacements per trajectory. The number of displacements per trajectory is determined by the length of the trajectory and temporal resolution. A large number of displacements per trajectory are easily obtainable

Single-Molecule Tracking

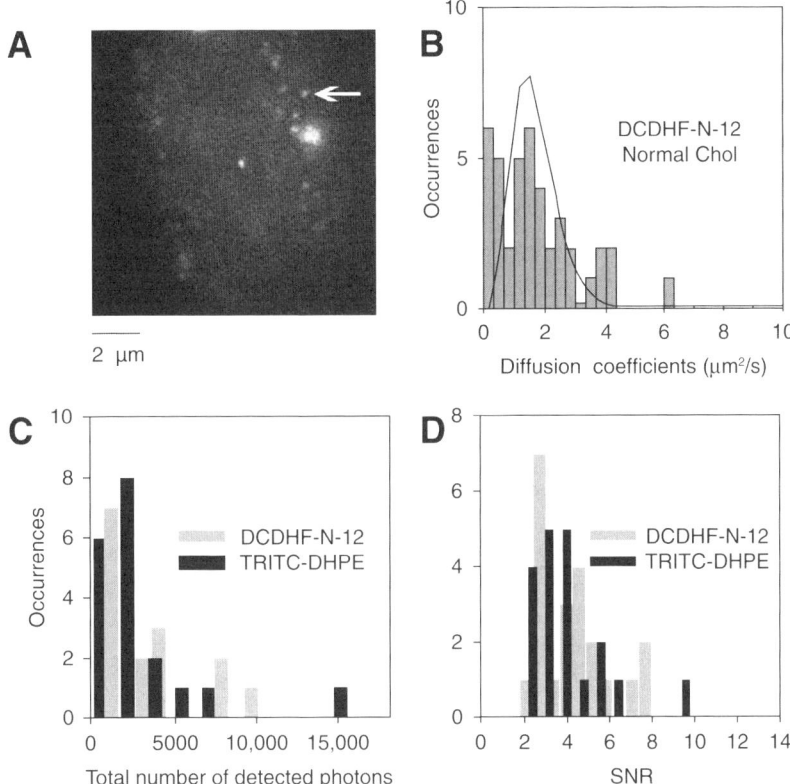

Fig. 9. Comparison of DCDHF-N-12 with a rhodamine derivative. (**A**) Representative cross-sectional fluorescence image of the CHO cell labeled with DCDHF-N-12. The white arrow points to a representative single-molecule of DCDHF-N-12 in the plasma membrane. (**B**) The diffusion of the DCDHF-N-12 molecules (excited using 532 nm) was observed at room temperature (22°C) with an integration time of 15.4 ms per frame. The trajectories were clipped to be 10 steps long (11 frames), and $D_{mean} = 1.3$ µm²/s from 41 trajectories. (**C**) Histogram of the total number of detected photons for DCDHF-N-12 (22 molecules) as compared with TRITC-DHPE (19 molecules). (**D**) Histogram of signal-to-noise ratios for DCDHF-N-12 (22 molecules) as compared with TRITC-DHPE (19 molecules) (modified with permission from Nishimura et al. *[31]*).

with beads and quantum dots because they stay on the membrane for several minutes and can be imaged with microsecond integration times. Experimental trajectories of single fluorophores are of a more limited length (up to ~3 s); this is the price paid for the noninvasive nature of the measurements with a single, 1-nm sized label. In general, it is difficult to capture representative probability distributions constructed from a few data points, especially when more than a single diffusing population is present. Fits to a small data set can induce artifacts.

Owing to the different on-times of the probes, at short time lags many trajectories contribute to the distribution whereas at long time lags distribution is determined by a few (e.g., one or two) long-lived trajectories. There is no reason to assume that those few trajectories behave as the mean of the initial population. The distribution at long time lags may or may not be the same as the behavior at short time lags. Thus, careful consideration of the sample size should be taken when determining maximum time lag of the analysis.

If too few displacements are used to calculate the experimental diffusion coefficient, the calculated value will differ from the real one (will be lower). Monte Carlo simulations of random walks should be used to determine the minimum number of displacements required for the fit. For example, random walk simulations suggest that for diffusion coefficients in the range 0.02–0.2 µm^2/s a minimum of ten displacements are necessary *(19)*.

7. *Data interpretation:*
   a. *Averaging over independent pairs vs averaging over all pairs:* the advantage of averaging over independent pairs is that each segment is an independent, nonredundant, and random walk. The disadvantage of averaging over independent pairs is that certain points are weighted more heavily. The disadvantage of averaging over all pairs is that pairs of points are highly correlated *(15,17)*. The advantage using all pairs, especially for short trajectories, is that each trajectory yields more displacements.
   b. *Mean square displacements:* any deviation from linearity of $\langle r^2 \rangle$ vs t is suggestive of deviations from Brownian motion owing to the entrapment in the domain with impermeable or permeable boundary, presence of immobile obstacles, and/or direct, cytoskeleton-mediated transport. When the motion is anomalous the diffusion coefficient varies with time and the slope of $\langle r^2(i\Delta t) \rangle$ ≈ $i\Delta t$ plotted over all $i\Delta t$ cannot be used to estimate the mean diffusion coefficient *(13)*. In order to calculate the unobstructed diffusion coefficient in a domain or in the presence of obstacles, only the initial, linear section of the $\langle r^2 \rangle$ vs $i\Delta t$ plot should be used. If the temporal resolution is too low, then the unobstructed diffusion coefficient cannot be determined.
   c. *Brownian or anomalous:* the shape of the $\langle r^2 \rangle$ vs t or D vs t plots can be used to determine the mode of diffusion that the probe is undergoing. In general, curving upwards is indication of directed motion, curving downwards suggests trapping in the domain and/or presence of immobile obstacles *(11–14)*. The simplest case is when the probe is inside a highly impermeable, static domain. At time lags before probe encounters the domain boundary the $\langle r^2 \rangle$ vs t is linear and D vs t has $\alpha = 1$. At time lags when displacements approach the domain diameter and the probe start encountering the boundary, its displacements will no longer increase and $\langle r^2 \rangle$ vs t will level off. The intercept of $\langle r^2 \rangle$ vs t plot with the ordinate can reveal the area of the domain. However, keep in mind that the maximum displacement is reached at $t \cong 4\langle r^2(\infty)\rangle/D$ *(12)*.

   If the molecule moves in the presence of immobile obstacles, at short time lags diffusion is Brownian and at longer time lags diffusion coefficient decreases.

As Saxton *(11)* has shown, the magnitude of the decrease and the shape of the D vs t plot depends on the concentration and size of the immobile obstacles. Some authors have reported methods of sensing non-Brownian motion based on the moment scaling spectrum *(33)*.

d. *Histograms of individual diffusion coefficients:* the probe can partition into different environments (reside inside or outside a domain, or be bound to another proteins). If the different environments influence the diffusion of the probe (e.g., viscosities differ), then this will be reflected in the values of the mean diffusion coefficients. In order to detect different populations of probes histograms of individual mean diffusion coefficients need to be constructed and compared with the expected probability distribution of diffusion coefficients *(17,19,34)*. However, if a molecule transitions between two diffusion coefficients, because the calculated diffusion coefficient is an average over the whole trajectory, one might not see two distinct populations depending on the fraction of time the molecule exhibits each diffusion coefficient.

Acknowledgment

We would like to thank Harden M. McConnell for instructive discussions.

References

1. Anderson, R. G. W. and Jacobson, K. (2002) A role for lipid shells in targeting proteins to caveolae, rafts and other lipid domains. *Science* **296**, 1821–1825.
2. McConnell, H. M. and Vrljic, M. (2003) Liquid-liquid immiscibility in membranes. *Ann. Rev. Biophys. Biomol. Struct.* **32**, 469–492.
3. Kenworthy, A. K., Nichols, B. J., Remmert, C. L., et al. (2004) Dynamics of putative raft-associated proteins at the cell surface. *J. Cell Biol.* **165**, 735–746.
4. Lagerholm, B. C., Weinreb, G. E., Jacobson, K., and Thompson, N. L. (2005) Detecting microdomains in intact cell membranes. *Annu. Rev. Phys. Chem.* **56**, 309–336.
5. Saffman, P. G. and Delbruck, M. (1975) Brownian motion in biological membranes *Proc. Nat. Acad. Sci. USA* **72**, 3111–3113.
6. Hughes, B. D., Pailthorpe, B. A., and White, L. R. (1981) The translational and rotational drag on a cylinder moving in a membrane. *J. Fluid Mech.* **110**, 349–372.
7. Rabinowitz, J. D., Vacchino, J. F., Beeson, C., and McConnell, H. M. (1998) Potentiometric measurement of intracellular redox activity. *J. Am. Chem. Soc.* **120**, 2464–2473.
8. Yildiz, A., Forkey, J. N., McKinney, S. A., Ha, T., Goldman, Y. E., and Selvin, P. R. (2003) Myosin V walks hand-over-hand: single fluorophore imaging with 1.5nm localization. *Science* **300**, 2061–2065.
9. Moerner, W. E. (1994) Fundamentals of single-molecule spectroscopy in solids. *J. Lumin.* **60–61**, 997–1002.
10. Moerner, W. E. and Fromm, D. P. (2003) Methods of single-molecule fluorescence spectroscopy and microscopy. *Rev. Sci. Instrum.* **74**, 3597–3619.

11. Saxton, M. (1994) Anomalous diffusion due to obstacles: a Monte Carlo study. *Biophys. J.* **66,** 394–401.
12. Saxton, M. J. (1993) Lateral diffusion in an archipelago. Single-particle diffusion. *Biophys. J.* **64,** 1766–1780.
13. Saxton, M. J. (1995) Single-particle tracking: effects of corrals, *Biophys. J.* **69,** 389–398.
14. Saxton, M. J. (1990) Lateral diffusion in a mixture of mobile and immobile particles. A Monte Carlo study. *Biophys. J.* **58,** 1303–1306.
15. Saxton, M. J. (1997) Single-particle tracking: the distribution of diffusion coefficients. *Biophys. J.* **72,** 1744–1753.
16. Schütz, G. J., Kada, G., Pastushenko, V. P., and Schindler, H. (2000) Properties of lipid microdomains in muscle cell membrane visualized by single molecule microscopy. *EMBO J.* **19,** 892–901.
17. Qian, H., Sheetz, M. P., and Elson, E. L. (1991) Single particle tracking. Analysis of diffusion and flow in two-dimensional systems. *Biophys. J.* **60,** 910–921.
18. Saxton, M. J. and Jacobson, K. (1997) Single-particle tracking: applications to membrane dynamics. *Annu. Rev. Biophys. Biomol. Struct.* **26,** 373–399.
19. Vrljic, M., Nishimura, S. Y., Brasselet, S., Moerner, W. E., and McConnell, H. M. (2002) Translational Diffusion of Individual Class II MHC Membrane Proteins in Cells. *Biophys. J.* **83,** 2681–2692.
20. Dietrich, C., Yang, B., Fujiwara, T., Kusumi, A., and Jacobson, K. (2002) Relationship of lipid rafts to transient confinement zones detected by single particle tracking. *Biophys. J.* **82,** 274–284.
21. Schütz, G. J., Schindler, H., and Schmidt, T. (1997) Single-molecule microscopy on model membranes reveals anomalous diffusion. *Biophys. J.* **73,** 1073–1080.
22. Smith, P. R., Morrison, I. E. G., Wilson, K. M., Fernandez, N., and Cherry, R. J. (1999) Anomalous diffusion of major histocompatibility complex class I molecules on HeLa cells determined by single particle tracking. *Biophys. J.* **76,** 3331–3344.
23. Feder, T. J., Brust-Mascher, I., Slattery, J. P., Baird, B., and Webb, W. W. (1996) Constrained diffusion of immobile fraction on cell surfaces: a new interpretation. *Biophys. J.* **70,** 2767–2773.
24. Saxton, M. J. (2001) Anomalous Subdifusion in Fluorescence Photobleaching Recovery: a Monte Carlo Study. *Biophys. J.* **81,** 2226–2240.
25. Simson, R., Sheets, E. D., and Jacobson, K. (1995) Detection of temporary lateral confinement of membrane proteins using single-particle tracking analysis. *Biophys. J.* **69,** 989–993.
26. Vrljic, M., Nishimura, S. Y., Moerner, W. E., and McConnell, H. M. (2005) Cholesterol depletion suppresses the translational diffusion of Class II Major Histocompatibility Complex proteins in the plasma membrane. *Biophys. J.* **88,** 334–347.
27. Murase, K., Fujiwara, T., Umemura, Y., et al. (2004) Ultrafine membrane compartments for molecular diffusion as revealed by single molecule techniques. *Biophys. J.* **86,** 4075–4093.

28. Willets, K. A., Callis, P. R., and Moerner, W. E. (2004) Experimental and theoretical investigations of environmentally sensitive single-molecule fluorophores. *J. Phys. Chem. B* **108,** 10,465–10,473.
29. Willets, K. A., Nishimura, S. Y., Schuck, P. J., Twieg, R. J., and Moerner, W. E. (2005) Nonlinear optical chromophores as nanoscale emitters for single-molecule spectroscopy. *Acc. Chem. Res.* **38,** 549–556.
30. Willets, K. A., Ostroverkhova, O., He, M., Twieg, R. J., and Moerner, W. E. (2003) New Fluorophores for Single-Molecule Spectroscopy. *J. Am. Chem. Soc.* **125,** 1174–1175.
31. Nishimura, S. Y., Lord, S. J., Klein, L. O., et al. (2006) Diffusion of Lipid-like Single-Molecule Fluorophores in the Cell Membrane. *J. Phys. Chem. B* **110,** 8151–8157.
32. Fujiwara, T., Ritchie, K., Murakoshi, H., Jacobson, K., and Kusumi, A. (2002) Phospholipids undergo hop diffusion in compartmentalized cell membrane. *J. Cell Biol.* **157,** 1071–1081.
33. Ewers, H., Smith, A. E., Sbalzarini, I. F., Lilie, H., Koumoutsakos, P., and Helenius, A. (2005) Single-particle tracking of murine polyoma virus-like particles on live cells and artificial membranes. *Proc. Nat. Acad. Sci. USA* **102,** 15,110–15,115.
34. Nishimura, S. Y., Vrljic, M., Klein, L. O., McConnell, H. M., and Moerner, W. E. (2006) Cholesterol depletion induces solid-like regions in the plasma membrane. *Biophys. J.* **90,** 927–938.

15

X-Ray Diffraction to Determine the Thickness of Raft and Nonraft Bilayers

Thomas J. McIntosh

Summary

Low-angle X-ray diffraction is a powerful method to analyze the structure of membrane bilayers. Specifically, the technique can be used to determine accurately the thickness of fully hydrated bilayers. Herein details are presented showing how this technique can measure the difference in thickness of bilayers in detergent-resistant membranes and detergent-soluble membranes extracted from model systems known to contain both raft and nonraft domains. The observed thickness difference may be critical in the sorting of transmembrane proteins between raft and nonraft bilayers.

Key Words: Bilayer thickness; detergent-resistant bilayers; electron density profiles; rafts; X-ray diffraction; hydrophobic matching.

1. Introduction

Rafts are thought to sequester specific lipids and proteins to small areas (microdomains) of plasma and Golgi membranes. As detailed in Chapter 1, because of the ability to sequester specific proteins and exclude others, rafts have been postulated to perform roles in a number of critical cellular processes. A key feature of rafts that may be important in sorting of membrane proteins relates to potential structural differences between raft bilayers and the surrounding nonraft bilayers. Owing to their high concentrations of cholesterol and sphingolipids with long, saturated hydrocarbon chains, rafts are thought to have thicker bilayers than the surrounding lipid matrix that contains unsaturated, shorter-chained phospholipids *(1)*. This structural feature is a factor to consider in protein sorting between rafts and surrounding membranes because of the effects of "hydrophobic matching" between the bilayer hydrocarbon thickness and the transmembrane domain length of resident proteins *(2–10)*. Thus, proteins with relatively long transmembrane hydrophobic regions might be expected to localize in the wide raft bilayers, whereas shorter transmembrane proteins should localize to the surrounding nondomain regions of the membrane *(11,12)*. However, some depictions of rafts *(13,14)* do not show any difference in thickness between rafts and the surrounding membrane, and it is critical to know as accurately as possible the hydrocarbon thickness of raft and nonraft bilayers for a thorough understanding of the energetics of sequestering of transmembrane proteins in the plane of the membrane *(11,12,15)*.

Low-angle X-ray diffraction provides one of the best methods to determine the thickness of unfixed, hydrated lipid bilayers. This chapter provides details and illustrations regarding how this method can be used to determine the relative bilayer and hydrocarbon thickness of raft and nonraft bilayers. Analysis shows that the hydrocarbon region of detergent-resistant raft bilayers is about 9 Å (25%) thicker than the hydrocarbon region of detergent-soluble nonraft bilayers *(16)*.

2. Materials

1. The lipids dioleoylphosphatidylcholine (DOPC) and bovine brain sphingomyelin (SM) from Avanti Polar Lipids (Alabaster, AL).
2. Cholesterol, cholesterol infinity reagent, benzidine, polyvinylpyrrolidone (PVP), thin-layer chromatography (TLC) plates, and Triton X-100 from Sigma Chemical Company (St. Louis, MO).
3. SM-2 Adsorbent BioBeads from BioRad (Hercules, CA).

3. Methods

3.1. Preparation of Raft-Containing Multilamellar Bilayers

Multilamellar lipid vesicles (MLVs) containing raft microdomains are made by the following procedure. An equimolar mixture of DOPC, SM, and cholesterol is first codissolved in chloroform or chloroform:methanol 3:1 (v:v). This lipid combination is used as it has been shown by microscopy to produce bilayers that contain rafts *(17–19)*. Next the solvent is removed by rotary evaporation and the dry lipid is hydrated with PVP solutions (0–40% PVP) made in either 25 mM KCl, 5 mM HEPES buffer (pH 7.4), or water. No difference has been observed in X-ray experiments or chemical analysis for samples prepared with this buffer or with water.

3.2. Detergent Extraction of Detergent-Resistant Membranes and Detergent-Soluble Membranes

Detergent-extraction procedures are similar to those of Ahmed et al. (1997) *(20)*. MLVs of 1:1:1 DOPC:SM:cholesterol in either water or buffer (total lipid concentration 3–4 mg/mL) are treated with 1% Triton X-100 for 30 min at 4°C and then centrifuged for 30 min at 4°C with an Eppendorf bench centrifuge (Eppendorf, Hamburg, Germany). The supernatant (containing detergent-soluble membranes [DSMs]) is removed and the pellet (containing detergent-resistant membranes [DRMs]) is resuspended in an equal volume of buffer or water.

3.3. Chemical Analysis of DRMs and DSMs

For the supernatant (DSMs) and the pellets (DRMs) the phospholipid content is determined by a standard phosphate assay *(21)*, the cholesterol content is determined using the Sigma infinity (cholesterol oxidase) assay, and the type of phospholipid is determined by TLC. TLC is performed using chloroform:methanol:ammonium hydroxide 65:25:4 (v:v:v) as the solvent. For most experiments iodine vapor is used to detect the lipid spots and lanes with DOPC, and SM controls are used to identify the location of the DOPC and SM spots. Benzidine reagent, which stains sphingolipids but not phosphatidylcholines *(22)*, can be used to verify the location of SM on the TLC plate.

To estimate the DOPC to SM ratio in the DRMs and DSMs, the ratios of the densities of the respective spots in the iodine-treated TLC plates are compared with those of control iodine-treated TLC plates containing lanes with a range of mole ratios of DOPC to SM. Relative densities of the DOPC and SM spots are determined by scanning the TLC plates with an AGFA T2500 Scanner (Agfa-Gervaert N. V., Mortsel, Belgium) and using NIH Image Version 1.61 (Bethesda, MD) to measure the area under each peak. For the 1:1:1 DOPC:SM:cholesterol it was found, as expected for raft and nonraft bilayers, that DRMs are enriched in SM and cholesterol, whereas DSMs are enriched in DOPC *(16)*.

3.4. X-Ray Diffraction Analysis

3.4.1. Specimen Preparation and Data Collection

The structures of the bilayers in DRMs and DSMs (*see* **Note 1**) are obtained by low-angle X-ray diffraction techniques from both unoriented MLV suspensions in PVP solutions and oriented multilayers in a range of relative humidity atmospheres controlled by incubation with saturated salt solutions *(23,24)*. To reduce the Triton X-100 concentration before X-ray diffraction analysis, the DSMs and DRMs are washed three times for 60 min with SM-2 BioBeads (100 mg/mL) that had been previously washed in water, 1 M acetic acid, methanol, and water. The PVP solutions and relative humidity atmospheres provide a wide range of osmotic pressures to systematically modify the water content of the specimens; PVP solutions from 0 to 40% PVP give osmotic pressures of $0-1 \times 10^7$ dyn/cm^2, whereas relative humidity atmospheres of 98 to 32% relative humidity give osmotic pressures of 1×10^7 dyn/cm^2 to 1×10^9 dyn/cm^2 *(25,26)*. Increasing osmotic pressure removes water from between adjacent bilayers in the multilamellar arrays, thereby changing the unit cell dimensions and the lamellar repeat period. By recording lamellar X-ray diffraction patterns for these different repeat periods one can obtain structure factors along the reciprocal space axis and effectively trace out the continuous Fourier transform of the bilayer *(16,23,24,26)*. This transform can be used to determine the phase angles for each X-ray reflection, permitting electron density profiles across the bilayer to be calculated.

Unoriented MLVs in PVP solutions of various concentrations are pelleted, sealed in glass X-ray capillary tubes, and mounted in a temperature-controlled specimen chamber in a point collimation X-ray camera (Unicam, Cambridge, UK). Oriented lipid multilayers are prepared by placing a drop of an aqueous suspension of DRMs or DSMs onto a curved glass substrate and drying it under a gentle stream of nitrogen. The lipid multilayers oriented on the glass substrate are mounted in a temperature-controlled constant humidity sample chamber on a line-focus custom-built line-focus (single mirror) X-ray camera *(23,24)* so that X-ray beam line tangentially grazes the edge of curved surface. The curve in the glass substrate allows a range of Bragg diffraction angles to be simultaneously accessed.

For both unoriented MLVs and oriented bilayers, X-ray patterns are recorded at ambient temperature on X-ray film. Densitometry provides the uncorrected intensity ($I[h]$) of each lamellar reflection h. Structure amplitudes ($F[h]$) are obtained by applying standard correction factors, which are different for oriented and unoriented specimens *(23,26)*. That is, for unoriented specimens $F(h)$ is proportional to $[h^2I(h)]^{1/2}$, whereas for oriented specimens $F(h)$ is proportional to $[hI(h)]^{1/2}$. The validity of these correction factors has been demonstrated by showing that the resulting electron density profiles are the same for unoriented and oriented bilayers of the same compositions at the same osmotic pressure *(23)*.

3.4.2. X-Ray Data Analysis and Phase Angle Determination

To obtain electron density profiles across the bilayer, a Fourier analysis of the X-ray diffraction patterns is performed. Phase angles ($\phi[h]$) for each diffraction order h are determined by using the osmotic stress experiments to trace out the continuous transform of the bilayer *(16,23,24,26)*. For these centrosymmetric multilamellar systems the analysis is simplified as the phase angle for each reflection can either be zero or 180, so that for each order h the structure factor, $\exp[i\phi(h)] \cdot F(h)$, is equal to $+/-F(h)$. For each multilamellar system possible con-tinuous transforms for each possible phase combination ($+/-F[h]$ for each order h), are calculated by use of the sampling theorem *(27)* for one particular data set. The phase combination that gives the transform that best matches the other structure factors is judged as the correct combination *(23,26,28)*. For incorrect phase angle combinations there is a large mismatch between some of the structure factors and the transform *(16,26,28,29)*. **Figure 1** shows structure factors and continuous Fourier transforms of DRMs and DSMs obtained from 1:1:1 DOPC:SM:cholesterol *(16)* using the correct phase combination. Note that for

X-Ray Diffraction

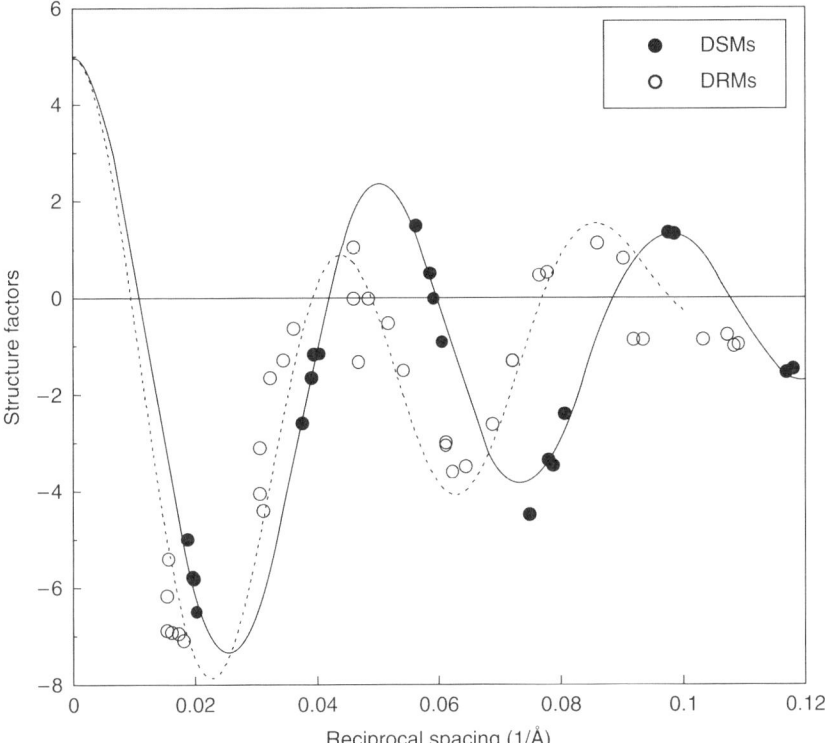

Fig. 1. Structure amplitudes for DSMs and DRMs obtained from 1:1:1 DOPC:SM: cholesterol bilayers. The solid and dotted lines are continuous transforms calculated with the sampling theorem for DSMs and DRMs, respectively. Note that for each system the continuous transform closely matches all of the structure amplitudes, indicating that the phase combination has been correctly assigned. Data were taken from *(16)*.

each system with this phase combination the continuous transform closely matches all of the structure factors.

3.4.3. Calculation of Electron Density Profiles

Electron density profiles across the bilayer are calculated from Fourier reconstructions using the X-ray structure factors

$$\rho(x) = (2/d) \sum \exp[i\phi(h)] \, F(h) \cos(2\pi x h/d) \tag{1}$$

where x is the distance from the center of the bilayer, d is the lamellar repeat period, $\phi(h)$ is the phase angle of order h, $F(h)$ is the X-ray structure amplitude for each order h, and the sum is over h.

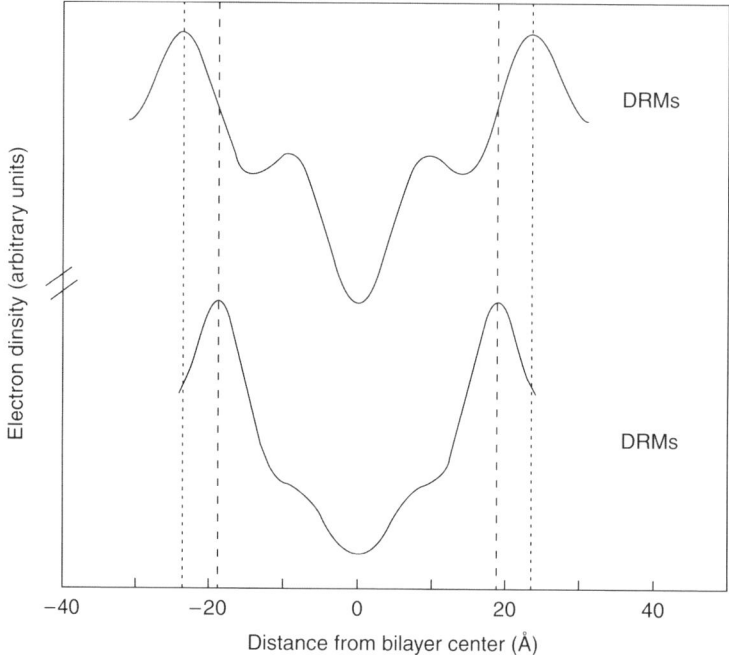

Fig. 2. Electron density profiles at a resolution of $d/2h_{max}$ ~7 Å for DRMs and DSMs obtained from 1:1:1 DOPC:SM:cholesterol bilayers. The vertical dotted and dashed lines are positioned at the headgroup peaks of DRMs and DSMs, respectively. Data were taken from **ref. 16**.

Figure 2 shows electron density profiles across DRM and DSM bilayers. For each profile the electron density peaks (at approx ±23 Å for DRMs and ±19 Å for DSMs) correspond to the position of the high-density phosphate headgroups, the low-density troughs in the geometric center of the profiles correspond to the localization of the low electron density lipid hydrocarbon terminal methyl groups, and the medium density regions between the terminal methyl trough and the headgroup peaks correspond to the localization of the methylene groups in the lipid hydrocarbon chains. The higher density for the hydrocarbon chain regions for DRMs compared with DSMs is because of the increased cholesterol content in the DRMs; the cholesterol rings have a higher electron density than the phospholipid hydrocarbon chains *(24,30)*. The most important and most easily measured feature of these profiles is the distance between the headgroup peaks (d_{pp}) across the bilayer (47 Å for DRMs and 38 Å for DSMs) (*see* **Note 2**).

3.4.4. Determination of Bilayer Thicknesses

The aforementioned values of d_{pp} can be used to estimate the bilayer and hydrocarbon core thicknesses of DRM and DSM bilayers (*see* **Note 3**).

As the lipid headgroup for phosphatidylcholine or SM is about 10 Å in width *(26,31,32)*, the hydrocarbon thickness is estimated as $d_{hc} = d_{pp} - 10$ Å. Thus for the 1:1:1 DOPC:SM:cholesterol system the hydrocarbon thickness is 37 Å for DRMs and 28 Å for DSMs.

The bilayer hydrocarbon thickness is important because, as noted in **Subheading 1.**, it is thought to be a factor in protein–lipid interactions because of the effects of "hydrophobic matching" between the bilayer hydrocarbon thickness and the length of protein transmembrane domains *(2,3,33)*. That is, because of the energetic cost of exposing either hydrocarbon or hydrophobic amino acids to water, the most energetically favorable interaction between a bilayer and a transmembrane protein occurs when the bilayer hydrocarbon thickness matches the length of the protein transmembrane domain *(15)*. This hydrophobic matching effect has been argued to be a mechanism for the sorting of proteins with different transmembrane domain lengths between the Golgi apparatus and the plasma membrane *(1,34)*. Given that the length of an α-helix is approx 1.5 Å per amino acid residue, the estimated 9 Å difference in thickness of DRMs and DSMs would correspond to a difference of about six amino acid residues in matching protein transmembrane domains *(12)*. As a point of reference, this six amino acid difference corresponds to the average difference in transmembrane domain length between resident Golgi proteins and typical plasma membrane proteins *(1,34)*.

4. Notes

1. It has been found *(16)* that small amounts of residual detergent in DSMs can adversely affect the quality of the diffraction patterns and thereby decrease the resolution. Therefore, for analysis of DSMs the author also used bilayers prepared from the observed composition of DSMs as obtained from the chemical analyses described in **Subheading 3.3.** However, detergent contamination was not a problem with DRMs; diffraction patterns from detergent-extracted DRM gave similar quality patterns with the same structure factors as those obtained from bilayers made up of the DRM lipid composition determined by chemical analysis *(16)*.

2. For comparisons of headgroup peak separations between bilayer systems it is important to make sure that each of the profiles is calculated at similar resolution ($d/2h_{max}$) because the exact position of the headgroup peak depends on the resolution. However, as long as similar resolution is used for each profile and the resolution is better than 7–8 Å, quite accurate estimates of the differences in bilayer thickness can be determined *(24,26)*.

3. The values of d_{pp} of course will depend on the composition of the original bilayer, and will vary with the hydrocarbon chain length and the degree of hydrocarbon chain unsaturation of the phosphatidylcholine enriched in the DSM. Thus d_{pp} is somewhat smaller for DSMs enriched in DOPC (which contains two unsaturated chains) than for bilayers containing phosphatidylcholines made up of mixtures of saturated and unsaturated hydrocarbon chains. For example, at the same resolution

d_{pp} was found to be 38 and 41 Å for bilayers of egg phosphatidylcholine and stearoyloleoylphosphatidylcholine, respectively *(26,35)*.

Acknowledgment

This work was supported by National Institute of Health Grant no. GM27278.

References

1. Bretscher, M. S. and Munro, S. (1993) Cholesterol and the Golgi apparatus. *Science* **261,** 1280–1281.
2. Mouritsen, O. G. and Bloom, M. (1984) Mattress model of lipid-protein interactions in membranes. *Biophys. J.* **46,** 141–153.
3. Killian, J. A. (1998) Hydrophobic mismatch between proteins and lipids in membranes. *Biochim. Biophys. Acta* **1376,** 401–416.
4. Dumas, F., Lebrun, M. C., and Tocanne, J. F. (1999) Is the protein/lipid hydrophobic matching principle relevant to membrane organization and functions? *FEBS Lett.* **458,** 271–277.
5. Harroun, T. A., Heller, W. T., Weiss, T. M., Yang, L., and Huang, H. W. (1999) Experimental evidence for hydrophobic matching and membrane-mediated interactions in lipid bilayers containing gramicidin. *Biophys. J.* **76,** 937–945.
6. Monne, M. and von Heijne, G. (2001) Effects of 'hydrophobic mismatch' on the location of transmembrane helices in the ER membrane. *FEBS Lett.* **496,** 96–100.
7. Petrache, H. I., Zuckerman, D. M., Sachs, J. N., Killian, J. A., Koeppe, R. E., and Woolf, T. B. (2002) Hydrophobic matching mechanism investigated by molecular dynamics simulations. *Langmuir* **18,** 1340–1351.
8. Ridder, A. N. J. A., van de Hoef, W., Stam, J., Kuhn, A., de Kruijff, B., and Killian, J. A. (2002) Importance of hydrophobic matching for spontaneous insertion of a single-spanning membrane protein. *Biochemistry* **41,** 4946–4952.
9. Caputo, G. A. and London, E. (2003) Cumulative Effects of Amino Acid Substitutions and Hydrophobic Mismatch upon the Transmembrane Stability and Conformation of Hydrophobic alpha-Helices. *Biochemistry* **42,** 3275–3285.
10. Hwang, T. C., Koeppe, R. E., II, and Andersen, O. S. (2003) Genistein can modulate channel function by a phosphorylation-independent mechanism: importance of hydrophobic mismatch and bilayer mechanics. *Biochemistry* **42,** 13,646–13,658.
11. van Duyl, B. Y., Rijkers, D. T., de Kruijff, B., and Killian, J. A. (2002) Influence of hydrophobic mismatch and palmitoylation on the association of trans-membrane alpha-helical peptides with detergent-resistant membranes. *FEBS Lett.* **523,** 79–84.
12. McIntosh, T. J., Vidal, A., and Simon, S. A. (2003) Sorting of lipids and transmembrane peptides between detergent-soluble bilayers and detergent-resistant rafts. *Biophys. J.* **85,** 1656–1666.
13. Brown, D. A. and London, E. (2000) Structure and function of sphingolipid- and cholesterol-rich membrane rafts. *J. Biol. Chem.* **275,** 17,221–17,224.
14. Simons, K. and Ikonen, E. (2000) How cells handle cholesterol. *Science* **290,** 1721–1726.

15. Lundbaek, J. A., Andersen, O. S., Werge, T., and Nielsen, C. (2003) Cholesterol-Induced Protein Sorting: An Analysis of Energetic Feasibility. *Biophys. J.* **84,** 2080–2089.
16. Gandhavadi, M., Allende, D., Vidal, A., Simon, S. A., and McIntosh, T. J. (2002) Structure, composition, and peptide binding properties of detergent soluble bilayers and detergent resistant rafts. *Biophys. J.* **82,** 1469–1482.
17. Dietrich, C., Bagatolli, L. A., Volovyk, Z. N., et al. (2001) Lipid rafts reconstituted in model membranes. *Biophys. J.* **80,** 1417–1428.
18. Samsonov, A. V., Mihalyov, I., and Cohen, F. S. (2001) Characterization of cholesterol-sphingomyelin domains and their dynamics in bilayer membranes. *Biophys. J.* **81,** 1486–1500.
19. Veatch, S. L. and Keller, S. L. (2003) Separation of liquid phases in giant vesicles of ternary mixtures of phospholipids and cholesterol. *Biophys. J.* **85,** 3074–3083.
20. Ahmed, S. N., Brown, D. A., and London, E. (1997) On the origin of sphingolipid/cholesterol-rich detergent-insoluble cell membranes: physiological concentrations of cholesterol and sphingolipid induce formation of a detergent-insoluble, liquid-ordered lipid phase in model membranes. *Biochemistry* **36,** 10,944–10,953.
21. Chen, P. S., Jr., Toribara, T. Y., and Warner, H. (1956) Microdetermination of phosphorous. *Anal. Chem.* **28,** 1756–1758.
22. Kates, M. (1972) Techniques of Lipidology. *Isolation, Analysis and Identification of Lipids*. North-Holland Publishing Company, Amsterdam.
23. McIntosh, T. J., Magid, A. D., and Simon, S. A. (1987) Steric repulsion between phosphatidylcholine bilayers. *Biochemistry* **26,** 7325–7332.
24. McIntosh, T. J., Magid, A. D., and Simon, S. A. (1989) Cholesterol modifies the short-range repulsive interactions between phosphatidylcholine membranes. *Biochemistry* **28,** 17–25.
25. Parsegian, V. A., Fuller, N., and Rand, R. P. (1979) Measured work of deformation and repulsion of lecithin bilayers. *Proc. Natl. Acad. Sci. USA* **76,** 2750–2754.
26. McIntosh, T. J. and Simon, S. A. (1986) The hydration force and bilayer deformation: a reevaluation. *Biochemistry* **25,** 4058–4066.
27. Shannon, C. E. (1949) Communication in the presence of noise. *Proc. Inst. Radio Eng. NY* **37,** 10–21.
28. McIntosh, T. J. and Holloway, P. W. (1987) Determination of the depth of bromine atoms in bilayers formed from bromolipid probes. *Biochemistry* **26,** 1783–1788.
29. McIntosh, T. J., Simon, S. A., Ellington, J. C., and Porter, N. A. (1984) A new structural model for mixed-chain phosphatidylcholine bilayers. *Biochemistry* **23,** 4038–4044.
30. McIntosh, T. J. (1978) The effect of cholesterol on the structure of phosphatidylcholine bilayers. *Biochim. Biophys. Acta* **513,** 43–58.
31. McIntosh, T. J., Simon, S. A., Needham, D., and Huang, C. -H. (1992) Interbilayer interactions between sphingomyelin and sphingomyelin:cholesterol bilayers. *Biochemistry* **31,** 2020–2024.

32. McIntosh, T. J., Simon, S. A., Needham, D., and Huang, C. -H. (1992) Structure and cohesive properties of sphingomyelin:cholesterol bilayers. *Biochemistry* **31,** 2012–2020.
33. Nielsen, C., Goulian, M., and Andersen, O. S. (1998) Energetics of inclusion-induced bilayer deformations. *Biophys. J.* **74,** 1966–1983.
34. Munro, S. (1995) An investigation of the role of transmembrane domains in Golgi protein retention. *EMBO J.* **14,** 4695–4704.
35. Allende, D., Simon, S. A., and McIntosh, T. J. (2005) Melittin-induced bilayer leakage depends on lipid material properties: evidence for toroidal pores. *Biophys. J.* **88,** 1828–1837.

16

Small-Angle Neutron Scattering to Detect Rafts and Lipid Domains

Jeremy Pencer, Thalia T. Mills, Norbert Kucerka, Mu-Ping Nieh, and John Katsaras

Summary

The detection and characterization of lateral heterogeneities or domains in lipid mixtures has attracted considerable interest, because of the roles that such domains may play in biological function. Studies on both model and cell membranes demonstrate that domains can be formed over a wide range of length scales, as small as nanometers in diameter up to microns. However, although the size and shape of micron-sized domains are readily visualized in freely suspended vesicles, by techniques such as fluorescence microscopy, imaging of nanometer-sized domains has thus far been performed only on substrate-supported membranes (through, e.g., atomic force microscopy), whereas additional evidence for nanodomains has depended on indirect detection (through, e.g., nuclear magnetic resonance or fluorescence resonance energy transfer). Small-angle neutron scattering (SANS) is a technique able to characterize structural features on nanometer length scales and can be used to probe freely suspended membranes. As such, SANS shows promise to characterize nanometer-sized domains in model membranes. The authors have recently demonstrated the efficacy of SANS to detect and characterize nanodomains in freely suspended mixed lipid vesicles.

Key Words: Cholesterol; lipid membranes; membrane domains; rafts; small-angle scattering; unilamellar vesicles.

1. Introduction

The identification of putative functional domains or "rafts" in cell membranes has stimulated considerable study of the lateral organization of both cell and model membranes, in part because of the possible roles of these domains in biological processes such as immune response, synaptic transmission, and viral infection *(1)*. A variety of techniques have been used in order to characterize lateral heterogeneities in both model and cell membranes, including fluorescence microscopy *(2)*, nuclear magnetic resonance *(2,3)*, electron paramagnetic resonance *(4)*, fluorescence correlation spectroscopy *(5)*, fluorescence resonance energy transfer *(6)*, single-molecule tracking *(7)*, atomic force microscopy *(8)*, near-field optical microscopy *(9)*, and small-angle neutron scattering (SANS) *(10)*. Of this variety of techniques, fluorescence microscopy appears to have gained the greatest popularity, whereas small-angle scattering has found the least use in characterizing membrane domains. The popularity of visually based techniques over scattering is likely a consequence of the relative complexity in the analysis and interpretation of scattering data. Nevertheless, despite these complexities, SANS has inherent advantage over visual techniques, in probing nanometer length scales, and yielding ensemble-averaged information. Additionally, SANS uses selective lipid deuteration as a method of contrast enhancement rather than the addition or inclusion of bulky fluorescent dyes or labels, which can perturb or affect the phase behavior observed by techniques such as, for example, fluorescence microscopy *(11)*.

The detection of domains using SANS depends on the use of contrast variation. First, contrast between domains of different composition is enhanced by the selective deuteration of a lipid that is known (or suspected) to preferentially partition into one phase (or type of domain) over the other. Second, the contrast between the vesicle and medium is minimized, through the appropriate ratio of H_2O to D_2O in the buffer medium. This minimum contrast condition occurs when the scattering length density (SLD) of the medium is equal to the mean SLD of the vesicle. When lateral segregation occurs, the domains or heterogeneities in the vesicles will now have SLD that are unequal to the mean SLD, resulting in contrast between these domains and the medium. As will be shown, both the angle dependence of the SANS curve, and its integrated intensity will contain information related to the size and composition of domains present on the vesicle.

2. Materials

2.1. Buffer Media

1. 99% Purity D_2O (Cambridge Scientific, Andover, MA).
2. Ultrapure H_2O.
3. Buffer media as required, with either NaOH or HCl for pH titration, as required.

2.2. Lipid Mixtures

1. Reagent grade chloroform (Sigma-Aldrich, St. Louis, MO).
2. Rotary evaporator (e.g., Buchi, Sigma-Aldrich) or pressurized inert gas (e.g., N_2, He, and Ar).
3. Phospho and/or sphingolipids (Avanti Polar Lipids, Alabaster, AL).
4. Cholesterol or other sterols (Sigma-Aldrich).

2.3. Rapid Solvent Exchange

1. Chloroform (e.g., Sigma-Aldrich).
2. Buffer media (e.g., Hepes, Sigma-Aldrich).
3. Vortexer (e.g., Vortexer Shaker Miniroto 115V, Fisher Scientific, Ottawa,ON).
4. Vacuum pump (e.g., Integrated Speedvac System, Fisher Scientific).

2.4. Extrusion

1. Nuclepore polycarbonate track-etch membranes (pore diameters 200, 100, and 50 nm).
2. Either a hand-held mini-extruder (e.g., Avanti Polar Lipids) or high-pressure extruder (Northern Lipids Inc., Vancouver, BC).
3. Pressurized inert gas (e.g., N_2, He, and Ar).

2.5. Small-Angle Scattering

1. 1-mm Path length quartz cuvets (Hellma USA, Plainview, NY).
2. Circulating water bath temperature control.
3. Multiple-position sample changer.

3. Methods

As discussed, the detection of lateral segregation in model membranes by neutron scattering relies on the labeling, by selective deuteration, of at least one lipid species in the lipid mixture. As will be shown, without deuteration, most lipids and sterols have very similar SLD, making them difficult to distinguish by neutron scattering. Furthermore, the detection of lateral segregation is optimal under contrast matching conditions, i.e., when the SLD of the medium equals the mean SLD of the vesicles. In order to obtain reproducible results, it is essential that unilamellar vesicles (ULVs) be prepared under conditions wherein the lipid components are homogeneously mixed. Should extrusion be performed on laterally segregated mixtures, it is likely that individual vesicles produced will have different proportions of the various lipids present. Homogeneous mixing of the lipid components during extrusion can be achieved by maintaining the lipid film at a temperature above the miscibility transition, wherein miscibility transitions are either known from the literature *(2)* or can be estimated from miscibility transitions for similar mixtures. As a general rule, complete mixing can be achieved by maintaining the mixture at a temperature about 10°C above the gel–fluid transition of the lipid component with the highest melting point.

3.1. Estimation of Contrast Matching Conditions

1. As will be shown, it is advantageous to prepare extruded vesicles at conditions close to their contrast match point. Contrast match conditions correspond to a situation wherein the SLD of the buffer or medium is equal to the mean SLD of the ULV, and can be achieved by varying the D_2O content in a $D_2O:H_2O$ buffer.

Table 1
Neutron Coherent Scattering Lengths for Atoms Commonly Found in Lipid Molecules

Atom	Scattering length (fm)
C	6.6460
H	−3.7390
D	6.671
N	9.36
O	5.803
P	5.13

From **ref. *12***.

2. The SLD (ρ) of a molecule can be calculated by:

$$\rho = (1/V)\sum b_i$$

where b_i is the scattering length of the *i*-th atom in the molecule and V corresponds to the molecular volume. The scattering lengths of all atoms and their isotopes can be found in a number of sources *(12)*. **Table 1** summarizes atomic scattering lengths for atoms commonly found in lipid molecules.

3. The mean vesicle SLD can be calculated similarly, as the sum of the scattering lengths of all the components, divided by the total vesicle volume. The mean vesicle SLD can also be expressed in terms of the sum of the SLD of the various lipids in the vesicle.

$$\bar{\rho} = \frac{\sum_j n_j b_j}{\sum_i n_j V_j} = \frac{\sum_j n_j V_j \rho_j}{\sum_i n_j V_j}$$

where n_j, b_j, V_j, and ρ_j are the molar fraction, scattering length, molecular volume, and SLD of lipids of type *j* making up the vesicle, whereas $\bar{\rho}$ is the mean vesicle SLD. **Table 2** lists neutron scattering lengths (*b*) molecular volumes (*V*) and SLD (ρ) of typical lipids.

3.2. Preparation of Lipid Mixtures by Rapid Solvent Exchange

1. A detailed discussion of the preparation of vesicles by rapid solvent exchange (RSE) can be found in **ref. *15***.
2. Dissolve the various lipids to desired concentrations (e.g., 10 mg/mL) in chloroform.
3. Mix the various lipid solutions in the desired proportions into a small vial.
4. Add the appropriate buffer or water solution to the lipid–chloroform solution.
5. Vortex the sample and pump under vacuum for approx 1 min.

Table 2
Neutron Scattering Lengths, Molecular Volumes, and Corresponding Scattering Length Densities of Common Lipids

Molecule	Chemical formula	b (fm)	V (Å^3)	SLD (fm/Å^3)
DPPC (20°C)	$C_{40}H_{80}NO_8P$	27.63	1144	0.024
DPPC (50°C)	$C_{40}H_{80}NO_8P$	27.63	1232	0.022
dDPPC (20°C)	$C_{40}H_{18}D_{62}NO_8P$	672.99	1144 (est.)	0.588
dDPPC (50°C)	$C_{40}H_{18}D_{62}NO_8P$	672.99	1232 (est.)	0.546
DOPC (30°C)	$C_{44}H_{84}NO_8P$	39.26	1303	0.030
Cholesterol	$C_{27}H_{46}O$	13.25	629	0.021
Water (25°C)	H_2O	−1.68	30	−0.056
Heavy water (25°C)	D_2O	19.15	29.9	0.64

Lipid and cholesterol volumes taken from **refs. *13* and *14***.
Data for H_2O and D_2O are also shown for comparison.
DPPC, 1,2-dipalmitoyl-sn-glycero-3-phosphocholine; dDPPC, 1,2-dipalmitoyl-d62-*sn*-glycero-3-phosphocholine.
For dDPPC, the molecular volumes given are estimated (est.) from corresponding values for DPPC.

3.3. Preparation of Lipid Mixtures by Film Deposition

1. Dissolve the various lipids to specific concentrations (e.g., 10 mg/mL) in chloroform.
2. Mix the various lipid solutions in the correct proportions either in a small vial, or if solvent is to be removed through rotary evaporation, in a round-bottom flask e.g. Sigma-Aldrich.
3. Remove solvent by evaporation, either by flowing a gentle stream of N_2 (or any other inert gas) over the lipid solution or by rotary evaporation.
4. Remove any remaining solvent by vacuum pumping for at least 2 h.
5. Heat lipid film to 60°C. Note that for lipids with melting transition temperatures higher than 50°C, higher temperatures may be required for complete mixing of lipid components.
6. Preheat buffer to 60°C.
7. Add buffer to lipid film and disperse by vortexing, making sure to maintain the lipid dispersion at approx 60°C.

3.4. Lipid Extrusion

1. Preheat extruder to 60°C.
2. Preheat two 250-µL Hamilton syringes (Hamilton Company, Reno, NV) to 60°C.
3. Assemble extruder with 200-nm pore diameter polycarbonate membranes.
4. Pass preheated buffer solution through extruder a minimum of three times.
5. Extrude lipid dispersion at least 11 times through a 200-nm pore diameter polycarbonate membrane. Note that extrusion through a hand-held extruder requires an odd number of passes through the extruder, to avoid cross-contamination with sample that has not been extruded.

6. Repeat **steps 2–4** using the extruded dispersion but now with a 100-nm pore diameter polycarbonate membrane, instead of the 200-nm pore diameter membrane.
7. Using the 100-nm pore diameter extruded dispersion, repeat **steps 2–4** using a 50-nm pore diameter polycarbonate membrane, extruding the dispersion at least 25 times.
8. It is important to make samples with a relatively high lipid:buffer ratio as neutron sources produce relatively weak beams for experiments, compared with X-ray sources. However, high lipid concentrations can make extrusion difficult resulting in higher loss of lipid on the polycarbonate membranes. Moreover, too concentrated a sample will also introduce added complications to the analysis of the SANS data, as the appropriate structure factor is needed to account for the inevitable intervesicle interactions. As a result, extruding lipid samples with concentrations more than 50 mg/mL is not recommended. After extrusion, samples should be diluted to yield final concentrations no higher than approx 20 mg/mL, in order to avoid intervesicle interactions.

3.5. Small-Angle Neutron Scattering

1. Inject the vesicle suspension into a quartz cuvet, or sample cell. Boron-free optical quartz cells are appropriate for this use, as boron is a neutron absorber. It is also very important to avoid introducing air bubbles into the sample, as air bubbles will also scatter neutrons, complicating data analysis.
2. For samples with less than about 80% D_2O content, a 1-mm path length cell is recommended to avoid multiple scattering from hydrogenated components.
3. The neutron wavelength (λ) and sample-to-detector distance used should be appropriately chosen such that the length scales probed span the ranges including the vesicle size and membrane thickness. For 30-nm radius vesicles, with a typical bilayer thickness (e.g., 4–5 nm), the optimum q ($4\pi \sin\theta/\lambda$, where θ is the scattering angle) range for SANS measurements is $0.003 < q < 0.3$ Å^{-1}. For example, the NG3 30-m SANS instrument at the National Institute of Standards and Technology Center for Neutron Research, this q range is achieved using a λ of 8 Å, and sample-to-detector distance of 13, 4, and 1.3 m *(16)*.
4. The sample holder and sample changer should be capable of being preheated or cooled, with the sample (or samples) in place to the desired temperature. Keep in mind that the equilibration of the sample may not occur at the same rate as temperature equilibration of the sample holder.
5. Typically, SANS measurements are performed using a two-dimensional detector utilizing commercial multiwire proportional counters (MWPC) (Model XYP 64 × 64–10, manufactured by (CERCA, Romans, France) of the type developed at the Institute Laue-Langevin (ILL). As vesicles in the scattering volume have no preferred orientation, the signal is radially averaged. Data are corrected for the detector response, empty cell scattering, and detector dark count. Then, the corrected data is put on the absolute scale using calibrated standards or based on the incident neutron flux, as described in **ref. *16***. Most neutron scattering facilities provide their own software that will automatically perform radial averaging and the various corrections to SANS data.

6. Hydrogen atoms have a large incoherent scattering cross-section. In practice, this means that any sample (including the buffer) that contains significant amounts of hydrogen will contribute to the incoherent background (i.e., an essentially flat, or q-independent baseline added to the signal of interest). As small-angle scattering curves typically decay very rapidly with increasing q, this flat background is most significant for data at large values of q. Thus, a good estimate for the incoherent background can be obtained from the value of the scattered intensity, measured at the highest q range of the experimentally obtained data. This value, corresponding to the incoherent background, is then subtracted from the experimental data.

3.6. Contrast Matching

1. The most reliable method for determining the contrast match point of vesicles is through contrast variation. Contrast variation relies on the use of a series of solvents having various proportions of H_2O and D_2O, and consequently, different SLDs.
2. First, prepare a stock ULV solution. The ULV should be made in a solution with an SLD close to or at the estimated contrast match point. This solution will then be used to prepare a series of ULV samples at different contrasts. Note that if the estimated contrast match point is within 20% of either 100% H_2O or D_2O, preparation of ULV in either pure H_2O or D_2O, respectively, is recommended, as, for example, a stock ULV solution in 90% D_2O could not be diluted to produce ULV in 100% D_2O.
3. Divide the ULV suspension into five equal volumes.
4. Prepare five stock solutions of different H_2O/D_2O ratios for dilution of ULV samples. Solutions should be prepared so that their dilution with the ULV samples yields five samples that are well spaced, with respect to their final D_2O (%) concentration. It is recommended that the H_2O/D_2O ratios of the stock solutions are determined such that dilution with the ULV suspensions yield final D_2O (%) concentration of $x \pm 20\%$ D_2O, $x \pm 10\%$ D_2O, and x, where x is the estimated contrast match point, in D_2O (%). If the estimated contrast match point is too close to either the 100% H_2O or D_2O conditions, then other contrasts (i.e., D_2O [%] solutions) may be used, as required.
5. Measure SANS curves for all five samples at 60°C. In order to properly evaluate the contrast match point, it is important that the SANS curves contain both a significant portion of the flat incoherent background (high q data) and low q, wherein the curve is insensitive to internal ULV structural variations. **Figure 1A** shows SANS curves taken for a single sample under various contrast conditions, before subtraction of the incoherent background.
6. Subtract the incoherent background from the SANS curves (as discussed earlier).
7. The zero-angle scattered intensity, $I(0)$, can be estimated from the intensity of the lowest value of q measured.
8. Plot $I(0)$ as a function of either the D_2O fraction or the SLD of the solution, ρ_s. $I(0)$, measured from a suspension of ULV is proportional to the square of the SLD contrast, $I(0) \sim (\bar{\rho} - \rho_s)^2$, where $\bar{\rho}$ is the mean vesicle SLD. Note that ρ_s is a linear function of D_2O (%). Thus, the minimum value of $I(0)$ yields the contrast

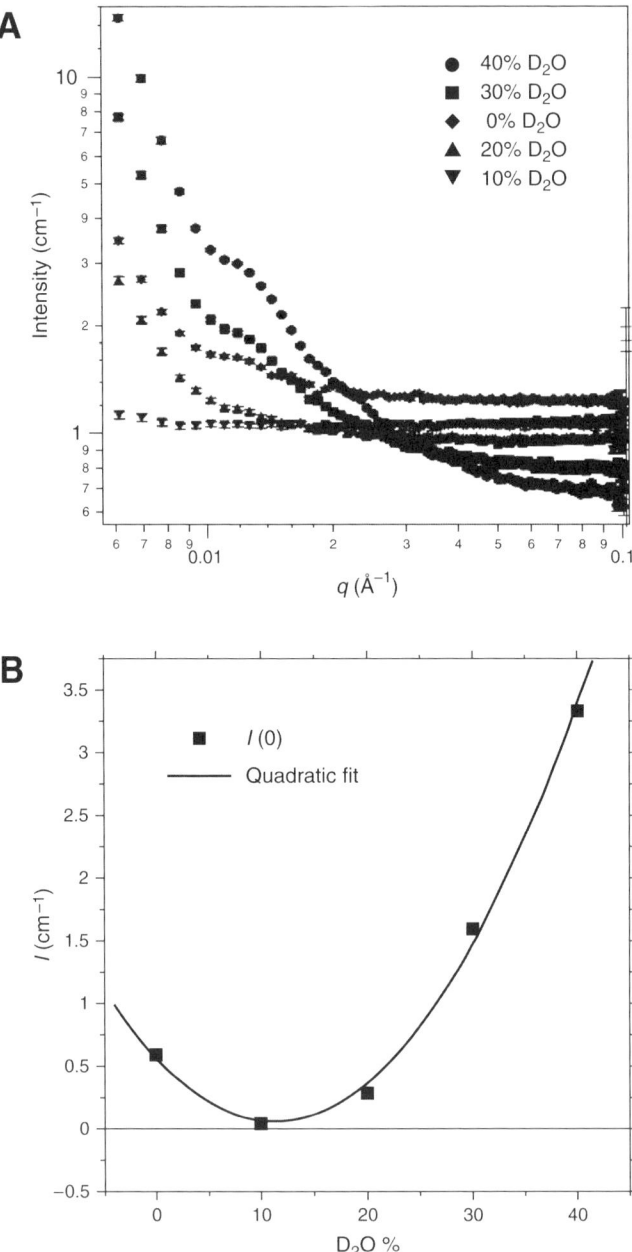

Fig. 1. (**A**) SANS curves plotted for 1:1:1 DOPC:DPPC:cholesterol ULV vs D_2O concentration in the buffer medium. (**B**) Zero-angle scattered intensity, $I(0)$, after subtraction of incoherent background, plotted as a function of D_2O fraction.

match point, or the vesicle's mean SLD, $\bar{\rho}$. A plot of $I(0)$ vs D_2O (%) is shown in **Fig. 1A**, along with a quadratic fit to the data.

3.7. Detection of Lateral Segregation

1. The scattering intensity from laterally heterogeneous vesicles can be expressed as a sum of three contributions:

$$I(q) = \left[F_{ave}(q) + F_{rad}(q) + F_{lat}(q) \right]^2$$

where $F_{ave}(q)$ is the scattering amplitude contribution from the mean vesicle SLD, $\bar{\rho}$, $F_{rad}(q)$ is the scattering amplitude contribution from radial fluctuations in the SLD (i.e., the difference between the acyl chain and headgroup SLD), and $F_{lat}(q)$ is the contribution to the scattered amplitude from lateral fluctuations in SLD (i.e., membrane domains). Typically, $F_{ave}(q)$ is the largest contribution to $I(q)$, except under contrast matching conditions, where $F_{ave}(q) = 0$. Thus, SANS measurements performed under contrast match conditions are optimal for measuring the contributions from $F_{rad}(q)$ and $F_{lat}(q)$.

2. Prepare a ULV sample at the experimentally determined contrast match point.

3. As there will be contributions to $I(q)$ from both F_{rad} and F_{lat}, it is recommended that at least one measurement be performed with ULV at high temperature, above the expected miscibility transition. This measurement will then show the contribution to $I(q)$ from $F_{rad}(q)$ alone, thus providing a baseline for possible temperature-dependent changes, owing to, for example, the onset of lateral segregation on cooling of the sample. **Figure 2B** shows $I(q)$ from laterally homogeneous vesicles at the contrast match point as a function of temperature. As can be observed from the figure, changes to $I(q)$, induced as a function of temperature, corresponding to changes in $F_{rad}(q)$, are small. Conversely, vesicles that become laterally heterogeneous on cooling show a dramatic change in $I(q)$ (shown in **Fig. 2A**), primarily because of the contribution from $F_{lat}(q)$.

4. Quantitative assessment of the degree of lateral segregation can be made by calculating the invariant, $Q = \int q^2 I(q) dq$. As discussed in **ref. 17**, the invariant (Q) can be expressed (in analogy to $I[q]$), as a sum of three contributions:

$$Q = Q_0 + Q_{rad} + Q_{lat}$$

where Q_0 is the invariant contribution from $\bar{\rho}$, Q_{rad} is the contribution owing to the difference between the acyl chain and headgroup SLD, and Q_{lat} is the contribution from membrane domains. Q_0 and Q_{rad} can be measured experimentally or estimated with reasonable accuracy using known component scattering lengths and molecular volumes **(Table 2)** from the following relationships:

$$Q_0 = 2\pi^2 V (\bar{\rho} - \rho_s)^2 \quad \text{and} \quad Q_{rad} = 2\pi^2 \left[V_{ac} (\rho_{ac} - \bar{\rho})^2 + V_h (\rho_h - \bar{\rho})^2 \right]$$

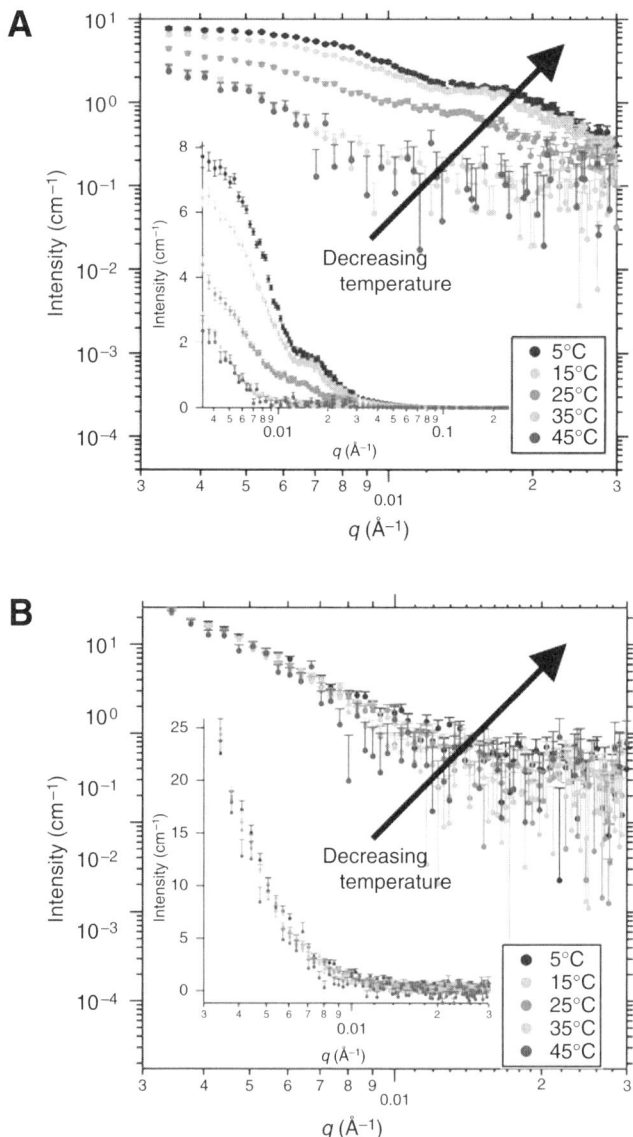

Fig. 2 (**A**) SANS curves plotted for 1:1:1 dDOPC:DPPC:cholesterol ULV vs temperature, under contrast matching conditions. (**B**) Similar curves plotted for 1:1:1 SOPC:dDPPC:cholesterol ULV vs temperature.

where Q_0 and Q_r are given with respect to a single vesicle, $V = V_{ac} + V_h$ is the total vesicle membrane volume, $\bar{\rho}$ is the vesicle mean SLD, ρ_s is the medium SLD, V_{ac} is the total membrane acyl chain region volume, ρ_{ac} is the mean acyl chain SLD, V_h is the total lipid headgroup region volume, and ρ_h is the mean headgroup SLD.

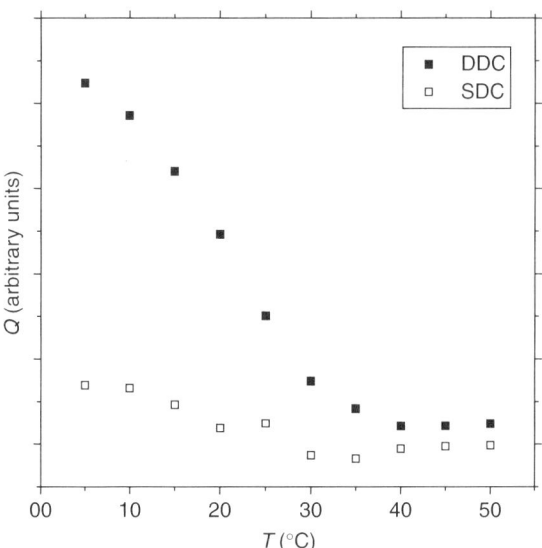

Fig. 3. The scattering invariant (Q) plotted vs temperature for 1:1:1 mixtures of DOPC:dDPPC:cholesterol (DDC) and SOPC:dDPPC:cholesterol (SDC). The data shown is normalized to lipid concentration.

5. Once Q has been calculated, $Q - Q_0 - Q_{rad}$ gives a quantitative indicator of the lateral segregation in the membrane. When the SLD of the lipid headgroups of the various components present in the membrane are the same then,

$$Q_f \sim a_1(1-a_1)(\rho_{1,ac} - \rho_{2,ac})$$

where a_1 is the relative area fraction of regions 1 and $\rho_{1,ac}$ and $\rho_{2,ac}$ are the SLD of the acyl chain portions of regions 1 and 2. Q_f increases as the difference in SLD of the two regions increase and reaches a maximum value when either region 1 or 2 make half of the ULV surface.

6. **Figure 3** shows plots of Q vs temperature evaluated for the scattering curves shown in **Fig. 2A,B**.

3.8. Model-Dependent Analysis

1. If information is known either about the domain composition or relative area, it is possible to model the scattering by calculating the form factor of heterogeneous membranes. For example, in **ref. *10***, form factors for laterally homogeneous and heterogeneous vesicles are calculated by a Monte Carlo method, as described in detail in **refs. *18,19***.
2. In **Fig. 4** experimental data is shown, along with a coarse-grained model, constructed on the basis of known domain compositions and surface areas *(2)*. Using this coarse-grained model in combination with known compositions and area fractions, the authors were able to use SANS to assess the size and number of domains present in ULV.

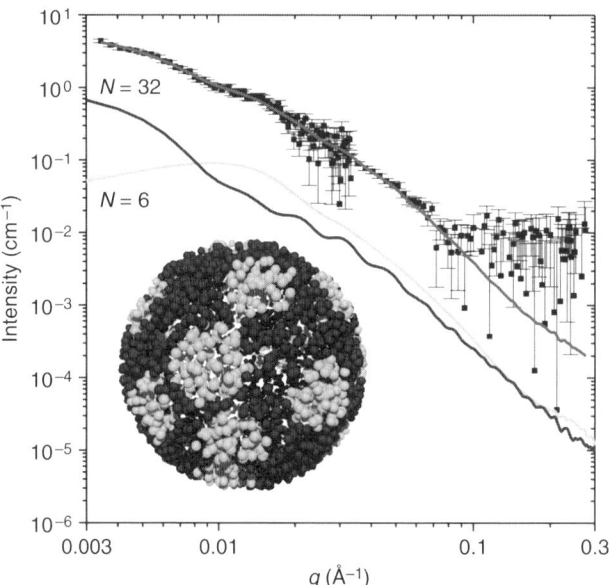

Fig. 4. SANS curve from DDC ULV at 25°C. The fit to the data, using coarse-grained modeling is shown (solid line) as well as curves calculated for ULV containing 6 or 32 domains. The inset shows a representative space-filling model of a vesicle with 32 domains.

4. Notes

1. There is some evidence in the literature that high membrane curvature may influence both the melting transition of single-component vesicles *(20)*, and the miscibility of lipids in multicomponent lipid mixtures *(21)*. Thus, the potential influence of membrane curvature should be considered when interpreting SANS data.
2. The RSE method produces MLV with low lamellarity (1–3 layers) compared with other methods, which produce MLV made up of hundreds of layers. Low lamellarity MLVs are easier to extrude through polycarbonate membrane pores, thus reducing the loss of lipid on the membrane. Thus, extrusion of vesicles prepared by the RSE method may increase yield and make the extrusion process easier, compared with extrusion of MLVs prepared by the hydration of lipid films.
3. The RSE method is also preferable to the film deposition method when using high sterol concentrations, because sterols such as cholesterol, might crystallize in dry lipid films above approx 45 mol% *(22)* or at much lower concentrations in the case of membranes containing polyunsaturated lipids *(23)*.
4. When lipid dispersions are extruded using a hand-held device, care should be taken to exert a slow and steady pressure on the syringe plungers during extrusion. If the lipid dispersion is pushed through the polycarbonate membrane of the extruder too quickly, there is a chance that the membrane will rupture, resulting in incomplete extrusion of the lipid dispersion.

5. For contrast matching, the zero-angle scattered intensity $I(0)$ can be estimated in several ways. As discussed, the intensity measured at the lowest scattering angle can be used to estimate $I(0)$. However, the scattering curves can be fit with the vesicle form factor and $I(0)$ can be determined as a fitting parameter, or the low-angle scattered intensity can be fit with a polynomial function and then extrapolated to obtain $I(0)$.

Acknowledgments

The authors thank G. W. Feigenson, R. M. Epand, D. L. Worcester, and S. Krueger for valuable discussions and advice. This work utilized facilities supported in part by the National Science Foundation under Agreement No. DMR-0454672. The authors acknowledge the support of the National Institute of Standards and Technology, US Department of Commerce, in providing the neutron research facilities used in this work. This work was also performed in part with support from the National Institutes of Health grant no. 1 R01 RR14812 and the Regents of the University of California through contributions from the Cold Neutrons for Biology and Technology research partnership.

References

1. Edidin, M. (2003) The state of lipid rafts: from model membranes to cells. *Annu. Rev. Biophys. Biomol. Struct.* **32,** 257–283.
2. Veatch, S. L., Polozov, I. V., Gawrisch, K., and Keller, S. L. (2004) Liquid domains in vesicles investigated by NMR and fluorescence microscopy. *Biophys. J.* **86,** 2910–2922.
3. Polozov, I. V. and Gawrisch, K. (2004) Domains in Binary SOPC/POPE Lipid Mixtures Studied by Pulsed Field Gradient 1H MAS NMR. *Biophys. J.* **87,** 1741–1751.
4. Chiang, Y.-W., Shimoyama, Y., Feigenson, G. W., and Freed, J. H. (2004) Dynamic Molecular Structure of DPPC-DLPC-Cholesterol Ternary Lipid System by Spin-Label Electron Spin Resonance. *Biophys. J.* **87,** 2483–2496.
5. Kahya, N., Brown, D. A., and Schwille, P. (2005) Raft partitioning and dynamic behavior of human placental alkaline phosphatase in giant unilamellar vesicles. *Biochemistry* **44,** 7479–7489.
6. Silvius, J. R. (2003) Fluorescence energy transfer reveals microdomain formation at physiological temperatures in lipid mixtures modeling the outer leaflet of the plasma membrane. *Biophys. J.* **85,** 1034–1045.
7. Kusumi, A., Nakada, C., Ritchie, K., et al. (2005) Paradigm shift of the plasma membrane concept from the two-dimensional continuum fluid to the partitioned fluid: high-speed single-molecule tracking of membrane molecules. *Annu. Rev. Biophys. Biomol. Struct.* **34,** 351–378.
8. Yuan, C., Furlong, J., Burgos, P., and Johnston, L. J. (2002) The Size of Lipid Rafts: An Atomic Force Microscopy Study of Ganglioside GM1 Domains in Sphingomyelin/DOPC/Cholesterol Membranes. *Biophys. J.* **82,** 2526–2535.

9. Ianoul, A., Burgos, P., Lu, Z., Taylor, R. S., and Johnston, L. J. (2003) Phase Separation and Interleaflet Coupling in Supported Phospholipid Bilayers Visualized by Near-Field Scanning Optical Microscopy in Aqueous Solution. *Langmuir* **19,** 9246–9254.
10. Pencer, J., Mills, T., Anghel, V., Krueger, S., Epand, R. M., and Katsaras, J. (2005) Detection of submicron-sized raft-like domains in membranes by small-angle neutron scattering. *Eur. Phys. J. E* **18,** 447–458.
11. Veatch, S. L., Leung, S. S., Hancock, R. E., and Thewalt, J. L. (2007) Fluorescent probes alter miscibility phase boundaries in ternary vesicles. *J. Chem. Phys, B.* **111,** 502–504.
12. Sears, V. F. (1992) Neutron Scattering Lengths and Cross Sections. *Neutron News* **3,** 26–37.
13. Nagle, J. F. and Tristram-Nagle, S. (2000) Structure of Lipid Bilayers. *Biochim. Biophys. Acta* **1469,** 159–195.
14. Knoll, W., Schmidt, G., Ibel, K., and Sackmann, E. (1985) Small-Angle Neutron Scattering Study of Lateral Phase Separation in Dimyristoylphosphatidylcholine-Cholesterol Mixed Membranes. *Biochemistry* **24,** 5240–5246.
15. Buboltz, J. T. and Feigenson, G. W. (1999) A novel strategy for the preparation of liposomes: rapid solvent exchange. *Biochim. Biophys. Acta* **1417,** 232–245.
16. Glinka, C., Barker, J., Hammouda, B., Krueger, S., Moyer, J. J., and Orts, W. J. (1998) The 30 m Small-Angle Neutron Scattering Instruments at the National Institute of Standards and Technology. *J. Appl. Cryst.* **31,** 430–441.
17. Pencer, J., Anghel, V. N. P., Kučerka, N., and Katsaras, J. (2006) Scattering from Laterally Heterogeneous Vesicles I: Model Independent Analysis *J. Appl. Cryst.* **39**.
18. Henderson, S. J. (1996) Monte Carlo Modeling of Small-Angle Scattering Data from Non-Interacting Homogeneous and Heterogeneous Particles in Solution. *Biophys. J.* **70,** 1618–1627.
19. Zhou, J., Deyhima, A., Krueger, S., and Greguricka, S. K. (2005) LORES: Low resolution shape program for the calculation of small angle scattering profiles for biological macromolecules in solution. *Comp. Phys. Commun.* **170,** 186–204.
20. Gruenewald, B., Stankowski, S., and Blume A. (1979) Curvature influence on the cooperativity and the phase transition enthalpy of lecithin vesicles. *FEBS Lett.* **102,** 227–229.
21. Brumm, T., Jørgensen, K., Mouritsen, O. G., and Bayerl, T. M. (1996) The effect of increasing membrane curvature on the phase transition and mixing behavior of a dimyristoyl-sn-glycero-3-phosphatidylcholine/distearoyl-sn-glycero-3-phosphatidylcholine lipid mixture as studied by fourier transform infrared spectroscopy and differential scanning calorimetry. *Biophys. J.* **70,** 1373–1379.
22. Huang, J., Buboltz, J. T., and Feigenson, G. W. (1999) Maximum solubility of cholesterol in phosphatidylcholine and phosphatidylethanolamine bilayers. *Biochim. Biophys. Acta* **1417,** 89–100.
23. Shaikh, S. R., Cherezov, V., Caffrey, M., et al. (2006) *J. Am. Chem. Soc.* **128,** 5375–5383.

17

Exploring Membrane Domains Using Native Membrane Sheets and Transmission Electron Microscopy

Bridget S. Wilson, Janet R. Pfeiffer, Mary Ann Raymond-Stintz, Diane Lidke, Nicholas Andrews, Jun Zhang, Wenxia Yin, Stanly Steinberg, and Janet M. Oliver

Summary

The flow of information in cells requires the constant remodeling of cell signaling and trafficking networks. To observe the remodeling events associated with activation of receptors on the cell surface, the authors have generated and analyzed high-resolution topographical maps of colloidal gold nanoprobes (3–10 nm) marking receptors, signaling proteins, and lipids in native membranes. The technology involves sandwiching of cells between glass cover slips and electron microscopy (EM) grids, followed by ripping. Membrane sheets on EM grids are fixed, labeled with functionalized nanoprobes, and imaged by transmission electron microscopy. Probe coordinates are extracted from digitized images and the distributions of the probes are analyzed with respect to each other and to membrane features like clathrin-coated pits, caveolae, and the cortical cytoskeleton.

Key Words: Immunoelectron microscopy; membrane sheets; rip-flips; microdomains; lipid rafts.

1. Introduction

Plasma membrane microdomains are sites for the interactions of proteins and lipids that mediate signaling, receptor internalization, and exocytosis. In this chapter, protocols for using transmission electron microscopy (TEM) are described to document the organization of gold-labeled membrane proteins and lipids in native membrane sheets. With these methods, the authors have followed the distribution and fate of a variety of cell surface receptors and their associated signaling molecules, including the high-affinity immunoglobulin (Ig)E receptor (FcεRI) *(1–3)*, the B-cell receptor *(4)*, the G-protein-coupled formyl peptide receptor *(5)*, and members of the ErbB family of growth factor receptors *(6)*. Hancock's group has led efforts to apply similar methods to the study of Ras proteins, specifically H-Ras and K-Ras isoforms *(7,8)* and the technique has recently been applied to evaluate the clustering behavior of viral-coat proteins *(9)*. The method is adapted from that of Sanan and Anderson *(10)* and is supported by spatial stastistics applications that improve quantitation and interpretation of digitized point coordinates.

2. Materials

2.1. Solutions (Prepare All Working Solutions on Day of Use)

1. HEPES buffer (50 mL): 25 mM HEPES, 25 mM KCl, and 2.5 mM magnesium acetate, pH 7.2.
2. Poly-L-lysine (stock solution): prepare 8 mg/mL poly-L-lysine hydrobromide (Sigma P-1524) in sterile water. Store in 50-µL aliquots at −20°C.
3. Poly-L-lysine (working solution): 50 µL stock poly-L-lysine in 950 µL dH$_2$O.
4. 2% PFA (4 mL): 0.5 mL 16% paraformaldehyde prepared in water, then added to 3.5 mL HEPES buffer or phosphate-buffered saline (PBS).
5. 2% Glutaraldehyde (10 mL): 2 mL 16% glutaraldehyde and 8 mL PBS6. 0.2 M Sodium cocadylate buffer (NaCacodylate-3H$_2$O) (0.5 L): 21.4 g (under hood) and bring up to 500 mL with double-distilled water. Bring to pH 7.2. Store at 4°C.
6. 1% Osmium (2 mL): 0.5 mL 4% osmium (Electron Microscopy Sciences [Hatfield, PA]), 1 mL 0.2 M cacodylate buffer, and 0.5 mL H$_2$O.
7. 1% Tannic acid (25 mL): 0.25 g tannic acid, 25 mL dH$_2$O, or 0.1 M NaCacodylate buffer. Shake well until acid goes into solution. Filter with a 0.2-µm Millipore filter disk.
8. 1% Uranyl acetate for rinse: dilute 2% uranyl acetate 1:1 in H$_2$O. Note that 2% uranyl acetate is a saturated solution and has a tendency to fall out of solution. Gentle heating is needed when making stock solution. Once made, keep in 60-mL syringe wrapped with foil. Use new 0.2-µm Millipore filter and filter small amount for staining, add to equal volume H$_2$O.

2.2. Tools and Setup

1. 15-mm Round glass cover slips (Electron Microscopy Sciences; cat. no. 72195-15).
2. Hexagonal, nickel grids no. 26431, 200 mesh (Ernest Fuller), formvar, and carbon coated.
3. Filter paper (Whatman's ashless, no. 42, 11 cm circles).
4. Cellulose acetate disks (Osmonics Inc.; cat. no. A02SP02500).
5. 37°C Plate warmer with lid, lining sides with damp paper towels to establish moist chamber.
6. Fume hood.
7. Two pair curved forceps (Electron Microscopy Sciences; cat. no. 72801-D).
8. Aspirating apparatus.
9. Divided Petri dishes.

3. Methods
3.1. Specimen Preparation

1. Prepare cells by plating onto sterile, clean glass cover slips. Culture 1–4 d, to 70–80% confluency.
2. On the day of the experiment, glow discharge grids and treat with poly-L-lysine.
   a. Grids (formvar and carbon coated) are glow discharged for 30 s.
   b. An aliquot of poly-L-lysine is thawed and diluted to 0.4 mg/mL. Grids are floated on poly-L-lysine drops for 30 min at room temperature in a covered 2-mL Petri dish lined with parafilm, with cell contact side down.
3. Transfer the grids facedown to distilled water for 10 s.
4. Using a sheet of filter paper, wick off excess water and place onto filter paper, treated side up.
5. At this stage, one has to have decided if one will "activate" the cells or subject the cells to other pretreatments (cholesterol depletion, pharmacological inhibitors, and so on). This step is not necessary if one is looking for steady-state distributions of membrane constituents. If one intends to activate:
   a. Remove cover slips with cells from culture dish and wick off excess medium. Wipe bottom of cover slip with Kimwipe and place on surface of plate warmer set at 37°C. Move quickly to keep cells from drying out. Add 100 µL of medium +/–ligand and incubate the cover slips for a defined incubation period (1–10 min); be sure to close the cover to keep chamber moist.
   b. During incubation, prepare cellulose acetate disks on glass or plastic surface. First place a drop of HEPES buffer on the surface, apply the disk, then aspirate off excess medium so that the disk does not move when tapped with the aspirator. Place two to three grids with the poly-L-lysine side up on the acetate disk.
   c. One can stop the reaction at the end of incubation by transferring to cold buffer. However, a light prefix at this stage is recommended, by adding 100 µL of 0.5% paraformaldehyde to the cover slip. The cover slip is then held in the moist chamber for 7 min at room temperature, followed by a rinse in HEPES buffer. Note that this light prefix has the distinct advantage that it helps to preserve the cortical cytoskeleton attached to the inner membrane, without rendering the cells too "stiff."

   If one intends to label with antibody and gold reagents to the *outside* of the cell, before making sheets, one can do so at this stage. In this case, it is advisable to substitute a stronger prefix (such as 2% paraformaldehyde) and to extend the incubation time (up to 30 min). Prepare 2% PFA solution in the slightly hypotonic HEPES buffer to facilitate a small amount of cell swelling and improved recovery of sheets *(11)*. Note that the use of stronger fix at this stage is important to prevent antibody-induced cluster artifacts during the exterior labeling but it tends to lower the yield and quality of sheets recovered (*see* **Notes 1–6**).
6. Remove excess buffer by the wicking method. Turn the cover slip so that cells face the grids and put the cover slip over grids. Place a cork or other stopper on top and

press for 1–20 s depending on cell type and confluency. The time/firmness of this step is established by trial and error.
7. Gently remove the stopper. Turn the cover slip over and remove grids by sliding to the edge of the cover slip. Transfer grids facedown onto droplets of 2% PFA. Hold for 10–30 min at room temperature.
8. Transfer grids to PBS in quadrant Petri dishes. As cells are now fixed, all further steps are performed at room temperature.
9. To label the cytoplasmic face of the sheet (*see* **Note 7**) with gold-conjugated probes with the traditional two-step protocol, prepare 100 µL droplets of primary antibody solutions. Keep things organized by using divided Petri dishes or labeled sheets of parafilm. Aim for saturating concentrations of the primary Ab (2–10 µg/mL in 0.1% bovine serum albumin [BSA]/PBS). If feasible, identical grids should be incubated with isotype matched, irrelevant IgG as a negative control. Multiple grids (two or three) can be placed facedown on each droplet of antibody solution for labeling. Best results are obtained with monoclonal antibodies or affinity purified polyclonal antibodies; labeling specificity should be confirmed by Western blotting or other immunochemical techniques. Incubate 30–60 min in moist chamber.
10. Rinse three times with PBS for 5 min.
11. Set up 100 µL droplets of secondary antibody conjugated to 5- or 10-nm colloidal gold particles in 0.1% BSA/PBS. Use the recommended dilution from manufacturer (usually 1:20). Place grids facedown on droplets and incubate for 30 min in the moist chamber.
12. Rinse 3 × 5 min in PBS, then postfix by incubating grids facedown on droplets of 2% glutaraldehyde for 10 min, again in the moist chamber.
13. Transfer to PBS; can hold overnight at this step if necessary.
14. Proceed with staining, using the droplet method. Note that uranyl acetate is radioactive and needs to be discarded in radioactive waste containers.
    a. Working in the hood, place small squares of parafilm on glass plate and secure to plate by rubbing gently. Add 100 µL droplets of 1% osmium to squares, transfer the grids facedown, and incubate for 10 min in the hood.
    b. Transfer the grids facedown to Petri dishes containing 0.1 *M* NaCacodylate. Rinse once for 5 min. Note that **steps 14a** and **b** can be omitted if osmium staining obscures gold label in membrane features such as coated pits.
    c. Transfer grids to a Petri dish containing H_2O. Rinse 2 × 5 min.
    d. Make new parafilm squares on glass and add droplets of freshly filtered tannic acid to each square. Transfer the grids facedown to these droplets and incubate 10 min.
    e. Transfer grids to a Petri dish containing ddH_2O. Rinse 3 × 5 min.
    f. Make new parafilm squares for filtered uranyl acetate. Add 100 µL droplets to each square. Transfer the grids facedown to the droplets and incubate 10 min.
15. Transfer the grids to ddH_2O, rinsing 2 × 5 min each. Wick off excess H_2O from the back of the grid (do not touch face) and air-dry. Store face-up in a labeled container lined with Whatman paper (Whatmanplc, Middlesex, UK).

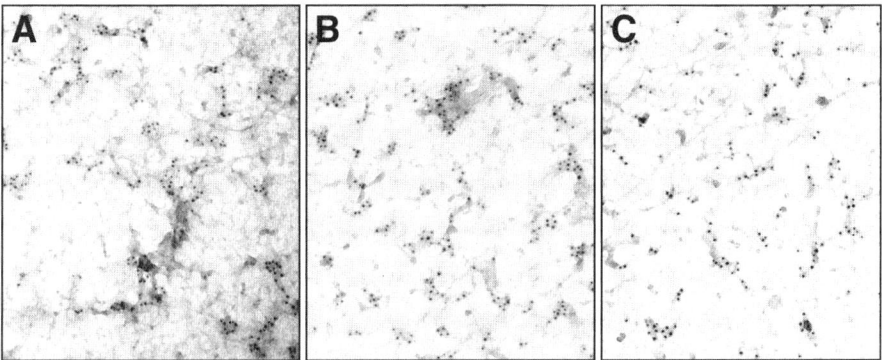

Fig. 1. The growth factor receptor, ErbB2, is clustered on the membranes of serum-starved SKBR3 breast cancer cells. Cells were prefixed with 0.5% PFA, then sheets were prepared on EM grids by the ripping technique. Sheets were fixed with 2% PFA for 10 min (**A**), 2% PFA for 30 min (**B**), and 4% PFA plus 0.1% glutaraldehyde for 10 min (**C**). After rinsing, sheets were labeled with antibodies to the cytoplasmic tail of ErbB2, followed by secondary antibodies conjugated to 5 nm gold.

3.2. Transmission Electron Microscopy

1. View specimens with TEM. If the TEM is equipped with high-resolution digital camera, acquire digital images. Alternatively, acquire photographic negatives and convert images to digital form using a high-resolution scanner (2000 dpi). **Figure 1** shows examples of membranes labeled from the inside with antibodies to the Erb2 cytoplasmic tail. The presence of small Erb2 clusters is typical of many membrane proteins observed on native sheets.
2. Fluorescence recovery after photobleaching (FRAP) can also be preformed on labeled membrane sheets. **Figure 2** shows the results of FRAP experiments performed on membrane sheets prepared from rat basophilic leukemia (RBL) cells whose FcεRI were primed with monovalent, Alexa 488-conjugated IgE. Sheets were prepared on glass cover slips, fixed for 10–30 min and placed on the inverted stage of a Zeiss LSM 510 microscope equipped with a 488-nm laser. A representative result is shown in **Figs. 2A–D**, wherein the sheet was fixed for 10 min with 2% PFA. The prebleach image is shown in (**A**), followed by the first image immediately postbleach of a stripe across the sheet (**B**). After 30 min (**C**), there is no recovery of the bleached area in the fixed membrane sheet. After 70 min, there is some generalized loss of fluorescence over the entire sheet because of repeated illumination but the bleached area still fails to recover (**D**). Plots in **Fig. 2E** show that four different fixation conditions (2% PFA for 10 min; 2% PFA for 30 min; 4% PFA for 10 min; and 4% PFA, 0.1% glutaraldehyde for 10 min) are equally effective at blocking recovery from photobleaching.
3. Single particle tracking experiments provide further support for the effectiveness of 2% PFA in stabilizing membrane proteins. In **Fig. 3**, RBL cells were labeled with

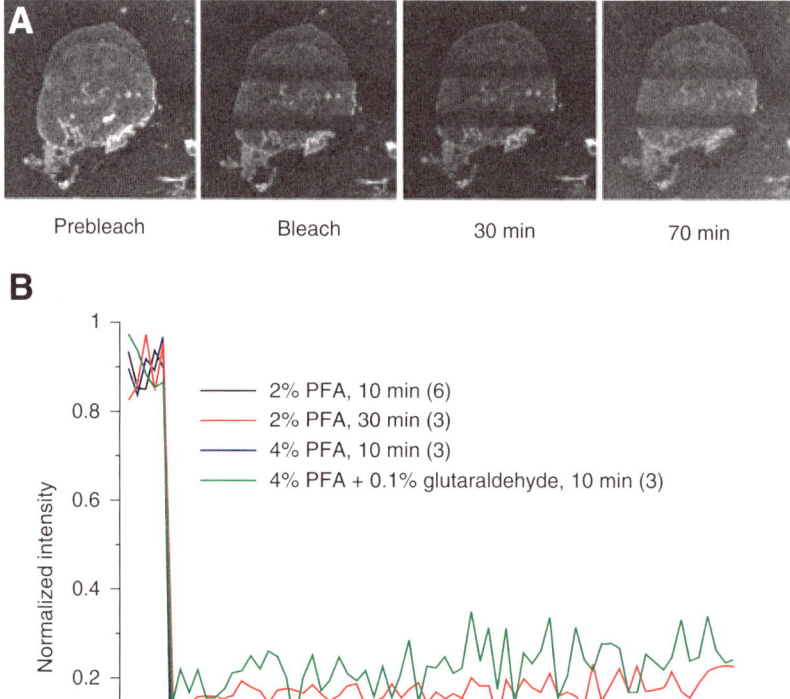

Fig. 2. Lack of photobleaching recovery in fixed membrane sheets. (**A**) Sheets were prepared from RBL cells labeled with IgE-Alexa 488. After ripping, the sample was fixed with 2% PFA for 10 min at room temperature. Rectangular regions were photobleached after frame 5 and the fluorescence intensity in that region monitored over time. (**B**) Sheets were prepared from IgE-Alexa 488-labeled cells and fixed under four different protocols (box). Plots summarize the fluoresence values in circular areas before bleaching and over 70 s of recovery after bleaching. Values are averaged for two to five experiments per condition and indicate a lack of recovery under all fixation protocols.

IgE–biotin conjugated to streptavidin-quantum dots (QDs). Careful titration of conjugation conditions, and subsequent biochemical analysis, confirmed that each IgE is monovalent and conjugated to a single QD. When used to prime FcεRI, IgE–QDs permit real-time tracking of individual receptors by fluorescence microscopy. Time series of IgE–QDs were collected at a frame rate of 33 Hz for 1000 frames using a sensitive emCCD camera (Andor iXon, Belfast, Northern Ireland). By plotting the

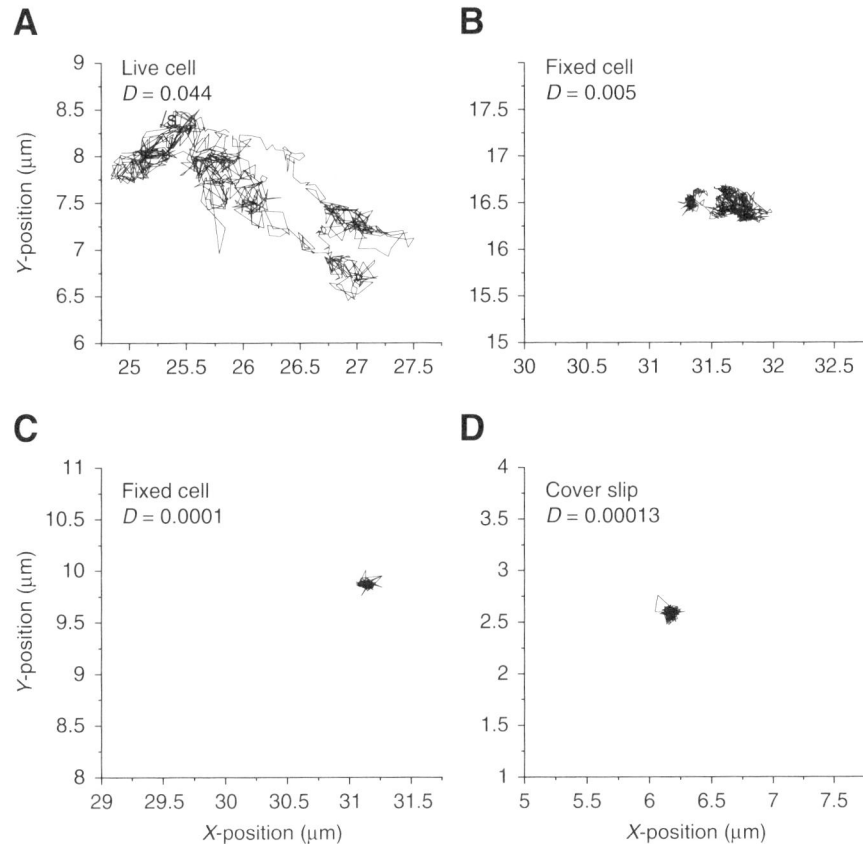

Fig. 3. Representative trajectories for monomeric IgE-QDs bound to IgE receptors on the surface of live (**A**) or fixed (**B,C**) RBL-2H3 cells. Cells in (**B,C**) were fixed for 30 min with 2% paraformaldehyde. The plot in (**D**) shows the tracking of a QD bound to the glass cover slip and serves as a control for signal-to-noise in the system.

mean-square displacement of each IgE–QD and fitting the linear portion, the diffusion coefficients (D) for single particles can be calculated. A typical trajectory of an individual IgE–QD-receptor complex on the surface of live RBL cells is shown in **Fig. 3A**. IgE–QDs on live cells had an average D of 0.043 ± 0.037 µm^2/s. In comparison, individual receptors on the surface of PFA fixed cells (**Fig. 3B,C**) are essentially immobile with an average D of 0.003 ± 0.008 µm^2/s, with less than 5% (6 out of 126) of spots tracked, maintaining a D similar to live cells ($D > 0.01$ µm^2/s). This low-diffusion coefficient is similar to QDs bound to glass (**Fig. 3D**), which provide an internal control for signal-to-noise and vibrational effects under the same measurement conditions.

4. Membrane sheets provide an unprecedented view of the topography of signaling. However, computational methods are essential to extract spatial coordinates of probes, to analyze and statistically validate the clustering and coclustering of these probes

and to integrate results between experiments in order to establish the relative spatial distributions of multiple different probes (*see* **Notes 8–12** for details and examples).

4. Notes

1. It is often desirable to label two or three membrane proteins on the same sheets. Obviously, care must be taken to use primary and secondary Abs from different species, to match the specificity of the gold-conjugated second antibodies to the species of the first, and to use gold particles of different sizes in double- and triple-labeling experiments. Because primary antibodies are raised in only a limited number of species, it is rarely possible to label more than two or three different proteins in one experiment.

2. Electron microscopy is, of course, an art as well as a science. In this technique, part of the art lies in applying just the right amount of pressure to yield a substantial harvest of clean membranes from a particular cell type. In addition, the thickness of the formvar coating the grid, the coverage with poly-L-lysine, the degree of wetness of both the cover slip and the cellulose acetate disk, and the speed of the removal of the grid with attached cell membranes from the cover slip, all affect the quality of the preparations. Even with practice, it still requires patience at the microscope to find good sheets.

3. The subject of fixation, and its application to the rigorous estimation of membrane raft and domain size, has been controversial. A thoughtful review has recently addressed both fixation chemistry and related literature *(12)*. Herein, results of a series of experiments that permit one to make an educated choice of the fixation conditions, balancing the need to maximize antigenicity for immunogold-labeling protocols with the need to minimize opportunities for lateral movement of membrane constituents during labeling procedures.

    The previous work used relatively short periods (7–10 min) of fixation with 2% PFA before labeling. The remarkable changes in FcεRI distributions following antigen treatment, accompanied by identification of distinct patterns for coclustering of receptors with signaling partners like Syk, Cbl, and others, and of receptor segregation from components of the Linker for activation of t-cells (LAT)-scaffolded secondary signaling domain, demonstrated the power and reproducibility of this approach *(1,2)*. Glycosyl phosphaticlylinsositol (GPI)-anchored Thy-1 molecules were shown to be distributed on fixed RBL cells in singlets and small clusters of up to four to five gold particles under saturating labeling conditions *(13)*. Using homo fluorescence resonance energy transfer (FRET) methods, Mayor and colleagues *(14)* estimated that perhaps 40% of GPI-anchored proteins on the surface of chinese homster ovary (CHO) cells are present in clusters of up to four. Remarkably, it was shown that another putative "lipid-raft" marker, GM1 ganglioside, is distributed essentially at random even when paraformaldehyde-fixed cells are labeled with multivalent reagents (pentameric cholera toxin–biotin plus avidin–gold) *(13)* (*see* **Fig. 7A** below). These results imply that, although gangliosides are not directly crosslinked by paraformaldehyde, their movement is severely restricted in even lightly fixed membranes—presumably by immobilized corrals or islands of membrane proteins and the associated cytoskeleton.

Figure 1A–C compares the observed cluster size for ErbB2, a member of the EGFR family of growth factor receptors, labeled with antibodies applied to the cytoplasmic face of membrane sheets that were prepared under different fixation protocols from SKBR3 breast cancer cells. In these experiments, cells were prefixed with 0.5% PFA before ripping, and then sheets were fixed for 10 min with 2% PFA **(Fig. 1A)**, for 30 min with 2% PFA **(Fig. 1B)**, or for 15 min with 4% PFA, 0.1% glutaraldehyde **(Fig. 1C)**. The average appearance of receptor clusters is similar under all of these conditions. Taken together with the FRAP and single-particle results **(Figs. 2** and **3)**, these results indicate the reliability of 2% PFA as a routine fixation protocol for the membrane sheet method.

4. The size of the gold particles matters. Most modern TEMs can reliably resolve gold particles as small as 2–3 nm in size. Practically speaking, most laboratories are limited by the gold reagents that are available commercially, with the broadest range of secondary reagents available in the range of 5–15 nm. Quality control, with respect to variability of gold particle diameter in a given preparation, can limit the usefulness of certain commercial preparations. This is particularly important when applied to double-labeling protocols, wherein the two sizes of gold must be unequivocally distinct. For experiments wherein a single species is labeled, the authors generally rely on 5-nm gold reagents. Compared with the estimated size of the IgG itself (around 15 nm), the probe is sufficiently small and labeling efficiency is generally very good. The 0.8- and 1.4-nm nanoprobe particles have been tested, but their use on membrane sheets was abandoned because of the need for silver enhancement. For double-labeling protocols, typically one species is labeled with 5 nm and the second species with 10 nm. It is important to note that labeling efficiency is always markedly poorer with the larger diameter probe, which is explained by steric hindrance and other factors. Typically the experiment is repeated, *reversing the size of the probes*, to obtain an accurate profile of cluster size and coclustering status. If local resources are available, an alternative is to make two preparations of colloidal gold in ones' own laboratory that can be resolved from each other in the smaller range useful for routine TEM (e.g., 2–3 and 5–6 nm). Samples of each preparation must be examined by TEM for lack of aggregates and for uniformity; this analysis must be performed both before and after conjugation to antibodies.

5. Preparation of gold probes using recombinant proteins. Gold probes can also be created using recombinant proteins capable of binding unique features in the plasma membrane. The authors have recently reported the use of gold conjugated to recombinant, monomeric perfringolysin O (PFO) as a probe for localizing cholesterol on the inner face of the plasma membrane *(15)*. The recombinant PFO was engineered in the laboratory of Arthur Johnson (Texas A & M University), then produced in *Escherichia coli* and purified in the laboratory. Routinely, PFO is adsorbed onto 5 nm colloidal gold (BBInternational, Cardiff, UK) using the manufacturer's instructions. It is also possible to conjugate ligands to gold particles, permitting both activation and tracking of receptors simultaneously. The biggest challenge is to titrate the conjugation conditions such that no more than one probe

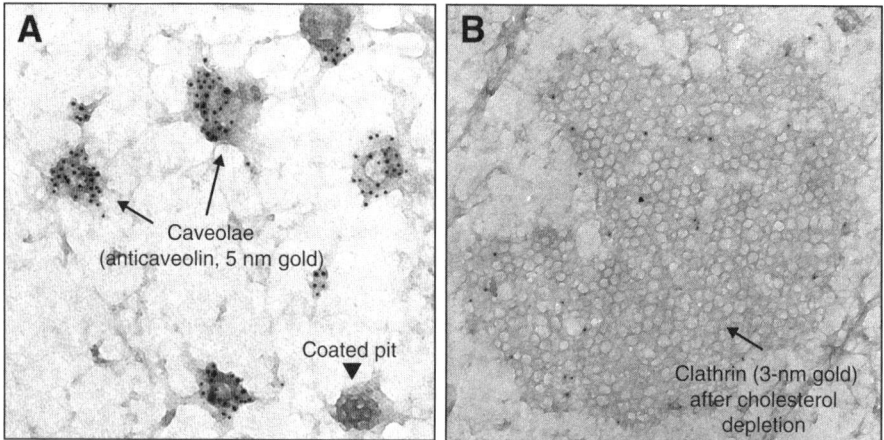

Fig. 4. Caveolae, clathrin-coated pits, and flat clathrin arrays are typical features of good membrane sheet preparations. **(A)** Sheets prepared from NIH-3T3 cells were labeled for caveolin (5 nm gold reagents, arrows) as a marker for caveolae. A coated pit is marked with an arrowhead. **(B)** NIH-3T3 cells were pretreated for 1 h with methyl-β-cyclodextrin to deplete membrane cholesterol, followed by preparation of sheets and immunogold labeling for clathrin. The flat clathrin array shown here is typical of cholesterol-depleted cells.

> molecule will be attached to each gold particle, unless (as in the case of the IgE receptor) the ligand is physiologically polyvalent.
6. Key structural features can be recognized in native membrane sheets. Successful membrane sheets will contain obvious features of the plasma membrane, such as coated pits. Sheets prepared from well-differentiated epithelial and endothelial cells will also be decorated with caveolae and caveolae-like structures. Examples of these are shown in **Fig. 4A**, wherein sheets were prepared from NIH-3T3 fibroblasts and labeled with monoclonal anticaveolin antibodies followed by antimouse secondary antibodies conjugated to 5 nm gold. Caveolae are abundant in the preparation, as indicated by dense anticaveolin labeling (arrows). For comparison, an unlabeled coated pit is observed in the lower right of the image (arrowhead). Membrane sheets often contain a mix of coated pits, recognizable by the distinctive "honey-comb" cage around the budding vesicle, as well as flat clathrin arrays with diameters from 60 to 80 nm. If the cells have been pretreated with cholesterol-depleting reagents, such as the methyl-β-cyclodextrin-treated fibroblast in **Fig. 4B**, flat clathrin arrays become dominant features of the sheets. Caveolae also flatten in severely cholesterol-depleted cells, although caveolin-labeling typically persists in clusters on flat membrane (not shown).
7. With practice, membrane sheets often retain important elements of the cortical cytoskeleton. The authors have documented the distribution of both filamentous actin (using commercial sources of biotinylated phalloidin, followed by avidin–gold) and

Exploring Membrane Domains 255

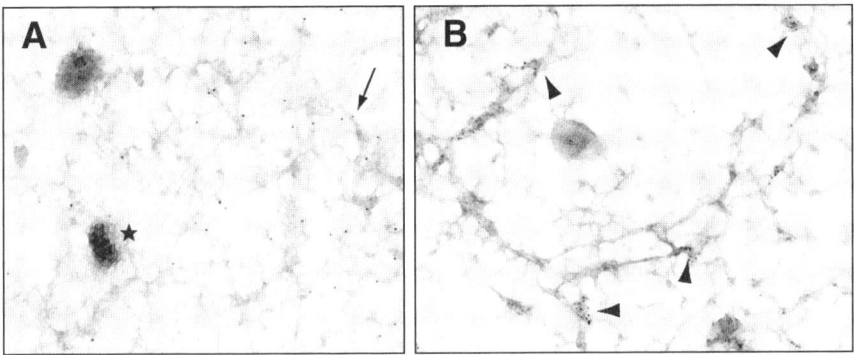

Fig. 5. The cytoskeleton can be observed and labeled on membrane sheets. Sheets were prepared from RBL-2H3 cells and labeled with biotin–phalloidin and 5 nm avidin–gold (**A**) or labeled with antimyosin antibodies plus antirabbit 5 nm gold (**B**). The star in (**A**) denotes a clathrin-coated pit.

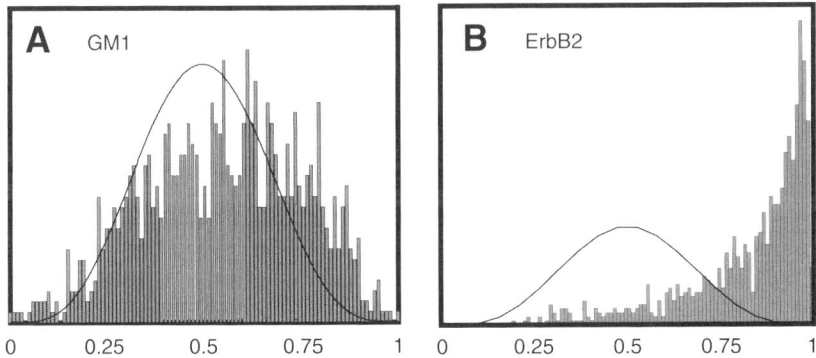

Fig. 6. (**A**) The Hopkins test for spatial randomness. Agreement of the data with the analytical curve (solid black line) indicates randomness, as in the analysis for GM1 ganglioside in fixed samples (**A**). In (**B**), a strong shift to the right indicates that the distribution of ErbB2 receptors is statistically nonrandom (i.e., clustered).

    myosin (using immunogold labeling) on membrane sheets prepared from RBL-2H3 cells *(2)*. Additional examples of these are shown in **Fig. 5A,B**.
8. The computational team has developed a suite of tools to evaluate the spatial point coordinates of gold particles, specifically focusing on determining (1) the degree of randomness/clustering of individual species and (2) the extent to which two different species colocalize. These tools have been refined over a 10-yr period *(13,16–17)* and both Matlab-format (Mathworks, Natick, Mass) and Windows friendly Cellspan versions of the statistical programs are available for download from http://cellpath.health. unm.edu/stmc/ (**Fig. 6A**).

9. The current data acquisition procedures use an ImageJ (http://rsb.info.nih.gov/ij/) plugin that automates the acquisition of gold particle data from digital images. To understand clustering, it is important to begin with micrographs containing a maximal area of the cell membrane. Therefore, gold particle distributions are obtained using the lowest magnification (around × 30,000) that still permits identification of the smallest gold particles. An algorithm based on cropping and thresholding is fast and over 95% accurate in producing a file of coordinates for larger (i.e., ≥10 nm) particles. The alternative filtering algorithm is more accurate for smaller gold particles (in the range of 5 nm). In addition, because no algorithm is completely accurate, each image should be visually checked with the user interface and corrected for missed particles and removal of false identifications *(17)*.

10. The Hopkins test, a statistical evaluation of clustering, which tests spatial randomness by comparing nearest neighbor distances from random points and randomly chosen gold particles, has proven simple and intuitive for analysis of single-size particle distribution. As described *(18)*, the expected distribution for random points is a bell-shaped curve centered about 0.5; this value is plotted as a smooth curve in both **Fig. 6B,C**. The histogram in **Fig. 6B** shows Hopkins values for the distribution of GM1 ganglioside when labeled on prefixed membranes *(13)*. The plot is marginally shifted to the right, indicating a near-random distribution in the absence of active crosslinking. In contrast, **Fig. 6C** shows the Hopkins result for ErbB2 receptors. Clusters of gold-labeled receptors in fixed membranes can be directly observed in **Fig. 1**. The histogram in **Fig. 6C** shows a marked shift to the right, confirming a high degree of clustering.

11. One important set of tools used to evaluate spatial distributions at the nanometer scale arose from work on a much different scale—that of tree patterns in ecology. The Ripley's *K* function was first described in 1976 *(19)* and later refined by Haase *(20)*. The Haase approach uses $L(t) - t$ for each distance t, where $L(t) = [K(t)/\pi]^{1/2}$, and $K(t)$ is the Ripley's *K* function. Because the transformation from *R* to *L* is nonlinear and the assumption of complete spatial randomness (CSR) is violated at large and small scales, it is important to assess *L* using Monte Carlo simulations. As its first application to EM data *(16)*, several groups have implemented Ripley's *K* function to analyze clustering of a single-molecular species labeled with gold particles *(8,20,21)*.

Most frequently, a variant of the Ripley's test that is particularly useful for defining spatial relationships between two different species is used *(13,17)*. Termed the Ripley's bivariate test, it specifically evaluates coclustering for statistical significance using Monte Carlo simulations. **Figure 7A,B** illustrate the use of Ripley's bivariate test, as demonstrated in **ref. *13***. The dotted lines give a 99% confidence envelope for CSR estimated from 100 Monte Carlo simulations. For the condition of GPI-anchored Thy-1 vs GM1 (**Fig. 7A**), the solid black $L(t) - t$ line is within the confidence envelope, indicating that the deviation from CSR is not statistically significant. In contrast, aggregation of Thy-1 by applying crosslinking reagents to live cells enhances the coclustering of Thy-1 with the palmitoylated transmembrane adaptor protein, LAT (**Fig. 7B**). Here, $L(t) - t$ rises above the confidence interval over a range of 30–200 nm, indicating a high degree

Exploring Membrane Domains 257

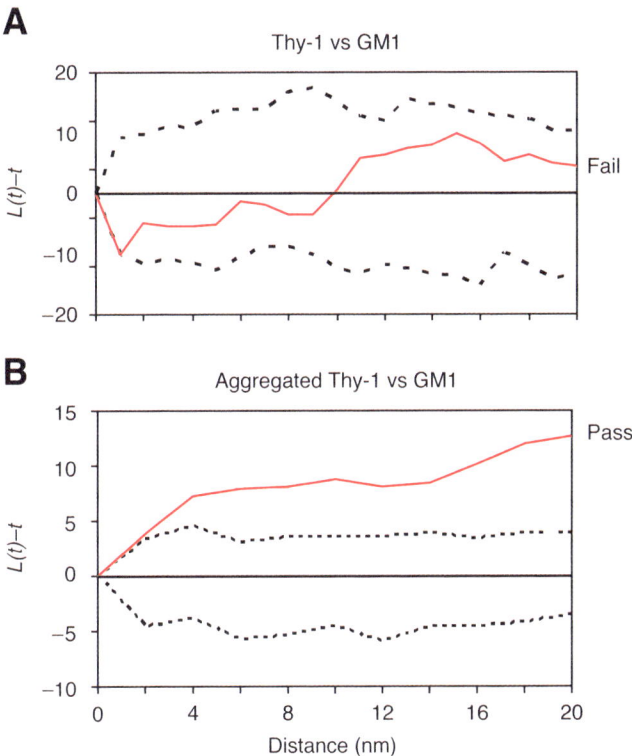

Fig. 7. Ripley's bivariate analysis. (**A**) Fixed samples were double-labeled for Thy-1 and GM1 using two different sizes of gold. This example *fails* the Ripley's test for coclustering, because the solid black line falls within the confidence interval represented by the dotted lines. (**B**) Cells were pretreated at 37°C with mouse anti-Thy-1 plus antimouse gold before ripping. Sheets were fixed and double-labeled to document the distribution of LAT. This example *passes* the Ripley's test for coclustering, because the solid black line rises above the confidence interval represented by the dotted lines.

of coclustering between the two species in this distance range. Prior et al. *(8)* have implemented a related bivariate K-function analysis in analyzing the distinct distributions of H-Ras and K-Ras clusters on fibroblast membrane sheets.

12. Clustering algorithms also have useful applications in analysis of particle distributions, particularly for estimating relative changes after cellular responses to ligand or other treatment conditions. **Figure 8** shows the outcome of a coclustering algorithm based on pairwise analysis of particles within a given small distance of each other, which is repetitively applied to evaluate relationships of *all* particles. This algorithm provides a method to estimate the percent of mixing for two species, as well as their relative cluster size, and is not sensitive to modest changes in the cutoff distance. The example shown here estimates

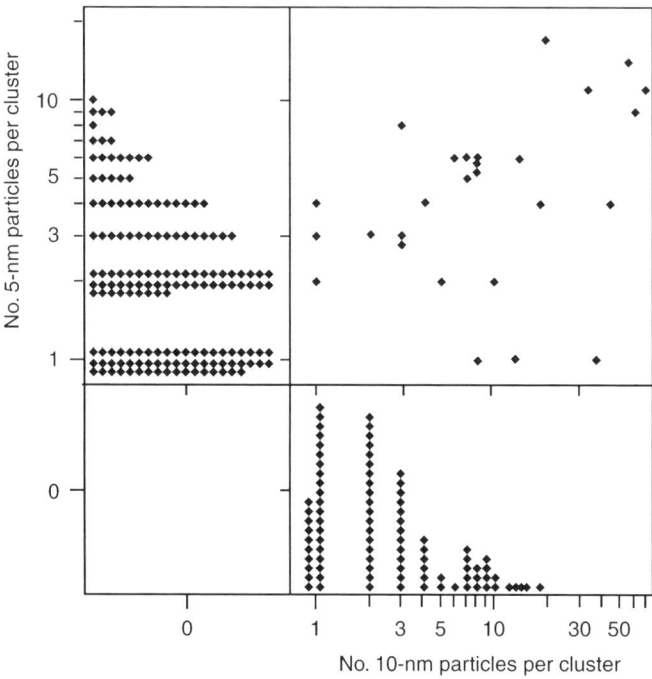

Fig. 8. Use of clustering algorithms. Clusters are identified by putting any two particles that are separated by less than a specified cutoff distance into the same group. In this micrograph, clusters of 5 and 10 nm particles are defined by a cutoff distance, $3d$, where d is the diameter of the particles. Clusters of only individual species are plotted in the upper left and lower right quadrants. Clusters that contain both species (mixed clusters) are plotted in the upper right quadrant; the percent coclustering can be calculated from values in this quadrant.

that 27% of species A (labeled with 5 nm particles), and 57% of species B (labeled with 10 nm particles) are present as coclusters. Because clustering algorithms can produce false positive results (particularly at high density), it is still important to use statistical analysis, such as the Ripley's bivariate test, to test that the coclustering is significant.

5. Conclusions

The membrane sheet technique has broad applications in membrane biology. It offers a "snap-shot" view of the plasma membrane at the time of preparation. By carefully balancing cell pretreatment strategies, fixation conditions, and probe design, one can achieve high-resolution topographic information with low risk of artifacts. It is strongly believed that this technique is best applied, not as a sole method of analysis, but in combination with other biochemical, biophysical, and microscopic techniques. For example, complementary immunoprecipitation studies can provide bulk information about the postranslational modifications of proteins in signaling cascades and evidence of protein–protein assemblies. Complementary live cell imaging applications can provide time resolution, and particularly when combined with biophysical approaches, potentially resolve questions about the diffusion characteristics and exchange rates of particles that can be shown at the TEM level to organize into clusters. Confocal microscopy and immunogold-scanning electron microscopy provide another important level of information—that of the three-dimensional spatial orientation of molecules on membrane projections, such as microvilli and membrane ruffles, as well as specific redistributions because of establishment of cell polarity and/or cell adhesion. These important cellular features are not apparent in the two-dimensional information obtained on membrane sheets.

The authors hope that other groups will find the membrane sheet method valuable in their research programs and will share—and extend—the goal of integrating knowledge of biochemical interactions involved in signaling pathways with knowledge of the topography of membrane-associated signaling proteins and lipids. The results should extend the understanding of how cells regulate the efficiency, specificity, amplitude, and duration of signaling and to identify new drug targets useful for the treatment of asthma, cancer, and other diseases.

Acknowledgments

This work was supported in part by NIH RO1 Grants to B. S. Wilson (A1051575 and CA119232) and J. M. Oliver (GM49814) and by NIH Grant P20 GM067594 that established the New Mexico Center for the Spatiotemporal Modeling of Cell Signaling. The authors thank Ryan Molecke and Diana Roberts for development of CellSpan, the spatial statistics package. Use of the electron microscopy facility at the university of New Mexico (UNM) School of Medicine and of the fluorescence microscopy facility at the UNM Cancer Research and Treatment Center is gratefully acknowledged. Microscopy equipment was funded in part through NIH S10 Grants RRI5734 and RR022493 (EM) and RR14668, RR19287, and RR016918 (fluorescence). Infrastructure support for fluorescence microscopy was from NCI Grants R24 CA88339 and P30 CA118100.

References

1. Wilson, B. S., Pfeiffer, J. R., and Oliver, J. M. (2000) Observing FcεRI signaling from the inside of the mast cell membrane. *J. Cell Biol.* **149,** 1131–1142.
2. Wilson, B. S., Pfeiffer, J. R., Surviladze, Z., Gaudet, E. A., and Oliver, J. M. (2001) High resolution mapping of mast cell membranes reveals primary and secondary domains of FcεRI and LAT. *J. Cell Biol.* **154,** 645–658.
3. Valna, P., Ledbduska, P., Draberova, L., et al. (2004) Negative regulatory role of the non-T cell activation linker in mast cell signaling. *J. Exp. Med.* **200,** 1001–1013.

4. Kim, J. -H., Cramer, L., Mueller, H., Wilson, B. S., and Vilen, B. (2005) Independent trafficking of Ig-α/β and μm is facilitated by dissociation of the BCR complex. *J. Immunol.* **175,** 147–154.
5. Xue, M., Prossnitz, E., Oliver, J. M., and Wilson, B. S. FPR and FcεRI occupy common signaling domains for localized crosstalk. *Mol. Biol. Cell* published on line 10.1091/mbc. E05-11-1073.
6. Yang, S., Raymond-Stinz, M., Hsieh, G., Oliver, J. M., and Wilson, B. S. Signaling domains of ErbB family members in SKBR3 breast cancer cells—an EM approach. *Submitted.*
7. Prior, I. A., Harding, A., Yan, J., Sluimer, J., Parton, R. G., and Hancock, J. F. (2001) GTP-dependent segregation of H-Ras from lipid rafts is required for biological activity. *Nat. Cell Biol.* **3,** 368–375.
8. Prior, I. A., Muncke, C., Parton, R. G., and Hancock, J. F. (2003) Direct visualization of Ras proteins in spatially distinct cell surface microdomains. *J. Cell Biol.* **160,** 165–170.
9. Hess, S. T., Kumar, M., Verma, A., Farrington, J., Kenworthy, A., and Zimmerberg, J. (2005) Quantitative electron microscopy and fluorescence spectroscopy of the membrane distribution of influenza hemagglutinin. *J. Cell Biol.* **169,** 965–976.
10. Sanan, D. A. and Anderson, R. G. W. (1991) Simultaneous visualization of LDL receptor distribution and clathrin lattices on membranes torn from the upper surface of cultured cells. *J. Histochem. Cytochem.* **39,** 1017–1024.
11. Heuser, J. (2000) The production of 'cell cortices' for light and electron microscopy. *Traffic* **1,** 545–552.
12. Kusumi, A. and Suzuki, K. (2005) Toward understanding the dynamics of membrane-raft-based molecular interactions. *Biochim. Biophys. Acta* **1746,** 234–251.
13. Wilson, B. S., Steinberg, S. L., Liederman, K., et al. (2004) Distinct distribution and behavior of raft markers in native membranes. *Mol. Biol. Cell* **15,** 2580–2592.
14. Sharma, P., Varma, R., Sarasij, R. C., et al. (2004) Nanoscale organization of multiple GPI-anchored proteins in living cell membranes. *Cell* **4,** 577–589.
15. Frankel, D. J., Pfeiffer, J., Oliver, J. M., Wilson, B. S., and Burns, A. R. (2006) Revealing the topography of cellular membrane domains by combined Atomic Force Microscopy/Fluorescence Imaging. *Biophys. J.* **90,** 2404–2413.
16. Sanders, M. L. (1996) Characterizing molecular aggregation on cell surfaces from spatial point patterns. PhD Thesis. University of New Mexico, Albuquerque, NM.
17. Zhang, J., Leiderman, K., Wilson, B. S., Oliver, J. M., and Steinberg, S. L. (2005) Characterizing the topography of membrane receptors and signaling molecules from spatial patterns obtained using nanometer-scale electron-dense probes and electron microscopy. *Micron* **37,** 14–34.
18. Jain, A. K. and Dubes, R. C. (1988) Algorithms for clustering data. *Advanced Reference Series.* Prentice-Hall, Englewoood Cliffs, NJ, 7632p.

19. Ripley, D. D. (1976) The second-order analysis of stationary process. *J. Appl. Prob.* **13,** 255–266.
20. Haase, P. (1995) Spatial pattern analysis in ecology based on Ripley's K–function: introduction and methods of edge correction. *J. Vegetation Sci.* **6,** 575–582.
21. Philimonenko, A. A., Janacek, J., and Hozak, P. (2000) Statistical evaluation of colocalisation patterns in immunogold labeling experiments. *J. Struct. Biol.* **132,** 201–210.

18

Atomic Force Microscopy of Lipid Domains in Supported Model Membranes

Alan R. Burns

Summary

Atomic force microscopy (AFM) has been a significant tool in the characterization of lipid domains in model membranes. With AFM, one can image the structure of membranes in a natural fluid environment with a lateral resolution that approaches 1 nm and vertical resolution of 0.1 nm. The AFM technique is discussed, with a special emphasis on imaging soft, compliant membranes that are supported on solid substrates such as glass or mica. In typical model membranes, lipid domains are formed by phase separation in multicomponent lipid mixtures and are observed by nm-level height differences owing to lipid packing. A general procedure for creating supported lipid bilayers through vesicle fusion is discussed.

Key Words: Atomic force microscopy; bilayer membranes; lipid domains; lipid rafts; scanning probes; vesicle fusion.

1. Introduction
1.1. Brief Review of Scanning Probe Applications in Lipid Domain Studies

Almost since its inception (1), atomic force microscopy (AFM) has been applied to biological samples and has been readily adapted to imaging soft materials in a fluid environment. Unlike transmission electron microscopy or fluorescence imaging, the nanoscale proximal probing technique of AFM intrinsically yields lateral resolution based on true height information; and thus, is able to directly map topographic features in membranes. Furthermore, it has the potential, through subtle mechanical interactions, to map changes in material composition or packing density that directly relate to membrane structure such as lipid domains. AFM thus bridges a very significant gap that is now present in light and electron microscope membrane imaging and characterization.

It has been well established in the imaging of biomembranes that AFM can provide undistorted, molecular-scale lateral (<2 nm) and height (<0.1 nm) resolution, owing in large part to sensitive feedback detection with small forces (<0.2 nN) and sharp probes (<20 nm radius) *(2–4)*. Although much of the imaging is still performed in "contact mode" (CM-AFM), wherein the probe tip is continuously scanned over the surface, comparable resolution is achieved through intermittent "tapping mode" or "amplitude modulation mode" (AM-AFM). Both techniques will be discussed.

AFM has been successfully used in numerous studies *(5–13)* to image domain structure in supported lipid monolayers and bilayers on the basis of topographic height differences between gel-phase (gel), liquid-ordered (L_o), and liquid-disordered domains (L_d). It is generally recognized that, for lipids of equal length, those height differences are directly related to the packing density of the phase-separated domains: densely packed gel domains > L_o domains > loosely packed L_d domains. Like unsupported membranes visualized by fluorescence imaging, the domain boundaries imaged by AFM can also reflect the domain morphology. Gel domains tend to be irregularly shaped, whereas L_o and L_d domains tend to be more rounded *(14)*.

Scanning probe microscopy has been largely responsible for the burgeoning nanotechnology, which spans material sciences from electronics to biology. Because of its versatility in imaging most materials under a wide variety of conditions, AFM has been the main workhorse. As such, new advances in AFM imaging and AFM-based nanoscale materials characterization techniques have given rise to new advances in nanotechnology. Because AFM is in a continuous state of refinement in instrumentation and software at the hands of commercial manufacturers and researchers, it is best to discuss only the basic techniques that should be in the repertoire of all AFM-imaging instruments. These techniques may differ in implementation from one instrument to the next, so no reference is given to instrument-specific or software-specific procedures.

1.2. Basic Principles of AFM Operation

The basic principles of AFM are depicted in **Fig. 1**. The overall concept is that a nanoscale probe "tip" at the end of a cantilever is scanned over the top surface of a sample (e.g., supported bilayer). The up and down motion of the probe tip is ideally a direct measurement of the sample topography or "height" in "Z," as a function of "lateral" position in "*X*" and "*Y*." With few exceptions, the displacement in Z of the tip is detected by the sensitive optical lever technique: a laser beam (usually a 670 nm diode laser) is focused on the end of the cantilever just above the tip, which reflects it to a photodiode that is divided into four separate quadrants. Z motion is detected by the relative change in signal of the sum of the upper two quadrants relative to the sum of the lower two quadrants.

AFM of Lipid Domains in Supported Model Membranes

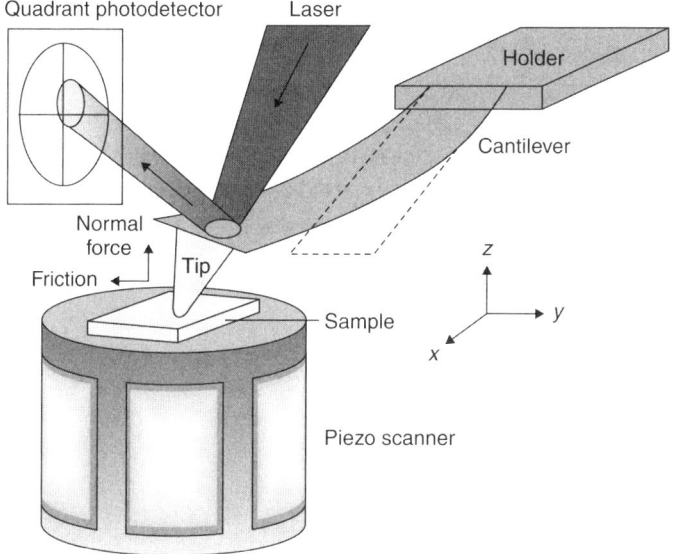

Fig. 1. Basic operation of AFM. A probe tip on the end of a cantilever is raster scanned in X and Y directions across a sample mounted on a piezo scanner (in this case the sample moves and the tip is stationary). A laser beam is reflected off the cantilever and is detected by a quadrant photodetector. Both vertical deflection (in Z) of the cantilever owing to normal forces, and torsional deflection owing to lateral forces are registered in the quadrant photodetector (*see* text for details).

This is the "normal force" signal. Torsional (side-to-side) motion of the cantilever is detected by the relative signal of the sum of the right quadrants relative to the sum of the left quadrants, and is caused by "lateral forces" owing to tip-induced friction. Because it is wished to eliminate potentially destructive lateral forces when imaging soft biological samples, this signal should always be maintained close to zero by minimizing the normal force load.

Scanning motion of the tip in X and Y over the sample is controlled by voltages applied to separate elements of a piezoelectric device or "piezo scanner." The piezo also controls the tip–sample interaction in Z. In some instruments, the sample is mounted on the scanner, and thus, the sample is moved relative to the tip (*see* **Fig. 1**); in others, the cantilever is mounted on the scanner and the sample is stationary. In some special cases *(5,6)*, Z motion is controlled by a separate piezo, and X and Y are controlled by another. When raster scanning in X and Y, one direction (X or Y) is chosen as the "fast" scan direction (usually in the range 1–10 Hz or lines/s), whereas the other is the "slow" scan direction. Maximum scan ranges for X and Y are usually <100 µm.

The Z motion of the piezo determines the amount of "normal force" applied to the sample at the probe tip. This is controlled by a feedback circuit that measures the normal force signal at the quadrant detector. The amount of voltage applied for Z piezo motion to maintain a constant force is directly related to the sample topography; and is thus processed as the "topography" or "height" signal. The "raw" photodiode signal is the "deflection" signal. The Z feedback circuit is engaged continuously when the probe is in contact with the sample, regardless whether stationary or scanning. All three piezo motions (X, Y, and Z) must be calibrated as discussed in **Subheading 3.2**.

The aforementioned description in which the probe tip is in continuous contact with the sample is commonly called CM-AFM. An alternative method, commonly called AM-AFM, involves feedback control of the amplitude of an oscillating (usually >8 kHz) probe tip that is only in intermittent contact with the sample. The up and down cantilever oscillation is driven by a small sinusoidal voltage applied to the Z piezo motion (or to a second, dedicated Z piezo). The modulated normal force signal is detected (and demodulated) with a phase-sensitive lock-in amplifier. This lock-in signal is maximum when the probe is "free" or out of contact and will be attenuated when intermittent contact occurs. The sample topography is thus mapped by applying voltages to the Z piezo required to maintain a set oscillation amplitude (e.g., 90% of free amplitude) that corresponds to soft intermittent contact. The main advantages of AM-AFM relative to CM-AFM are: (1) it eliminates lateral forces as it is in intermittent contact; and (2) lock-in detection is immune to thermal and piezo ("DC") drifts that must be constantly checked in CM-AFM. The main disadvantages are that it is a more complicated technique that requires more initial effort, and it can, at times, have slightly more noise.

The phase-shift between the drive signal and the amplitude signal in AM-AFM can also be detected and captured through simultaneous "phase images." The phase-shift is sensitive to material properties such as elasticity, viscoelasticity, and probe-sample adhesion *(15)*, and may reveal details in lateral structure absent in the topographic "height" images. However, contrast is difficult to achieve in phase imaging of soft biological materials because the imaging forces are kept to a minimum.

Photos of typical cantilevers and probe tips are shown in **Fig. 2**. Both the probe and cantilever are microfabricated from a single piece of silicon nitride or silicon. Typical sharp probe tips have a 10–20 nm radius of curvature; "very sharp" probes can have a radius less than 5 nm. In these special cases (not discussed herein) when it is important to know the tip radii, samples with atomically sharp features such as $SrTiO_3$ crystal planes *(16)* are used to indirectly trace the probe tip. Cantilevers come in rectangular "diving board" or "triangular" shapes. Regardless of the shape, the length and thickness of the cantilever is of critical importance,

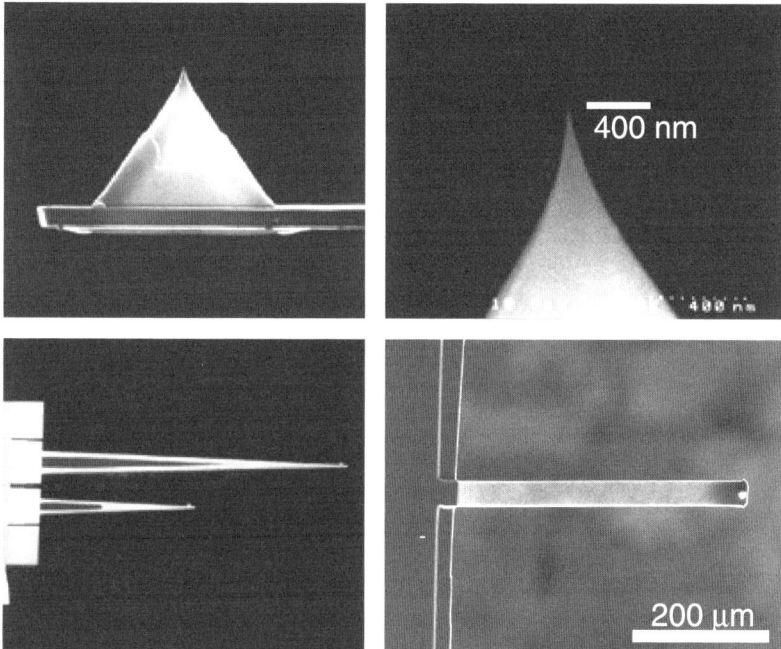

Fig. 2. Examples of AFM probe tips (upper) and cantilevers (lower). In the lower right is a rectangular cantilever, and in the lower left are two triangular cantilever of different lengths.

because together they determine the "stiffness" or force (spring) constant k. As will be discussed in detail next, when imaging delicate membranes in CM-AFM, soft, compliant cantilevers having a force constant $k < 0.1$ N/m are required. Stiffer cantilevers can be used for AM-AFM ($k < 1$ N/m).

2. Materials
2.1. Supported Model Lipid Bilayers
1. Lipids, commercially available from Avanti Polar Lipids (Alabaster, AL), Invitrogen-Molecular Probes (Eugene, OR), or Sigma Chemical Co. (St. Louis, MO). Kept at −20°C.
2. High-performance liquid chromatography grade chloroform and methanol.
3. Phosphate-buffered saline (PBS buffer): 150 mM NaCl, 50 mM sodium phosphate, 1.5 mM NaN$_3$, pH 7.4.
4. Vortexer (e.g., Thermolyne, Dubuque, IA).
5. Extruder (e.g., Northern Lipids, Vancouver, Canada) with 100 nm pore membranes (e.g., Whatman Nuclepore Track-Etch available from VWR Inc. [Westchester, PA] or Fisher Scientific [Pittsburgh, PA].
6. Argon.

7. Ultrapure water (e.g., 18 MΩ-cm from Barnstead Nanopure filter system, Dubuque, IA).
8. Dynamic light scattering apparatus (e.g., Protein Solutions, High Wycomb, UK).
9. 25-mm Diameter glass cover slips or mica (*see* **Note 1**).
10. H_2SO_4 and H_2O_2.
11. Leiden cover slip dish (e.g., Harvard Apparatus, Holliston, MA).
12. 70°C oven.
13. Vacuum desiccator *without the desiccant* (e.g., VWR or Fisher Scientific).

2.2. AFM Apparatus

1. AFM equipped with electronic controllers, cantilever holders suitable for fluid imaging, fluid cell (dictated by AFM/sample geometry), and software.
2. Air table for vibration isolation (e.g., Technical Manufacturing Corp., Peabody, MA, or Newport Corp., Irvine, CA).
3. Silicon nitride or silicon cantilevers with nominal force constants <0.1 N/m for CM-AFM or <0.5 N/m for AM-AFM (e.g., Veeco Metrology, Santa Barbara, CA; NT-MDT Mikromasch, Portland, OR).
4. Ultraviolet ozone cleaner (e.g., UVOCS Inc., Montgomeryville, PA).
5. Calibration samples (e.g., Ted Pella, Redding, CA; NT-MDT Mikromasch).

3. Methods
3.1. Preparation of Supported Lipid Bilayers Through Vesicle Fusion

There are two commonly used techniques for the preparation of supported lipid bilayers for AFM imaging. The first involves the use of a Langmuir-Blodgett trough to form monolayers on a water surface that are subsequently transferred to a substrate, one monolayer at a time. Langmuir-Blodgett deposition *(7,13,17)* requires special equipment (a computer-controlled trough), involves two separate depositions to form a bilayer, and often exposes the lipid monolayers or bilayers to air. Vesicle fusion, pioneered by H. McConnells' group *(18)*, is the alternative technique. It is far more convenient than Langmuir-Blodgett deposition, starts with preformed bilayers (vesicles), and is accomplished entirely under fluid conditions, (thus, never exposing the supported bilayer to air). Vesicle fusion is the method of choice for this author and will be discussed next.

As shown schematically in **Fig. 3**, vesicle fusion involves the adsorption, rupture, and fusion of lipid vesicles into a continuous supported bilayer. The "isolated" process shown in **Fig. 3** is one of many that have been proposed in various biophysical models *(19–21)*. There are many variations of the vesicle fusion technique involving choice of substrates (usually glass or mica, *see* **Note 1**), temperature, ionic strength of solution, and vesicle (liposome) preparation. It has been found to work best for vesicles larger than 100 nm in diameter *(21)*. Thus, 100 nm pores should be used with an extruder and the vesicle size distribution checked with light scattering (*see* **Fig. 4**). A general procedure based on samples mounted

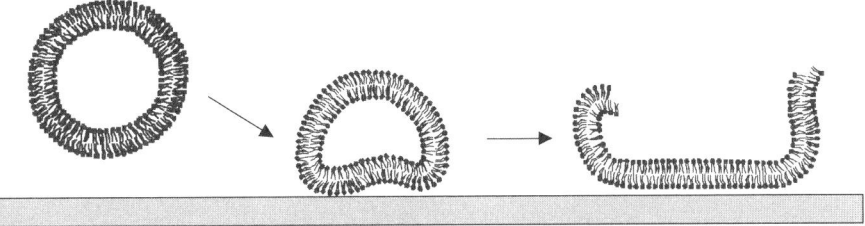

Fig. 3. Highly schematic depiction of vesicle fusion on a solid substrate (e.g., glass or mica). From left to right: A single 100 nm vesicle adsorbs to the substrate, deforms due spreading, and finally ruptures. The ruptured vesicle will fuse with others to form a complete bilayer. This is just one of several proposed mechanisms (*see* text). A 1–2 nm water layer separates the lipid bilayer from the substrate.

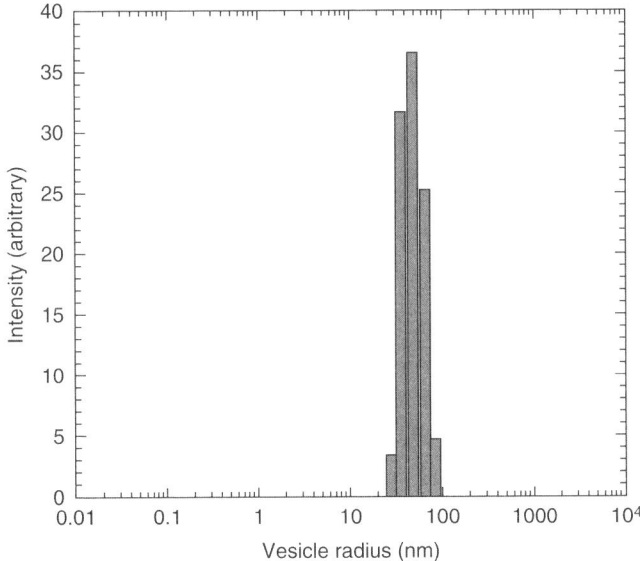

Fig. 4. Example of light scattering histogram that shows size distribution of vesicles after extrusion.

in a fluid-containing Leiden cover slip dish (*see* **Fig. 5**) is outlined next. The reader is encouraged to examine the other methods of vesicle fusion *(8–12,20)*.

3.1.1. Vesicle Preparation

1. Make stock solutions of lipids in high-performance liquid chromatography grade chloroform:methanol 3:1 (v:v) (*see* **Note 2**). Typical concentrations are 85 mg/mL. Store at 4°C under nitrogen in well-sealed flask. Make fresh stock solutions after about 30 d.

Fig. 5. Cover slip dish (Harvard Apparatus). It accommodates 25-mm glass cover slips (or mica) and is sealed by O-rings.

2. Measure accurate amounts of stock solutions (a calibrated glass syringe works well) into a clean test tube to eventually make 3 mL of a 3 mM solution (*see* **step 5** next) of lipids.
3. Place tube on rotary evaporator and dry down while keeping solution at 40–50°C with a water bath.
4. Store dried lipids under vacuum at room temperature until needed (e.g., vacuum desiccator *without the desiccant*). Vent with clean nitrogen or argon.
5. Resuspend dried lipids in 3 mL PBS buffer to make 3 mM solution.
6. Vortex for several seconds.
7. Freeze in 2-propanol/crushed dry ice and thaw in 60°C water bath. Repeat 10 times.
8. Assemble clean extruder, using 100 nm pore membrane, and argon gas lines (500 psi).
9. Adjust water jacket to 60°C, and using PBS buffer to prime system, extrude while checking for leaks.
10. Load lipid sample and extrude 10 times.
11. Check for size distribution with light scattering apparatus. A representative size distribution is shown in **Fig. 4**.
12. Store at room temperature for short periods (1 d), or 4°C for long periods.
13. Disassemble extruder and thoroughly clean and rinse parts in nanopure water.

3.1.2. Vesicle Fusion on Substrate

1. Clean glass cover slips in (7:3) H_2SO_4:H_2O_2 (*caution: this is potentially explosive when reacting with organics*), rinse thoroughly in distilled water and ultrapure water, and store under ultrapure water (18 MΩ-cm).

2. Just before use, dry cover slips under a stream of pure, dry nitrogen and mount in a Leiden cover slip dish. The dish shown in **Fig. 5** accommodates 25 mm cover slips. Check for leaks using nanopure water.
3. If mica is used, mount freshly cleaved mica in cover slip dish (*see* **Note 1**).
4. Warm vesicle suspension in 60°C water bath for several minutes.
5. Add 0.8 mL PBS buffer to assembled cover slip dish, followed by 0.2 mL of vesicle suspension.
6. Cover and incubate in 60–70°C oven for 1–2 h, making sure no significant evaporation occurs.
7. Cool slowly (1–2 h) to room temperature.
8. Rinse carefully and thoroughly with PBS buffer. Never expose bilayer to air.
9. Mount on AFM.

3.2. AFM Calibration Procedures

The following two sections are not daily procedures, but are important steps to AFM characterization of lipid rafts that are done to maintain accuracy.

3.2.1. Calibration of X, Y, and Z Piezo Motion

The piezoelectric motion in an AFM must be calibrated regularly (at least once a year) because the X, Y, or Z displacement "sensitivity" (nm/V) can change with age. The sensitivities for each direction of motion are stored in the AFM software and are used to convert raw data into images. The most common procedure is to image commercially available calibration grids or gratings and to use those accurate dimensions to readjust the stored sensitivities. Because piezos also exhibit nonlinearities over large ranges, it is recommended that one use calibration grids/gratings that correlate as much as possible with common feature sizes of interest. For lipid domains, this would correspond to lateral dimensions of 0.1–5 µm and heights <50 nm. Usually X and Y are done together **(Fig. 6A)** using a sample for lateral calibration, and Z is done separately using a sample for depth calibration **(Fig. 6B)**. An example of a line profile for depth calibration is shown in **Fig. 6C**. With these measurements, a new sensitivity calibration $C_{i,new}$ ($i = X, Y,$ or Z) is given by:

$$C_{i,new} = \frac{C_{i,old}}{\left(D_{measured}/D_{actual}\right)} (\text{nm/V})$$

where $D_{measured}$ is the displacement of the piezo, D_{actual} is the specified, calibrated feature size, and $C_{i,old}$ is the old sensitivity measured previously.

3.2.2. Calibration of Cantilever Force Constant

Unlike protein unfolding experiments or cell nanomechanics, very precise, accurate force measurements are not to be concerned with. However, in CM-AFM

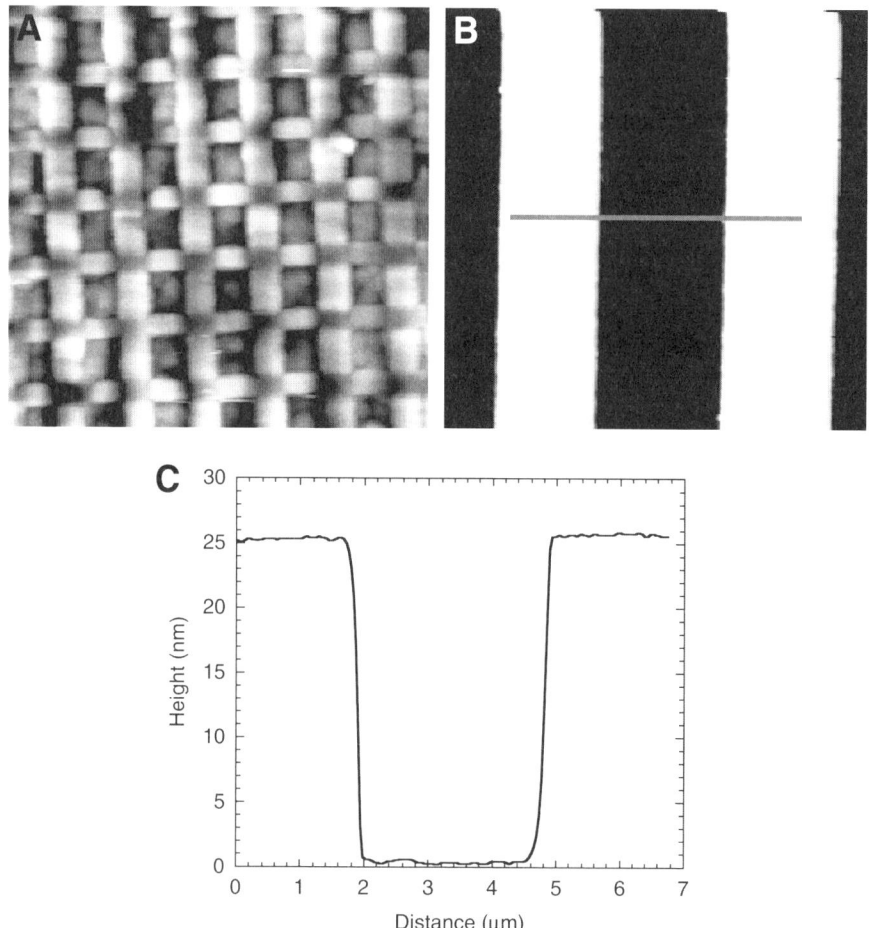

Fig. 6. Images of commercially available grids and gratings for piezo calibration. (**A**) A square grid with 463 nm periodicity in both X and Y (Ted Pella). (**B**) A grating with 25.5 nm etched pits (NT-MDT Mikromasch). (**C**) Cursor plot of cross-section (gray line) of grating shown in **B**.

the concern is with reproducibly applying small forces (loads) on soft, compliant lipid bilayer samples. Cantilevers come with a "nominal" force (spring) constant according to their fabrication dimensions; long, thin cantilevers are more compliant than short, thick ones. Because most cantilevers are shipped as part of a microfabricated batch that is fairly uniform, the force constant for a few random ones should be checked. There are many calibration procedures *(22–24)*. A simple procedure, based on thermal noise *(24)*, is briefly discussed herein.

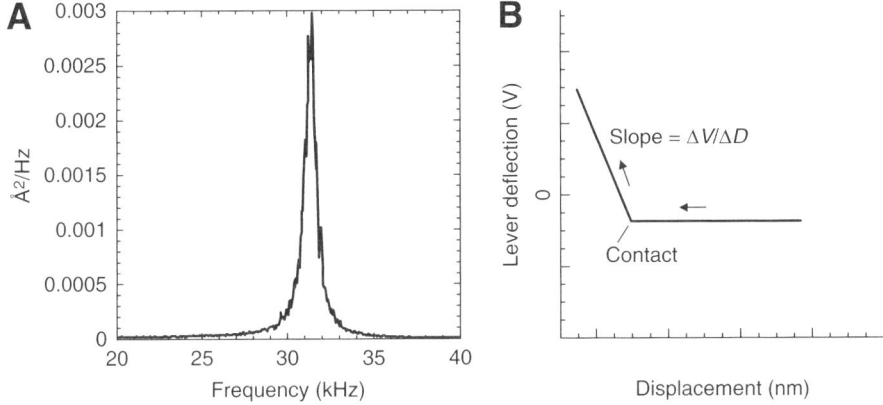

Fig. 7. Data used for calibration of cantilever force (spring) constant k. **(A)** Power spectrum of cantilever motion (Z deflection) in air at room temperature. The peak is the natural resonance frequency in air. The area under the peak is integrated to obtain k. This requires determination of the cantilever deflection sensitivity (V/nm) as shown in **(B)**. Based on **ref. *24***.

The force constant k can be expressed as a function of the mean-square of random cantilever deflection fluctuations d at temperature T:

$$k = \frac{k_B T}{\langle d^2 \rangle}$$

where k_B is Boltzmann's constant. One can measure the deflection fluctuations of a freely oscillating cantilever in air near its natural resonance frequency. This is most easily acquired in the frequency domain as shown in **Fig. 7A**, where $\langle d^2 \rangle$ is equal to the area under the Lorentzian curve (*see* **Note 3**). However, before one integrates the data, one has to know the cantilever deflection sensitivity (in V/nm) as detected by the quadrant photodiode. The deflection sensitivity is easily determined by measuring the slope of the voltage change during deflection vs displacement after cantilever contact with a substrate (assuming the Z piezo has been calibrated), as shown schematically in **Fig. 7B**.

Once the force constant k is known, it should be entered in the software. Then one can convert the cantilever deflection (voltage) vs displacement curves into force profiles as shown in **Fig. 8.**, which are used to characterize all of the forces experienced by the cantilever on approach to the surface (A) and while withdrawing (F).

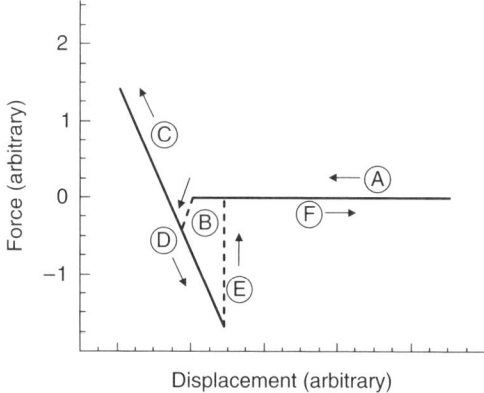

Fig. 8. Schematic representation of force profile experienced by cantilever. **(A)** approach, **(B)** jump-to-contact, **(C)** repulsive deflection on compression, **(D)** attractive deflection on withdrawal followed by, **(E)** pull-off force, and **(F)** retraction. Best results in imaging soft lipid bilayers are obtained by minimizing the long-range forces that can be responsible for jump-to-contact in **B** and pull-off forces in **E**.

3.3. Minimizing Long-Range Forces With Electrostatic Shielding

Long-range forces must be minimized in order that short-range forces be used in force feedback. It is through short-range forces (i.e., van der Waals, hydrophobic) that high-resolution imaging under well-controlled small loads (<0.2 nN) is possible *(25)*. Particularly noteworthy in **Fig. 8** are the "jump-to-contact" force (shown as arrow B) and the "pull-off" force (shown as arrow E). Both can, owing to long-range adhesive and electrostatic forces, overcome the spring constant of the cantilever. They need to be minimized (or shielded) as much as possible in imaging lipid bilayers. Electrostatic shielding can be implemented by using buffers that have an ionic concentration in the range 100–300 m*M* (e.g., PBS buffer in **Step 3** in **Subsection 2.1.**) *(25)*. An example of an actual force curve that is routinely observed for supported lipid bilayers in PBS buffer is shown in **Fig. 9**. One can see that the jump-to-contact and pull-off forces are absent, the approach and withdrawal are indistinguishable, and a small load setpoint is easily achievable.

3.4. Cantilever Preparation and Sample Mounting

1. Select CM-AFM (<0.1 N/m nominal) or AM-AFM (<1 N/m nominal) cantilever from Gel-Pak 4® holder (*see* **Note 4**).
2. Place cantilever on clean glass slide inside ultraviolet ozone cleaner and run cleaner for 15 min.
3. Retrieve cantilever and load onto fluid cell cantilever holder provided by AFM manufacturer.

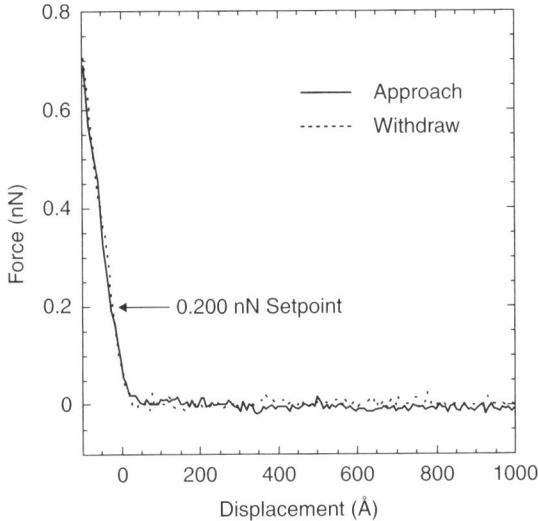

Fig. 9. Typical force profile obtained in CM with sample in PBS buffer that provides electrostatic shielding. These force profiles are also used to check for proper CM-AFM feedback setpoint (e.g., 0.2 nN) at zero displacement.

4. Mount sample (e.g., cover slip dish, fluid cell) and cantilever holder on AFM and turn on instrument.
5. Lower cantilever into fluid so that it is fully immersed.
6. Align laser onto end of cantilever using available optical tools provided by AFM manufacturer.
7. Center reflected spot on quadrant detector so that the normal force and lateral force signals are close to zero.
8. Let cantilever and sample thermally equilibrate (with laser and AFM electronics on) for an hour (*see* **Note 5**).

3.5. "CM-AFM" Imaging

1. After thermal drift has settled down, adjust laser spot on quadrant detector so that the normal force reading is slightly negative, corresponding to a small attractive force.
2. With a setpoint of zero and the software in CM, approach the surface using software control and bring cantilever into contact with the surface. The lever signal will go up to zero and the feedback circuit will take over.
3. Acquire a force vs displacement profile as depicted in **Fig. 9** and adjust feedback setpoint and/or signal bias to insure feedback at 200 pN or less. Notice in **Fig. 9** that the setpoint is at zero displacement. This procedure is highly specific to the AFM software; however, the physics is the same for all instruments and observing the force profile is critical in order to make sure that the load on the sample will cause no distortions and still provide good images. Because of cantilever and/or piezo drift, this step must be performed repeatedly all day (especially before any

image is acquired). A force profile is a good way to check for false engagement (*see* **Note 6**) on approach.
4. Optimize gains by looking at "height (topography)" signal in oscilloscope mode or perform repeated line traces (slow scan off) at a reasonably large scan size (e.g., 10 μm at <3 Hz). Depending on the AFM control electronics, the gains should be increased gradually until feedback oscillations occur in the line traces (*see* **Note 7**). These oscillations can sometimes be more visible in the "deflection" signal. Reduce gains until oscillations disappear.
5. Using line traces, minimize "slope" of both fast and slow scan directions. This procedure will depend on software/control electronics.
6. Before taking the first image, the vibration/acoustic background should be checked once (at the beginning of the session). Check software for instrument-specific procedure, or using the same spectrum analyzer used to calibrate force constants (**Subheading 3.2.2.**), acquire a low-frequency (0–250 Hz) spectrum of the "height (topography)" signal while in feedback, *but not scanning*. Most air tables eliminate noise >100 Hz, but not all noise <100 Hz. As shown in **Fig. 10**, this spectrum reflects the noise level in Z, which should be less than 1Å (0.1 nm) at all frequencies. If a large peak at 60 Hz (and its harmonics) is observed, alternating current (AC) electrical-mechanical noise ("hum") has coupled into the vibration isolation. This is usually caused by AC devices sitting on the same table as the AFM. Other vibration noise sources can be nearby fans, elevators, hallway traffic, and so on. It pays to locate the AFM in a quite, acoustically baffled environment.
7. Begin acquiring images for both "height (topography)" and "deflection" signals. Always check force profiles (**step 3** aforementioned) and adjust load before acquiring each image.
8. The scan rate can be increased for smaller, high-resolution images (e.g., up to 10 Hz for 100 nm scans). However, if the scan rate is too fast, the piezo scanner will cause ringing in the images.

3.6. AM-AFM Imaging

1. Follow **steps 1–2** in **Subsection 3.5.** in order to "find" surface of sample using CM. However, do not remain in contact; retract cantilever a few μm, staying close to sample.
2. Go into AM-AFM mode (software specific) and tune cantilever to find resonance peak. This procedure will vary with software. Often the strongest resonance is 8 KHz because of coupling with the fluid cell. This resonance can be used successfully.
3. Adjust the oscillation frequency to be slightly lower (few 100 Hz) than the resonance peak and adjust setpoint to be about 90% of oscillation amplitude.
4. Approach surface and make sure feedback is engaged.
5. Acquire a force profile to make sure the setpoint reflects 90% of amplitude and there is no false engagement (*see* **Note 6**). *See* **Fig. 11** for a sample AM-AFM force profile (for clarity, only approach is shown). Notice in **Fig. 11** that the output of the lock-in amplifier (which detects the oscillation amplitude) decreases as soon as intermittent contact is made. Fortunately, because AM-AFM uses a phase-sensitive

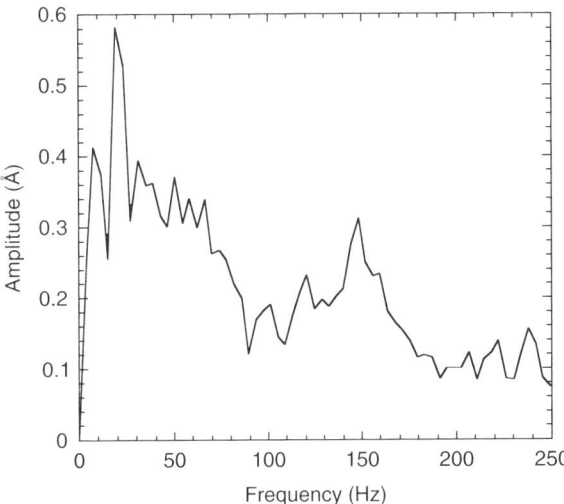

Fig. 10. Spectrum of vibrational noise experienced by stationary cantilever in contact with sample and under feedback.

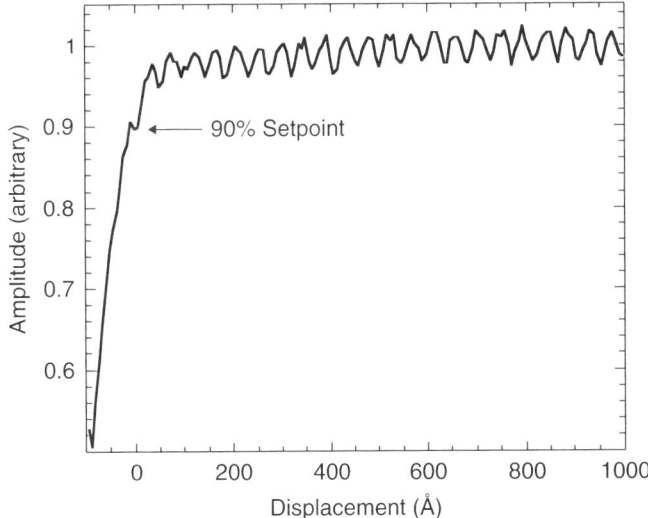

Fig. 11. Typical force profile obtained in AM-AFM mode, showing amplitude attenuation after contact and 90% setpoint. (Oscillations on approach are from excessive gain on lock-in and are not present during feedback.)

modulation signal, it is not subject to the DC drifts that affect CM-AFM. Thus, AM-AFM does not require constant checking of the force profile before imaging. However, it should be checked whenever the cantilever is withdrawn for some reason and brought back to the sample surface.

6. Perform **steps 4–6** in **Subsection 3.5**.
7. Begin acquiring images both for "height (topography)" and "phase-shift." Phase-shift images are usually featureless for uniform lipid bilayers. However, they can confirm defects in the bilayers that appear in the topography. AM-AFM scan rates should be kept slower (1–3 Hz) than CM-AFM.
8. If images are not sharp owing to poor tracking, adjust setpoint to slightly increase amplitude damping. Conversely, check for any sign of tip-induced distortion in the images owing to excessive damping or amplitude. If this is the case, retract cantilever and reduce damping and/or amplitude.

3.7. Image Analysis

Image acquisition and analysis is essentially the same for both CM-AFM or AM-AFM. While acquiring images, careful control of the feedback ('tracking") and scanning parameters prevent many artifacts (**Step 2** below). All commercial AFM instruments come with software to analyze and process the images. The basic routines such as imaging "flattening" (**Step 3** below) are essentially the same and should be well covered. More advanced techniques such as bandpass filtering, FFT analysis, or image statistics are useful, but not usually required for viewing and characterizing lipid domains. Remember, the images are *data* and should be processed with caution.

1. It is always a good practice to record and store the "raw data" DC level images, even if the AC level is displayed during acquisition. *Copies* of the DC level images can then be processed later on (**Step 3** below) to remove slow thermal and piezo drift and unwanted piezo "slope" that was not fully adjusted in **Step 5** of **Subsection 3.5**.
2. Images that have artifacts are for the most part useless and should be discarded. As noted earlier in **Subsections 3.5.** and **3.6.**, some artifacts such as blurred images can arise from improper feedback control because of inadequate gain or improper setpoints, whereas noise (oscillations) are because of excessive gain or piezo ringing. Tip-induced artifacts are more difficult to assess. Impurities can stick to the probe tip and suddenly reduce resolution; if that is the case, a new cantilever is required. A common problem that can be spotted quickly is double-tip artifacts. An example is shown in **Fig. 12**. One can see that each feature in the image seems to be duplicated at the same offset. That is because the AFM probe physically has two points (asperities) at the end of the tip, each of which give rise to a topographic signal under feedback control.
3. Owing to drifts noted earlier, the raw DC images almost always require "flattening." This process does not alter the content of the images, but removes slow changes (drift) in the DC background that occurs during acquisition; visually, drift shows up as darkened or lightened background in the image (*see* **Fig. 13A**). To make the image background appear uniform from top to bottom, the mean value of each scan line in the fast scan direction (e.g., X) is subtracted, i.e., adjusted to make the DC level uniform. The average value of the entire data page is kept constant. Similarly,

AFM of Lipid Domains in Supported Model Membranes

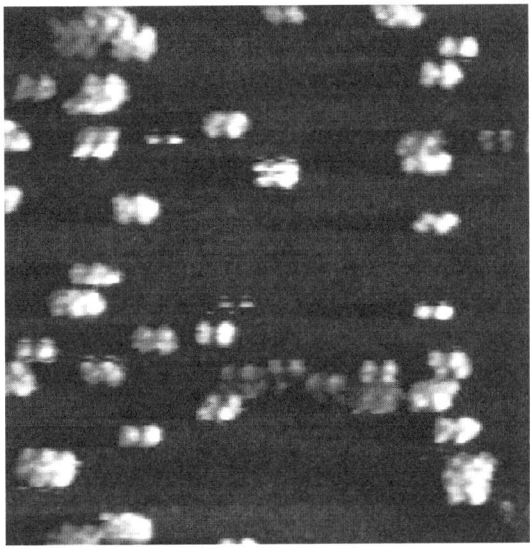

Fig. 12. Example of double-tip artifact in topography image. Bar = 200 nm.

to make the image appear uniform from side-to-side, the mean value of each scan line in the slow scan direction (e.g., Y) is subtracted. This can remove piezo ringing. Unwanted slope in X or Y can be removed by fitting each line to a linear equation. A typical image of lipid domains that has been "flattened" (no other procedure) is shown in **Fig. 13B**. Finally, if a tube scanner is used, then often large scans will exhibit a "bow" in the image owing to the arc traced out by the piezo. All commercial AFMs using tube scanners have a routine to flatten the bowing by fitting the background to a two-dimensional parabola or hyperboloid.

4. An example of a "cursor plot" or cross-section to measure lipid domain height differences is shown in **Fig. 13C** for the dashed line drawn in **Fig. 13B**. Statistical histograms (not shown) can also be used to examine average heights.

4. Notes

1. Mica is the most common substrate for AFM because it can be freshly cleaved with a razor blade. The freshly cleaved surface is atomically flat and free of ambient surface contamination. However, mica is not an intrinsically clean material; it has many surface contaminants such as K^+ and other impurities that give it color. It can be highly charged after cleaving, which could effect its interaction with lipid bilayers. Finally, it is not ideal for combined fluorescence imaging (*5,6*), because the crystal planes scatter light and the impurities can cause fluorescence background.
2. Solvents can be pure chloroform or other ratios with methanol. Lipids are often shipped in pure chloroform. Some lipids, such as GM1 ganglioside, will not dissolve in chloroform, but will dissolve in ethanol.

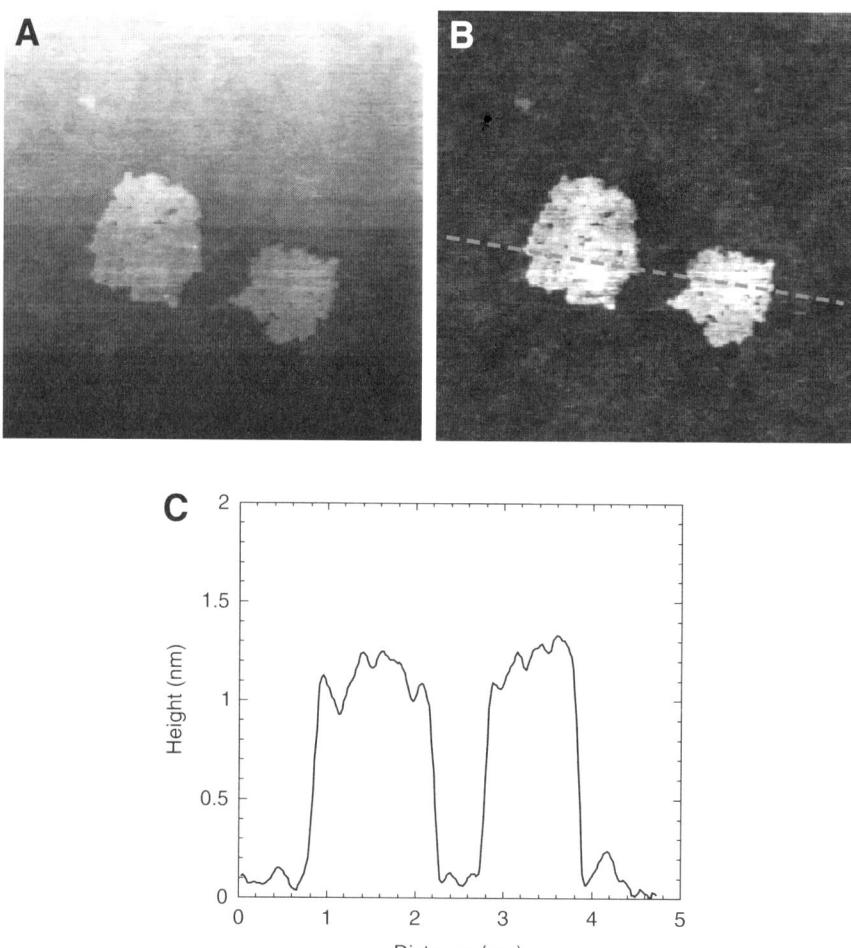

Fig. 13. Example of "flattening" AFM topography (height) image (TM-AFM) of 1,2-dipalmitoyl-*sn*-glycero-3-phosphocholine (DPPC) gel domains observed in fluid phase 1,2-dioleoyl-*sn*-glycero-3-phosphocholine (DOPC) (3:1 DOPC/DPPC, bar = 1 μm). **(A)** DC "raw" image shows Z piezo drift **(B)** Flattened image removes DC drift (*see* text), no other processing was performed. **(C)** Cursor plot of dashed line in **B** reveals the 1.1 ± 0.1 nm height difference of DPPC domains relative to DOPC.

3. This data was acquired with the Fourier transform spectrum analyzer provided in the AFM software. If not available, a stand-alone spectrum analyzer would be required.
4. Individual cantilevers are most often shipped in Gel-Pak 4® containers (plastic boxes with silicone bottoms). These are ideal for cantilever storage and transport and should be kept when emptied.

5. The reflective gold coating on the top side of cantilevers has a different thermal coefficient of expansion than silicon nitride or silicon. This can cause the cantilever to bend before it gets fully equilibrated.
6. False engagement occurs on approach when the feedback loop falsely senses the setpoint and stops the approach before contact. This can occur owing to noise or adjusting the quadrant photodiode signal too close to the setpoint.
7. Control electronics vary in available gains. If integral and proportional gains are available, then maximize the integral gain first and adjust the proportional gain second (about 2X integral gain). If there is one adjustable gain and a time constant available, then maximize gain at high time constant, followed by reduction in time constant.

Acknowledgment

This work was supported in part by the Division of Materials Science, Office of Basic Energy Sciences, United States Department of Energy. Sandia is a multiprogram laboratory operated by Sandia Corporation, a Lockheed Martin Company, for the United States Department of Energy under contract DE-AC04-94AL85000.

References

1. Binning, G., Quate, C., and Gerber, C. (1986) Atomic force microscope. *Phys. Rev. Lett.* **56,** 930–933.
2. Czajkowsky, D. and Shao, Z. (1998) Submolecular resolution of single macromolecules with atomic force microscopy. *FEBS Lett.* **430,** 51–54.
3. Müller, D., Fotiadis, D., and Engel, A. (1998) Mapping flexible protein domains at subnanometer resolution with the atomic force microscope. *FEBS Lett.* **430,** 105–111.
4. Müller, D., Heymann, J., Oesterhelt, F., et al. (2000) Atomic force microscopy of native purple membrane. *Biochim. Biophys. Acta* **1460,** 27–38.
5. Burns, A. (2003) Domain structure in model membrane bilayers investigated by simultaneous atomic force microscopy and fluorescence imaging. *Langmuir* **19,** 8358–8363.
6. Burns, A., Frankel, D., and Buranda, T. (2005) Local mobility in lipid domains of supported bilayers characterized by atomic force microscopy and fluorescence correlation spectroscopy. *Biophys. J.* **89,** 1081–1093.
7. Dufrêne, Y., Barger, W., Green, J. -B., and Lee, G. (1997) Nanometer-scale surface properties of mixed phospholipid monolayers and bilayers. *Langmuir* **13,** 4779–4784.
8. Giocondi, M. -C., Boichot, S., Plénat, T., and Le Grimellec, C. (2004) Structural diversity of sphingomyelin microdomains. *Ultramicroscopy* **100,** 135–143.
9. Giocondi, M. -C., Vié, V., Lesniewska, E., Milhiet, P. -E., Zinke-Allmang, M., and Le Grimellec, C. (2001) Phase topology and growth of single domains in lipid bilayers. *Langmuir* **17,** 1653–1659.
10. Lin, W. -C., Blanchette, C., Ratto, T., and Longo, M. (2006) Lipid asymmetry in DLPC/DSPC-supported lipid bilayers: a combined AFM fluorescence microscopy study. *Biophys. J.* **90,** 228–237.

11. Ratto, T. and Longo, M. (2002) Obstructed diffusion in phase-separated supported lipid bilayers: A combined atomic force microscopy and fluorescence recovery after photobleaching approach. *Biophys. J.* **83,** 3380–3392.
12. Rinia, H., Snel, M., van der Eerden, J., and Kruijff, B. (2001) Visualizing detergent resistant domains in model membranes with atomic force microscopy. *FEBS Lett.* **501,** 92–96.
13. Yuan, C. and Johnston, L. J. (2001) Atomic force microscopy studies of ganglioside GM1 domains in phosphatidylcholine and phosphatidylcholine/cholesterol bilayers. *Biophys. J.* **81,** 1059–1069.
14. Veatch, S. L. and Keller, S. L. (2002) Organization in lipid membranes containing cholesterol. *Phys. Rev. Lett.* **89,** 268,101.
15. García, R. and Peréz, R. (2002) Dynamic atomic force microscopy methods. *Surf. Sci. Rep.* **47,** 197–301.
16. Carpick, R., Agraït, N., Ogletree, D., and Salmeron, M. (1996) Variation of the interfacial shear strength and adhesion of a nanometer-sized contact. *Langmuir* **12,** 3334–3340.
17. Roberts, G. (1990) in *Langmuir-Blodgett Films*, (Roberts, G., ed.), Plenum press, New York, pp. 317–411.
18. Watts, T., Brian, A., Kappler, J., Marrack, P., and McConnell, H. (1984) Antigen presentation by supported planar bilayers containing affinity-purified I-Ad. *Proc. Natl. Acad. Sci. USA* **81,** 7564–7568.
19. Boxer, S. (2000) Molecular transport and organization in supported lipid membranes. *Curr. Opin. Chem. Biol.* **4,** 704–709.
20. Johnson, J., Ha, T., Chu, S., and Boxer, S. (2002) Early steps of supported bilayer formation probed by single vesicle fluorescence assays. *Biophys. J.* **83,** 3371–3379.
21. Reviakine, I. and Brisson, A. (2000) Formation of supported phospholipid bilayers from unilamellar vesicles investigated by atomic force microscopy. *Langmuir* **16,** 1806–1815.
22. Carpick, R. and Salmeron, M. (1997) Scratching the surface: fundamental investigations of tribology with atomic force microscopy. *Chem. Rev.* **97,** 1163–1194.
23. Costa, K. (2006) Imaging and probing cell mechanical properties with the atomic force microscope, in *Cell Imaging Techniques: Methods and Protocols, Methods in Molecular Biology, vol. 319,* (Taatjes, D. and Mossman, B., eds.), Humana Press, Totowa, NJ, pp. 331–361.
24. Hutter, J. L. and Bechhoefer, J. (1993) Calibration of atomic force microscope tips. *Rev. Sci. Instrum.* **64,** 1868–1873.
25. Müller, D., Fotiadis, D., Scheuring, S., Müller, S., and Engel, A. (1999) Electrostatically balanced subnanometer imaging of biological specimens by atomic force microscope. *Biophys. J.* **76,** 1101–1111.

19

Atomistic and Coarse-Grained Computer Simulations of Raft-Like Lipid Mixtures

Sagar A. Pandit and H. Larry Scott

Summary

Computer modeling can provide insights into the existence, structure, size, and thermodynamic stability of localized raft-like regions in membranes. However, the challenges in the construction and simulation of accurate models of heterogeneous membranes are great. The primary obstacle in modeling the lateral organization within a membrane is the relatively slow lateral diffusion rate for lipid molecules. Microsecond or longer time-scales are needed to fully model the formation and stability of a raft in a membra ne. Atomistic simulations currently are not able to reach this scale, but they do provide quantitative information on the intermolecular forces and correlations that are involved in lateral organization. In this chapter, the steps needed to carry out and analyze atomistic simulations of hydrated lipid bilayers having heterogeneous composition are outlined. It is then shown how the data from a molecular dynamics simulation can be used to construct a coarse-grained model for the heterogeneous bilayer that can predict the lateral organization and stability of rafts at up to millisecond time-scales.

Key Words: Cholesterol; dioleoyl phosphatidylcholine; dipalmitoyl phosphatidylcholine; lipid bilayer; mean field theory; molecular dynamics; sphingomyelin.

1. Introduction

As discussed in other chapters, in this volume there is abundant evidence gathered in recent years suggesting that cellular plasma membranes contain small domains of different lipid composition. Of particular interest are nano-size membrane fragments insoluble in detergent. These fragments are called detergent-resistant membrane domains (DRM) or "rafts" *(1–4)*. In a biological context there is much evidence that nanometer-sized domains, presumably similar in structure to DRMs, are important membrane structural components in signal transduction *(5–8)*, protein transport *(9–11)*, and sorting of membrane components *(12–15)*. There is also evidence for rafts functioning as sites for the binding and transport into the cell of several pathogens and toxins, including the human immunodeficiency virus-1 and the prion protein PrP^{sc} *(16,17)*.

At lower temperatures much of the sphingolipid and cholesterol (Chol) components of mammalian cell membranes can be isolated in DRM fragments *(18,19)*. Data from fluorescence polarization measurements of liposomes incorporating diphenylhexatriene show that the extracted DRM domains are in a liquid-ordered (L_o) phase *(20)*. In the L_o phase the lipid chains are highly ordered but whole lipid molecules have rotational and lateral diffusion coefficients comparable with those in the liquid-crystalline (L_α) phase. Consequently, a major focus of raft research involves studies of bilayer systems in the L_o phase *(18)*. These systems typically consist of mixtures of saturated phospholipids, sphingolipids, and Chol. Reitveld and Simons *(2)* suggested that "rafts" are L_o phase domains dispersed in a L_α phase bilayer. A variety of experimental techniques have been used to study the properties of rafts or related L_o phase domains in simple model membrane systems. The model membrane systems are generally multilamellar vesicles, monolayers, or giant unilamellar vesicle systems. Techniques used typically include fluorescence microscopy *(21–23)*, single-particle tracking *(24)*, differential-scanning calorimetry, X-ray diffraction *(25)*, and atomic force microscopy (AFM) *(26–28)*. A comprehensive review of experimental studies of rafts and DRMs *(29)* concludes that actual rafts in membranes may be very small (10 nm in diameter) and that the intermolecular interactions, which drive the formation and stability of these nanodomains is still largely unknown.

Computational modeling offers a potential way to gain insights into the nature of the lateral organization in membranes that is not easily obtained in experiments. Simulation studies are broadly divided into two classes; atomistic simulations and simplified model simulations. At its most basic level, computational modeling is done by atomistic molecular dynamics (MD) or Monte Carlo simulations. In recent years, numerous MD simulations have been run on saturated phospholipid and Chol systems to investigate atomic level interactions and properties of bilayers in the L_o phase *(30–37)*. Most of these simulations focus on the detailed interaction of Chol with the saturated phospholipid and the clustering behavior leading to the L_o phase. There have been relatively few simulation and modeling attempts that directly focus on the issues related to the separation of L_o and L_α phases in bilayer systems *(36,38,39)*. The reason for the lack of such studies is twofold; first, to observe formation of a nanometer-size domain requires a system as large as the domain plus surrounding bilayer, which can mean at least thousands of lipids, plus waters of hydration. Second, and even more seriously, very long time simulations are needed to allow forsufficient lateral reorganization to occur through diffusion. Typically, the times should be in the millisecond range, i.e., four orders of magnitude longer than the longest atomistic MD simulations accessible to current hardware and software systems.

The approach described in this review is to first use MD simulations to model the interactions between lipids and Chol in mixed lipid bilayers at an atomic level. Then the MD data are used as input to construct a mean field-based coarse-grained model that can serve as a predictive platform for the lateral organization of the membranes.

2. Materials

2.1. Lipids Simulated

1. Dipalmitoyl phosphatidyl choline (DPPC).
2. Dioleoyl phosphatidyl choline (DOPC).
3. 1-Palmitoyl, 2-oelyphosphatidylcholine (POPC).
4. 18:0 Sphingomyelin (SM).
5. Phosphatidylinositol (PI).
6. Chol.

2.2. Software Tools

1. GAUSSIAN http://www.gaussian.com *(40)* for the calculation of forcefield parameters.
2. There are several popular software packages that may be used for lipid bilayers MD simulations. GROMACS open source software http://www.gromacs.org *(41–43)* is the program used for MD simulations by the group. Other popular MD simulation software includes NAMD (http://www.ks.uiuc.edu/Research/namd/) and the commercial software CHARMM (http://www.charmm.org/). All these software packages rely on message passing interface-based parallelism, so they can be easily installed and used on wide variety of parallel computers ranging from Symmetric Multiprocessing Platforms SMP to small and large-scale clusters.
3. The algorithm for Configurational Bias Monte Carlo may be found in Ref. 44. A widely used Monte Carlo Software Package that implements this algorithm and is freely available is TOWHEE (http://towhee.sourceforge.net).
4. *Analysis code and tools:* the GROMACS package provides many programs for the calculation of properties of simulated systems. In addition several analysis programs have been developed. These programs are written in "C" language and use GROMACS libraries. Examples include code for the calculation of X-ray form factors, and bilayer Voronoi tessellation construction.

2.3. Hardware Resources

1. *Linux clusters:* a Linux cluster is an inexpensive and convenient solution that can deliver relatively low-scale performance owing to a common communication bottleneck unless installed with a low-latency performance bandwidth communication device. This can lead to poor scaling of a simulation with increasing numbers of processors, thereby reducing the advantage that liux clusters offer. GROMACS

does not scale well with the number of processors (*see* **Notes 1–5** for details) so a small size cluster usually provides adequate computing backbone.

2. *Supercomputing centers:* simulations can be run at a supercomputing center. Performance of the simulation depends on the exact facilities available at the center and the queuing and resource allocation policies of the center.
3. *Large-scale grids:* grids are not a particularly useful computing backbone for atomistic MD simulations owing, again, to communication bottlenecks. However, if one wishes to accumulate statistical data for, for example, domain structure and stability, then multiple independent runs are an excellent use of a grid environment.

3. Methods
3.1. Atomistic MD Methods

In MD, snap shots of the system are generated by starting with initial positions and velocities of all the atoms in the system and integrating Newton's equations of motion for all these atoms. However, in doing so one must make critical assumptions. One basic assumption is that the dynamical time-scales for electronic degrees of freedom are several orders of magnitude faster than the motion of the individual atomic nuclei. Hence, it is possible to approximate the effects of electronic motions by averages when defining bonded and nonbonded interactions. A second assumption is that all the nonbonded interactions are represented by pairwise interactions. **Equation 1** shows a typical potential energy form that incorporates these assumptions, and is used in most of the classical MD software packages.

$$V_{total} = \sum_{bond} K_b (r-r_0)^2 + \sum_{angle} K_\theta (\theta-\theta_0)^2 + \sum_{improper} K_\Phi (\Phi-\Phi_0)^2 \\ + \sum_{dihedral} K_\phi [1+\cos(n\phi-\phi_0)] + \sum_{\substack{nonbonded \\ pair}} \left\{ \frac{q_i q_j}{r_{ij}} + \left[\frac{C_{ij}^{(12)}}{r_{ij}^{12}} - \frac{C_{ij}^{(6)}}{r_{ij}^6} \right] \right\}. \quad (1)$$

The sum runs over (in order) bonds, bond angles, improper and proper dihedrals, and all pairs of atoms that are on different molecules or are separated by more than four interatomic bonds on a molecule. The dihedral function in **Eq. 1** represents the energy of a connected set of four consecutive atoms on a molecule. The parameters in each term in the sum must be determined as described next. The negative gradient of this potential provides the force used in the solution of Newton's equations. In the implementation of MD simulations of a mixed or raft-like lipid bilayer, through the above equation, the following steps are required:

3.1.1. Force Field Parameters

To obtain correct dynamics it is necessary to specify the set of parameters in **Eq. 1**. These parameters are called *force field parameters*. In principle it is necessary to determine these parameters by independent calculations and simulations that determine the values that provide the best fit to independent experimental data. Many groups have already calculated force fields for lipids and Chol *(45–47)*. However, if a simulation of a new lipid or a Chol analog such as ergosterol is planned, then the following force field parameters should be calculated.

1. Bond-length and -angle parameters.
2. Partial charges.
3. The non-bonded parameters C^{12}_{ij} and C^{12}_{ij} in **Eqa.1**.
4. Dihedral parameters.

3.1.2. Atomistic MD Simulation Setup

When setting up an MD simulation with GROMACS, a number of decisions must be made and stored in an input file and in the molecular topology files (GROMACS manual http://www.gromacs.org/documentation/reference_3.3/online.html):

1. *Hydrogen atoms:* a decision must be made regarding the inclusion of hydrogen atoms. Simulations that include all hydrogen atoms explicitly are referred as *all-atom simulations*. Usually all-atom simulations are more accurate, but they impose a huge penalty in time. This penalty arises from the large number of additional degrees of freedom associated with the hydrogen atoms. Also, explicit hydrogens in the system restrict the integration time step because hydrogen atom is an order of magnitude lighter than the other atoms in the simulation. As an alternative, one can use a *united-atom model*, in which apart from hydroxyl, amide, and water hydrogen atoms, all hydrogens are combined with the atom to which they are connected, forming a larger 'pseudoatom'. For example, CH_2 and CH_3 become pseudoatoms with larger interaction radii than abare carbon atom. Usually for lipid systems, the united-atom simulations are preferred because they strike a balance between accuracy and the simulation speed.
2. *Boundary conditions:* a simulated system is very small compared with the corresponding experimental system. In such a small system the boundaries can have significant effect on the physical properties of simulated systems. The usual choice is periodic boundary conditions, whereby the system is replicated indefinitely in all the directions.
3. *Ensemble choice:* the propagation of a simulated lipid bilayer, by MD integration of Newtonian equations over time, samples the states on a constant energy hypersurface in the phase space of the system. This gives rise to a simulated system with fixed number of particles (N), fixed volume (V), and fixed energy (E). Therefore, an NVE simulation produces samples from the microcanonical ensemble. To produce samples from different ensembles, the system can be coupled to a heat bath at constant temperature (NVT ensemble) and also can be coupled to pistons at constant temperature and pressure (NPT ensemble). Hybrid ensembles such as, constant pressure perpendicular to the membrane and constant surface tension within the membrane (NγT), or constant surface area (NAPT) can be constructed.

For a detailed review of coupling techniques *see* Frenkel and Smit *(44)*. The NPT, NγT, and NAPT ensembles are most frequently used in bilayer simulations (*see* **Notes 6** and **7** regarding isotropic and semi-isotropic pressure couplings).

4. *Lennard–Jones cutoff:* in MD software the bonded interactions are efficiently implemented as static lists. The nonbonded interactions, such as Lennard–Jones and electrostatics interactions, cannot be implemented as static lists because one needs to consider all possible pairs of atoms in these interactions. With periodic boundaries these sums have to incorporate the effect of atoms from periodic images. However, if the interaction strength diminishes with distance faster than r^{-3} (*see* **Note 7** for details) then one can use a cutoff method whereby the potential is calculated only up to cutoff radius from central atom and considered to be zero beyond cutoff. Simulation software usually allows the user to select this cutoff value. A conservative choice is 1.8 nm cutoff, which introduces approx 3×10^{-4} kJ/mol error in typical CH_3–CH_3 interaction at cutoff distance (although many researchers use a much shorter cutoff of 1 nm to improve simulation efficiency giving rise to an error of approx 10^{-2} kJ/mol in a typical CH_3–CH_3 interaction).

5. *Electrostatics:* the long range electrostatic forces present a challenge. One can simply truncate the electrostatic interaction, but this has been shown to produce artifacts *(48)* as it diminishes with distance only as r^{-1}. The situation becomes even more challenging with periodic boundaries as electrostatic interaction sum becomes conditionally convergent and requires special care in summing. A preferred choice for performing electrostatics sums in this situation is the Ewald summation algorithm, for example, *see* **ref. 44**. In software implementations the Ewald sum algorithm is further improved using smooth particle mesh technique *(49)*. The Ewald sum algorithm requires parameters such as real space cutoff, number of Fourier space vector, and so on (for details about choosing these parameters *see* **Note 7**).

3.1.3. Running a Simulation

1. *Construction of a bilayer for simulation:* in simulations, a small patch of a bilayer self-assembles in a lipid water solution in few tens of nanoseconds *(50)* but usually simulations are performed starting with preassembled bilayers. Programs have been developed to construct bilayers with various lipid compositions under various conditions. For the simulations of raft-like bilayers one needs to perform simulations with several different possible initial states, such as preformed domain or random distribution of the lipid components in the system (*see* **Notes 5** and **8**).

2. *Equilibration:* the equilibration time depends on the size and chemical composition of the simulation, and on the initial state of the system. Typically, equilibration is examined by calculating time correlation functions of relevant physical quantities of the system, for example, in protein systems one can observe autocorrelation function (and correlation times) for Ramachandran angles. In lipid systems usually area per lipid and chain order parameters are closely monitored. Apart from physical properties, thermodynamic properties such as temperature and pressure are also closely monitored during equilibration process.

3. *Diffusion and simulation time-scales:* the duration of an MD simulation run is determined by the slowest degrees of freedom that one wants to study. In raft-like lipid simulations this time is determined by the diffusion coefficient of the lipid constituents. Lateral diffusion coefficients for the lipids in the L_α phase, determined from experiment, are of the order of approx 5×10^{-12} m^2 *(51)*. This value along with Einstein's relation gives a root mean square displacement of approx 20 Å for a single lipid in 200 ns (*see* **Note 9**). So, if the organization is investigated on the length scale of 20 Å then, typical simulation runs should be several hundreds of nanoseconds. On a small size linux cluster a GROMACS MD simulation consisting of around 200 lipids with adequate amount of water (*see* **Notes 2–4**, and **9**) can achieve about a half nanosecond per day. Hence, one needs at least few months of wall clock simulation time to observe organization on 20 Å organization.

3.1.4. Analysis of Simulation Results

MD simulations: in general the strategy for the analysis of simulation data is to calculate properties of the systems that can be measured experimentally. If simulations agree with experimental data, then calculations of other properties represent predictions of the simulation.

1. *Area per lipid and Voronoi tessellation:* area per lipid is one of the key properties of lipid systems. In experiments it is deduced from a model fitted to the electron density *(52,53)*. In the simulation of pure systems the average area per molecule is generally calculated by taking the ratio of twice the area of the simulation cell to the total number of lipid molecules. However, in mixtures this crude method cannot give areas for each molecular species separately. The problem of calculating the correct area per lipid in Chol–DPPC mixtures has been addressed by Hofsass et al. *(35)* and Chiu et al. *(34)*. Hofsasβ et al. resolved this issue by considering the volumes of the constituent molecules and writing the average thickness of the bilayer in terms of the simulation cell volume and area. Chiu et al. performed several simulations with varying concentration of Chol and observed a linear relation between the area per molecule and the concentration from which the area per DPPC and the area per Chol could be calculated. An alternative method for calculating area per molecule in a binary lipid mixture was proposed by Nagle and Edholm *(54)*. The method is based on the use of the partial specific area and requires simulations at several different stoichiometries to extract the partial specific area for each component. However, for ternary or higher mixtures of molecules neither of these methods can be used. One method that can be used is based on Voronoi tessellation of the projected coordinates of certain key atoms in each lipid onto some plane parallel to the bilayer surface. **Figure 1** shows a Voronoi tessellation for the simulation of ref. *(55)*, and *see* **Note 10** regarding this technique.
2. *Thickness of the bilayer:* to measure the thickness in reliable way, the simulation setup is first compared with a typical AFM experimental setup, for example, *see* **ref**. *26*. In the AFM experiments the thickness is measured with respect to a flat surface on which the bilayer is supported. As there is no such flat reference

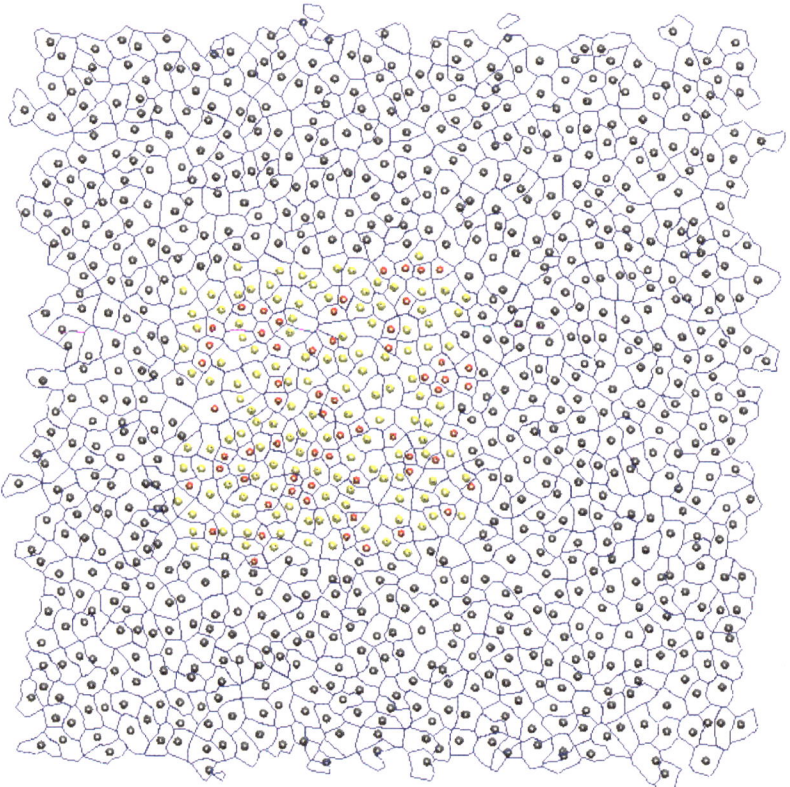

Fig. 1. Voronoi tessellation of domain system consisting of SM, Chol, and DOPC (reprinted from **ref. 52**, with permission).

surface, an algorithm proposed by Pandit et al. *(56)* was used. This algorithm gives a surface-to-point correlation function. For each phosphorus in the top leaflet, first, the phosphorus in the lower leaflet that is approximately below it, is identified. This is achieved by

a. Tessellating the lower leaflet into Voronoi polygons.
b. Projecting coordinates of phosphorus from top leaflet on to this tessellated surface.
c. Identifying the polygon in which the projected coordinates fall. This procedure identifies a transbilayer "neighbor" for each lipid in the top leaflet.

With such identification, the distance of phosphorus in the top leaflet is defined with respect to the surface defined by the phosphorus atoms in the lower leaflet, as the normal distance between phosphorus atoms from two leaflets, which are "vertical neighbors" of each other. This allows to calculate the densities of phosphorus atoms of lipids in one leaflet with respect to the surface defined by the phosphorus atoms in the other leaflet. From the peak of the density distribution, one obtains the phosphorus-to-phosphorus thickness of the bilayer.

3. *Electron density and form factor:* small angle X-ray scattering is a quantitative tool for the structural study of lipid bilayer membranes *(52,53)*. Small angle X-ray scattering patterns reveal the form factors of the scattered X-rays. Although it is possible to calculate electron density profiles from form factors, this effort depends on models for the unknown X-ray phase factors. Therefore, MD simulations should calculate form factors predicted by the simulations for direct comparison with experimental data. In these calculations the simulation system is divided in slices along the bilayer normal and the electron density is calculated by binning the electron counts into slices. Because the electron density is symmetric around the center of the bilayer plane, one can calculate the form factor by expanding the electron density as a cosine transform. In simulations this is done by numerically evaluating the following cosine transform:

$$F(q) = \int_{-L/2}^{L/2} [\rho_e(z) - \rho_e^w] \cos(qz) dz$$

where L is the box length along z-direction, $\rho_e(z)$ is the electron density profile obtained from simulation, and ρ_e^w is the electron density of the bulk water in the simulations. The integration is evaluated using either the 1/3 or the 3/8 Simpson's rule. **Figure 2** shows the electron density and the corresponding form factor for a bilayer made up of quaternary mixture of POPC, SM, Chol, and PI.

4. *Lipid chain order parameters:* the ordering of hydrocarbon tails is determined in nuclear magnetic resonance experiments with deuterated lipid chains by measuring the quadrupolar splitting, associated with the deuterium order parameters for the C–D bonds in the hydrocarbon chains. The order parameter tensor (S), is defined as

$$S_{ab} = \frac{1}{2} < 3\cos(\theta_a)\cos(\theta_b) - \delta_{ab} > \quad a,b = x,y,z$$

where θ_a is the angle made by a-th molecular axis with the bilayer normal and δ_{ab} is the Kronecker δ. In the simulations, with the united-atom force field, the order parameter for saturated and unsaturated carbons S_{CD} can be determined using the following relations,

$$-S_{CD}^{sat} = \frac{2}{3} S_{xx} + \frac{1}{3} S_{yy}$$

$$-S_{CD}^{unsat} = \frac{1}{4} S_{zz} + \frac{3}{4} S_{yy} \pm \frac{\sqrt{3}}{2} S_{yz} \qquad (2)$$

respectively, *(57)*. The order parameter profiles can be calculated separately for each type of molecule in the simulation. **Figure 3** shows the order parameter profiles of POPC, SM, and PI in a simulation.

5. *Pair correlation functions:* pair correlation functions are crucial to the understanding of the structural properties of a raft-like bilayer. If properly sampled, the calculated correlation functions can give information about change in free energy

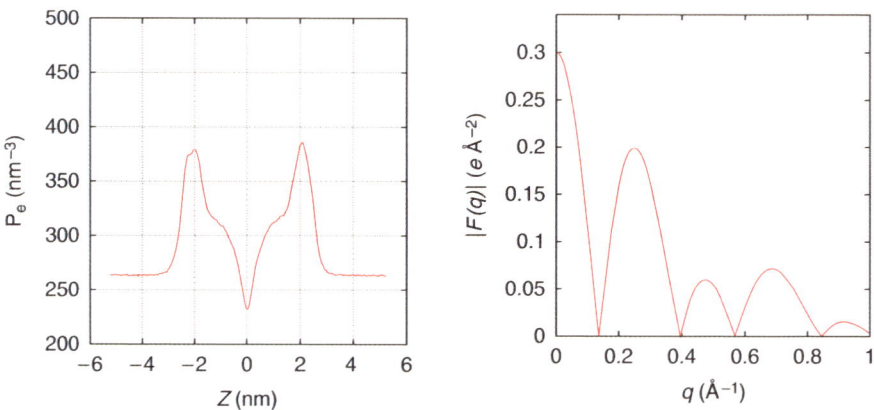

Fig. 2. Electron density profile (**left**) and the corresponding form factor (**right**) for system consisting of POPC, SM, Chol, and PI.

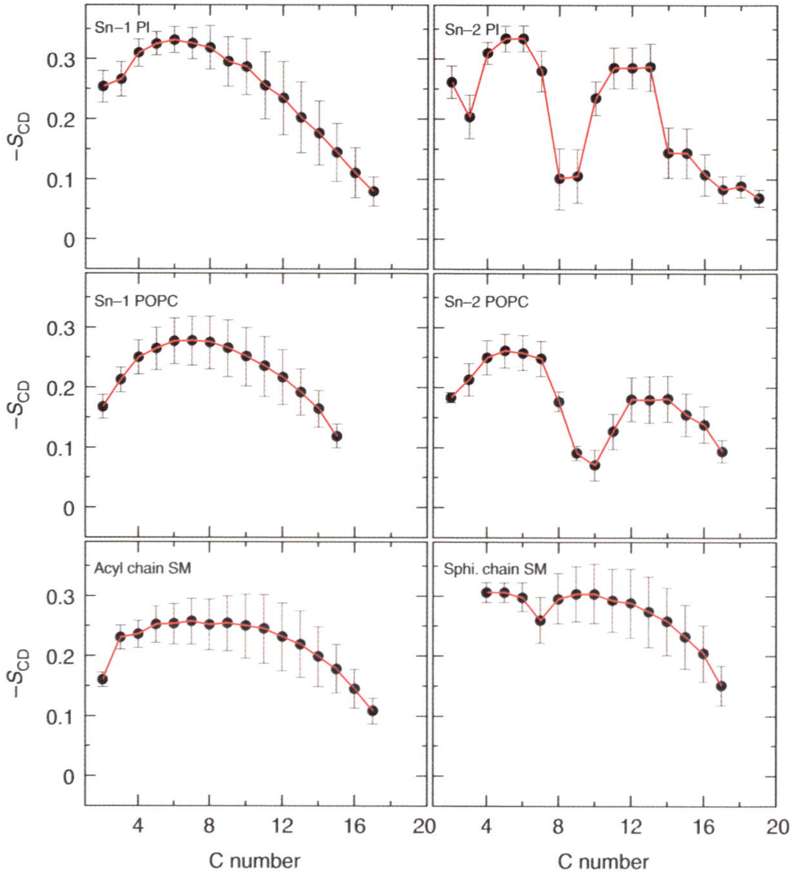

Fig. 3. Order parameters of POPC, SM, and PI chains.

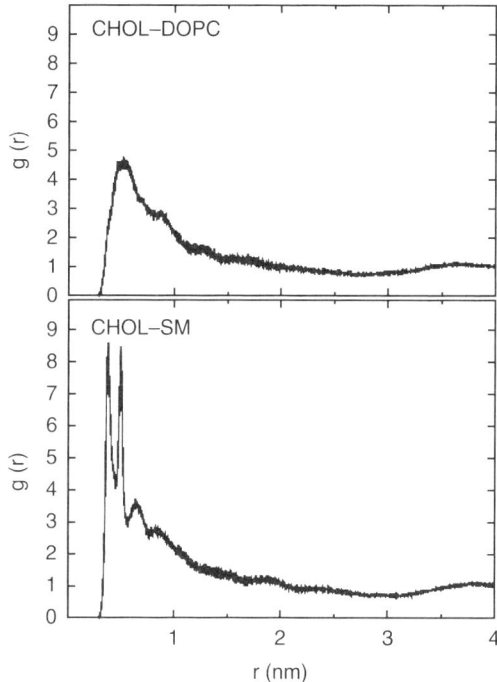

Fig. 4. RDF of SM–Chol and DOPC–Chol in ternary mixture system (reprinted from **ref. 59** with permission).

with the configuration of two particles as reaction coordinate. Most commonly, the pair correlation functions are averaged over the solid angles giving only the radial dependence. These are denoted as *radial distribution functions* (RDF). The RDF is defined as

$$g(r) = \frac{N(r)}{4\pi r^2 \rho \delta r} \tag{3}$$

where $N(r)$ is the number of atoms in the shell between r and $r + \delta r$ around the central atoms, ρ is the number density of atoms, taken as the ratio of the number of atoms to the volume of the simulation cell. In a ternary mixture simulation of DOPC, SM, and Chol, the RDF between Chol–SM and Chol–DOPC is used to investigate the role of Chol in the formation of raft-like domains (*see* **Fig. 4**). **Figure 4** shows that the first coordination shell of SM, with respect to Chol. The first coordination shell is defined by the location of the sharp peak in the RDF. In Chol–SM case the first shell consists of two peaks, indicating two possible binding locations for SM, whereas for Chol–DOPC, RDF has only one peak. However, it turns out that the number of SM or DOPC molecules in the first shell of Chol are the same. This information can be obtained by integrating RDF up to first coordination shell. This suggests that, even if there are two binding locations for SM, the coordination number of Chol with SM and DOPC is nearly the same, so that Chol does not show specific binding preference for SM or DOPC. Recent fluorescence spectroscopy and differential calorimetric studies performed on mixtures of PC and SM and Chol also indicate a lack of specific interaction between SM and Chol *(58)*. Hence, the RDFs by themselves do not provide direct insight into the role of Chol in domain formation.

The Chol molecule has one flat face (the α-face) and one face that is rough because of protruding methyl groups (the β-face). As Chol lies primarily in the hydrocarbon region of the bilayer, it is reasonable to question whether this specific design of the Chol molecule plays any role in promoting domain formation. For correlations involving two variables, orientation and position, one can define a bivariate correlation function $g(r, \varphi)$ between one selected backbone carbon atoms of DOPC and SM molecules, respectively, and the oxygen atom of Chol, defined by

$$g(r,\varphi) = \frac{N(r,\varphi)}{2\pi r \rho \delta r \delta \varphi}$$

where the distance r and ρ are defined as in RDFs, the angle φ is the angle made by the distance vector with respect to the positive x-axis of the Chol body coordinate frame (*see* **Fig. 5**), and $N(r, \varphi)$ is the number of the selected lipid carbon atoms in an area element $r\delta r\delta\varphi$ at the point (r, φ) from the oxygen of Chol.

Figure 6 shows $g(r, \varphi)$ for SM and DOPC averaged over last 150 ns of the trajectory. The figure clearly shows that, on the simulation time-scale, SM is associated with the α-face of Chol. On the other hand DOPC does not show a preference for either face of Chol. The preferential arrangement of SM around Chol indicates that on the simulation time-scales, the Chol molecules tend to locate at the interface between the SM and the DOPC regions of the membrane, with the α-face of the Chol molecule interacting most strongly with SM chain. Such an arrangement again should be entropically favored because the saturated SM chains pack well around the α-face and more disordered unsaturated DOPC chains should pack well around the β-face.

6. *The definition of a domain and domain identification:* although the chain order and the thickness of the bilayer in simulations can be calculated *(55)*, the large error bars on these quantities do not provide a well-resolved temporal description of the lateral organization of the raft-like bilayer system (*see* **Note 11**). A more robust criterion can be based on the projected polygonal areas of the molecules. As a rule of thumb, higher chain order gives rise to smaller areas per molecule. Hence, if one considers all molecules having their polygonal area below a certain cutoff value to be "ordered" and find connected aggregates based on polygon neighbors, then the domains based on this chosen cutoff value can be identified as connected sets of molecules with areas less than the cutoff. Such domains are called A-domains where A is the cutoff value used to identify the domains. Because the choice of a particular area cutoff does not uniquely characterize a nanoscopic domain, A-domains for various cutoff values of A are examined *(59)*.

Computer Simulations of Raft-Like Lipid Mixtures

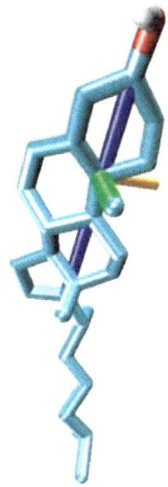

Fig. 5. Body center coordinate system for Chol, z-axis is represented by blue, x-axis is represented by green, and y-axis is represented by orange cylinders.

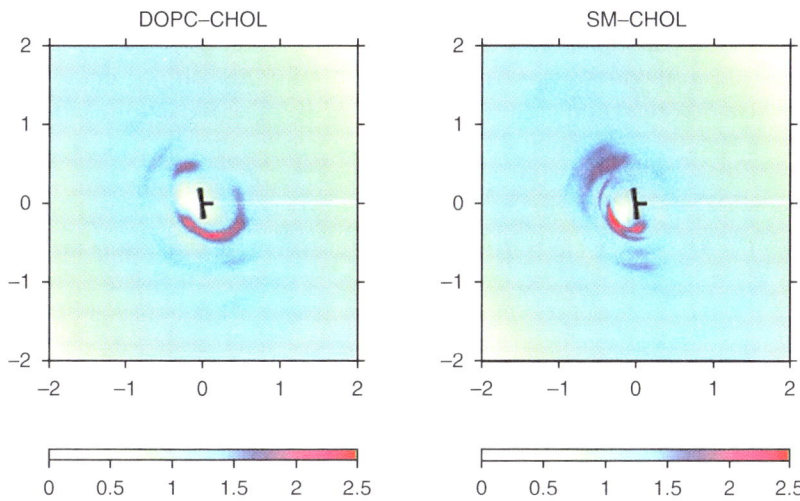

Fig. 6. Bivariate correlation function between SM–Chol and DOPC–Chol (reprinted from **ref. 59**, with permission).

3.2. Coarse-Grain Models

3.2.1. Self Consistent Mean Field Theory Models

Although the details of the construction of equilibrium self-consistent mean field theory (SCMFT) models are given in the chapter by Schick in this volume, our group has developed an SCMFT-based method that allows for the simulation of the time evolution toward equilibrium. The complete details of the model are described in a recent publication *(60)*. An outline of the overall strategy for this approach is:

1. Construct a mapping from an atomistic lipid bilayer onto a two-dimensional (2D) field based on the average chain order parameters or another molecular structural property.
2. Embed Chol or other molecules that are to interact with the lipid field onto the 2D field.
3. Construct a model based on statistical mechanics for the thermodynamic evolution of the field.
4. Use Langevin dynamics for the translation and rotation of the embedded objects.
5. Calculate all interactions and other parameters for the model from atomistic MD simulations.

3.2.2. Pseudoatom Models

In this approach lipid and water molecules are represented by chains of pseudoatoms. Each pseudoatom may actually represent several atoms from an atomistic simulation. After the pseudoatom is defined one follows the same basic procedure as in atomistic MD simulations. Because of the reduced number of df, the simulations can typically be several orders of magnitude larger in size and time-scale, compared with atomistic MD. The model of Marrink et al. *(61)* is an excellent example of this process.

3.2.3. Analysis of Simulation Results: Coarse-Grain Models

1. *SCMFT-MD models:* although many atomic level details are lost in the mapping to an SCMFT model one can monitor distributions of molecules through 2D RDF's. In some cases it is possible to extract some 3D information from the 2D simulation. Because this type of modeling is based on statistical mechanics ideas, thermodynamic quantities like free energy and heat capacities are directly accessible. By using the model described in **Section 3.2.1**, one could investigate the lateral organization induced by Chol and compare the results with the DSC experimental curves.
2. *Coarse-grained models based on pseudoatoms:* the analysis of this type of model basically follows the analysis of atomistic MD simulations with appropriate reinterpretations owing to atoms (*see* **Notes** *12* and *13*).

4. Notes

1. *Scalability of the simulations:* not all MD code packages are equally scalable on parallel computers. In general, GROMACS and NAMD perform better in parallel environment than CHARMM. The choice of GROMACS and NAMD depends on the system size, the model used for electrostatics, and personal preference.

2. *Resource considerations:* if a dedicated linux cluster is available then one can run accurate simulations of bilayers with more than 1000 lipids, plus waters of hydration. If one is using a shared computing resource, larger simulations may sit in queue, making them less efficient in terms of wall clock throughput.
3. *System size and scalability:* if Particle Mesh Ewald Summation is used for the calculation of electrostatic interactions, then scalability is limited. For a system of 100–200 lipids (50–100/leaflet), the performance of a linux cluster diminishes more than 8–12 nodes. For much large systems scalability is improved. But the tradeoff is that these simulations still take longer per nanosecond to run. In general, larger is better, if one has the resources.
4. *Waters of hydration:* in general one should include sufficient water molecules so that there is a layer of bulk-like water above the surface of the membrane. Without this layer, there can be indirect interactions between the polar groups on the two opposing leaflets owing to periodic boundaries. Typically, it is found that at least 30 waters per lipid (including Chol) are needed.
5. *Selection of the initial state:* the initial setup of the simulation state must be done carefully. There should not be any artificial, local, or global structures in the initial state, as these may produce unphysical correlations that may not disappear even after a very long simulation. Initial states with large potential energy embedded between some of the molecules will "blow up" owing to large repulsive force.
6. *Neutral charge groups:* simulation programs parse the atoms in a molecule into "charge groups" for the calculation of electrostatic interactions. These charge groups must be neutral to avoid unwanted charge–charge correlations. Additionally, the charge groups should not be excessively large, as this will also cause artifacts.
7. *Pressure coupling:* for the simulation of a planar lipid bilayer embedded in a 3D bath of water, the two axes in the plane of the bilayer are to be distinguished from the 3D, normal to the bilayer. Most MD code packages implement a constant pressure simulation by allowing the dimensions of the three axes to change during the simulation, in response to an applied pressure (this is implemented in several different ways that are beyond the scope of this chapter to describe). The user must choose whether to allow the three sides of the simulation box to be changed isotropically, or to decouple the three sides of the box. For a bilayer the best approach is to decouple the normal dimension, but couple the 2D of the box parallel to the bilayer plane (this avoids unphysical changes in the shape of the membrane in the box).
8. *Force field parameters:* simulations can be extremely sensitive to the details of the force field parameters. In the simulations of raft-like mixtures this is especially important. Force field parameters should be tested by running simulations of one-component bilayers that consist the lipids to be used, and compare carefully with experimental data.
9. *Diffusion constant:* for the calculation of diffusion constants, the system must have progressed to the point whereby a plot of the mean square deviation vs time is linear over a large number of time steps. If this is not the case the Einstein relation will not be valid.

10. *Voronoi tessellation:* tessellations based only on molecular centers of mass will cause artifacts. A Voronoi tessellation based on the centers of mass of lipids was proposed by several authors, for example, *see* **refs. *48* and *62–65*.** However, for any system with a mixture of molecules of different sizes, this method may overestimate the areas of smaller molecules and underestimate the areas of larger ones.
11. *Raft identification:* identification of larger raft-like structures from locally organized small domains is especially difficult. The issues to be addressed include the lifetime of the proposed domains, the detailed molecular interactions in the domains, and the structure of the lipid matrix around the domain.
12. *Coarse graining with pseudoatoms:* care must be taken in using simulations based on pseudoatoms to predict the structure and properties of rafts. The coarse-graining step can significantly affect the delicate balance of forces that are responsible for domain formation. For this reason, pseudoatoms based simulations should be closely tested with full atomistic simulations and to experimental data.
13. *Coarse graining using SCMFT:* the input of MD simulation data into mean-field model is subject to large uncertainties. This mapping should be based on RDF and intermolecular forces averaged over very large simulations in order to minimize statistical errors.

Acknowledgments

SAP thanks Prof. Ananth Grama for financial support under NSF Grant no. DMR0427540 and HLS thanks Prof. Shankar Subramaniam for hospitality at the San Diego Supercomputing center.

References

1. Simons, K. and Ikonen, E. (1997) Functional rafts in cell membranes. *Nature* **387,** 569–572.
2. Reitveld, A. and Simons, K. (1998) The differential miscibility of lipids as the basis for the formation of functional membrane rafts. *Biochim. Biophys. Acta* **1376,** 467–479.
3. Pralle, A., Keller, P., Florin, E. L., et al. (2000) Sphingolipid-cholesterol rafts diffuse as small entities in the plasma membrane of mammalian cells. *J. Cell Biol.* **148,** 997–1007.
4. Jacobson, K. and Dietrich, C. (1999) Looking at lipid rafts? *Trend Cell Biol.* **9,** 87–91.
5. Manes, S., Mira, E., Gomez-Moulton, C., et al. (1999) Membrane raft microdomains mediate front-rear polarity in migrating cells. *EMBO J.* **18,** 6211–6220.
6. Aman, M. J. and Ravichandran, K. S. (2001) A requirement for lipid rafts in B cell receptor induced Ca^{2+} flux. *Curr. Biol.* **10,** 393–396.
7. Xavier, R., Brennan, T., Li, Q., et al. (1998) Membrane compartmentation is required for efficient T cell activation. *Immunity* **6,** 723–732.

8. Kawabuchi, M., Satomi, Y., Takao, T., et al. (2000) Transmembrane phosphoprotein cbp regulates the activities of scr-family tyrosine kinases. *Nature* **404,** 999–1002.
9. Rozelle, A. L., Machesky, L. M., Yamamoto, M., et al. (2000) Phosphatidylinositol 4,5-biphosphate induces actin-based movement of raft-enriched vesicles through WASP-Arp2/3. *Curr. Biol.* **6,** 311–320.
10. Cheong, K. H., Zachetti, D., Schneeberger, E., and Simons, K. (1999) VIP17/MAL, a lipid raft-associated protein, is involved in apical transport in MDCK cells. *Proc. Natl. Acad. Sci. USA* **96,** 6241–6248.
11. Viola, A., Schroeder, S., Sakakibara, Y., and Lanzavecchia, A. (1999) T Lymphocyte costimulation mediated by reorganization of membrane microdomains. *Science* **283,** 680–682.
12. Manie, S. N., Debreyne, S., Vincent, S., and Gerlier, D. (2000) Measles virus structural components are enriched into lipid raft microdomains: a potential cellular location for virus assembly. *J. Virol.* **74,** 305–311.
13. Harder, T., Scheiffele, P., Verkade, P., and Simoms, K. (1998) Lipid domain structure of the plasma membrane revealed by patching of membrane components. *J. Cell Biol.* **141,** 929–942.
14. Sönnichsen, B., Nielsen, E., Reitdorf, J., et al. (2000) Distinct membrane domains on endosomes in the recycling pathway visualized by multicolor imaging of Rab4, Rab5, and Rab1. *J. Cell Biol.* **149,** 901–914.
15. Zerial, M. and McBride, H. (2001) Rab proteins as membrane organizers. *Nat. Rev. Mol. Cell Biol.* **107,** 17.
16. Fantini, J., Garmy, N., Mahfoud, R., and Yahi, N. (2002) Lipid rafts: structure, function, and role in HIV, Alzheimers, and prion diseases. *Exp. Rev. Mol. Med.* **20,** 1–21.
17. Brügger, B., Glass, B., Haberkurt, P., et al. (2006) The hiv lipidome: A raft with an unusual composition. *Proc. Natl. Acad. Sci. USA* **103(8),** 2641–2646.
18. "Brown D. A. and Rox, J. K. (1992) Sorting of GPI–Anchored proteins to glycolipid–enriched membrane subdomains during transport to the apical cell surface. *Cell* **68,** 533–544.
19. Brown, D. A. and London, E. (1998) Functions of lipid rafts in biological membranes. *Annu. Rev. Cell Dev. Biol.* **14,** 111–136.
20. Schroeder, R., London, E., and Brown, D. A. (1994) Interactions between saturated acyl chains confer detergent resistance on lipids and GPI–anchored proteins: GPI–anchored proteins in liposomes and cells show similar behavior. *Proc. Natl. Acad. Sci. USA* **91,** 12,130–12,134.
21. Dietrich, C., Volovyk, Z. N., Levi, M., Thompson, N. L., and Jacobson, K. (2001) Partitioning of Thy-1, GM1, and cross-linked phospholipids analogs into lipid rafts reconstituted in supported model membrane monolayers. *Proc. Natl. Acad. Sci. USA* **98,** 10,642–10,647.
22. Samsonov, A. V., Mihalyov, I., and Cohen, F. S. (2001) Characterization of cholesterol-sphigomyelin domains and their dynamics in bilayer membranes. *Biophys. J.* **81,** 1486–1500.

23. Veatch, S. L. and Keller, S. L. (2002) Organization in lipid membranes containing cholesterol. *Phys. Rev. Lett.* **89,** 268101-1–268101-4.
24. Dietrich, C., Yang, V., Fuiwara, T., Kusumi, A., and Jacobson, K. (2002) Relationship of lipid rafts to transient confinement zones detected by single particle tracking. *Biophys. J.* **82,** 244–284.
25. Gandhavadi, M., Allende, D., Vidal, A., Simom, S. A., and McIntosh, T. J. (2002) Structure, composition, and peptide binding properties of detergent soluble bilayers and detergent resistant rafts. *Biophys. J.* **82,** 1469–1482.
26. Rinia, H. A., Snell, M. M., van der Eerden, J. P., and de Kruijff, B. (2001) Visualizing detergent resistant domains in model membranes with atomic force microscopy. *FEBS Lett.* **501,** 92–96.
27. Yuan, C., Furlong, J., Burgos, P., and Johnston, L. J. (2002) The size of lipid rafts: an atomic force microscopy study of ganglioside GM1 domains in sphingomyelin/DOPC/cholesterol. *Biophys. J.* **82,** 2526–2535.
28. Lawrence, J. C., Saslowsky, D. E., Edwardson, J. M., and Henderson, R. M. (2003) Real–time analysis of the effects of cholesterol on lipid raft behavior using atomic force microscopy. *Biophys. J.* **84,** 1827–1832.
29. Edidin, M K. (2003) The state of lipid rafts: From model membranes to cells. *Annu. Rev. Biophys. Biomol. Struct.* **32,** 257–283.
30. Tu, K., Klein, M. L., and Tobias, D. J. (1998) Constant–pressure molecular dynamics investigation of cholesterol effects in a dipalmitoylphosphatidylcholine bilayer. *Biophys. J.* **75,** 2147–2156.
31. Smondyrev, A. M. and Berkowitz, M. L. (1999) Structure of dipalmitoylphosphatidylcholine/cholesterol bilayer at low and high cholesterol concentrations: Molecular dynamics simulation. *Biophys. J.* **77,** 2075–2089.
32. Pasenkiewicz-Gierula, M., Róg, T., Kitamura, K., and Kusumi, A. (2000) Cholesterol effects on the phosphatidylcholine bilayer polar region: A molecular simulation study. *Biophys. J.* **78(3),** 1376–1389.
33. Róg, T. and Pasenkiewicz-Gierula, M. (2001) Cholesterol effects on the phosphatidylcholine bilayer nonpolar region: A molecular simulation study. *Biophys. J.* **81(4),** 2190–2202.
34. Chiu, S. W., Jakobsson, E., Mashl, R. J., and Scott, H. L. (2002) Cholesterol–induced modifications in lipid bilayers: A simulation study. *Biophys. J.* **83(4),** 1842–1853.
35. Hofsasβ, C., Lindahl, E., and Edholm, O. (2003) Molecular dynamics simulations of phospholipid bilayers with cholesterol. *Biophys. J.* **84(4),** 2192–2206.
36. Pandit, S. A., Bostick, D. L., and Berkowitz, M. L. (2004) Complexation of phosphatidylcholine lipids with cholesterol. *Biophys. J.* **86(3),** 1345–1356.
37. Scott, H. L. (2002) Modeling the lipid component of membranes. *Curr. Opin. Struct. Biol.* **12(4),** 495–502.
38. Huang, J., Swanson, J. E., Dibble, A. R. G., Hinderliter, A. K., and Feigenson, G. W. (1993) Nonideal mixing of Phosphatidylserine and Phosphatidylcholine in fluid lamellar phase. *Biophys. J.* **64,** 413–425.
39. Anderson, T. G. and McConnell, H. M. (2001) Condensed complexes and the calorimetry of cholesterol–phospholipid bilayers. *Biophys. J.* **81(5),** 2774–2785.

40. Frisch, M. J., Trucks, G. W., Schlegel, H. B., et al. (1998) Gaussian 98 (Revision A.11.4, Gaussian, Inc. Pittsburgh PA).
41. Berendsen, H. J. C., van der Spoel, D., and van Drunen, R. (1995) Gromacs: A message-passing parallel molecular dynamics implementation. *Comp. Phys. Comm.* **91**, 43–56.
42. Lindahl, E., Hess, B., and van der Spoel, D. (2001) Gromacs 3.0: A package for molecular simulation and trajectory analysis. *J. Mol. Mod.* **7**, 306–317
43. van der Spoel, D., van Buuren, A. R., Apol, E., et. al. *Gromacs User Manual Version 3.0.* Nijenborgh 4, 9747 AG Grongen, The Netherlands.
44. Frenkel, D. and Smit, B. (2002) *Understanding Molecular Simulation From Algorithm to Applications*, after Academic Press: "San Diego".
45. Chiu, S. W., Clark, M. M., Jakobsson, E., Subramaniam, S., and Scott, H. L. (1999) Optimization of hydrocarbon chain interaction parameters: Application to the simulation of fluid phase lipid bilayers. *J. Phys. Chem. B* **103**, 6323–6327.
46. Berger, O., Edholm, O., and Jahnig, F. (1997) Molecular dynamics simulations of a fluid bilayer of dipalmitoylphosphatidylcholine at full hydration, constant pressure, and constant temperature. *Biophys. J.* **72**, 2002–2013.
47. Schlenkrich, M., Brickmann, J., Mackerell, A., and Karplus, M. (1996) *Biological Membrane: A Molecular Perspective from Computation and Experiment*, chapter An Empirical Potential Energy Function for Phospholipids: Criteria for Parameter Optimization and Applications, Birkhäuser, Boston, pp. 31–82.
48. Michael, P., Mikko, K., Marja, H., Emma, F., Lindqvist, P., and IIpo, V. (2003) Molecular dynamics simulations of lipid bilayers: Major artifacts due to truncating electrostatic interactions. *Biophys. J.* **84**, 3636.
49. Essmann, U., Perera, L., Berkowitz, M. L., Tom Darden, Hsing Lee, and Lee G. Pedersen. (1995) A smooth particle mesh Ewald method. *J. Chem. Phys.* **103(19)**, 8577–8593.
50. Marrink, S. J., Lindahl, E., Edholm, O., and Mark, A. E. (2001) Simulation of the spontaneous aggregation of phospholipids into bilayers. *J. Am. Chem. Soc.* **123**, 8638–8639.
51. Filippov, A., Orädd, G., and Lindblom, G. (2003) The effect of cholesterol on the lateral diffusion of phospholipids in oriented bilayers. *Biophys. J.* **84(5)**, 3079–3086.
52. Nagle, J. F. and Tristram-Nagle, S. (2000) Structure of lipid bilayers. *Biochim. Biophys. Act.* **1469**, 159–195.
53. Nagle, J. F. and Tristram-Nagle, S. (2000) Lipid bilayer structure. *Curr. Opin. Struct. Biol.* **10**, 474–480.
54. Edholm, O. and Nagle, J. F. (2005) Areas of molecules in membranes consisting of mixtures. *Biophys. J.* **89**, 1827–1832.
55. Pandit, S. A., Vasudevan, S., Chiu, S. W., Mashl, R. J., Jakobsson, E, and Scott, H. L. (2004) Sphingomyelin–cholesterol domains in phospholipid membranes: Atomistic simulation. *Biophys. J.* **87(2)**, 1092–1100.
56. Pandit, S. A., Bostick, D. and Berkowitz, M. L. (2003) An algorithm to describe molecular scale rugged surfaces and its application to the study of a water/lipid bilayer interface. *J. Chem. Phys.* **119(4)**, 2199–2205.

57. Douliez, J. P., Léonard, A., and Dufourc, E. J. (1995) Restatement of order parameters in biomembranes: Calculation of C-C bond order parameters for C-D quadrupolar splittings. *Biophys. J.* **68,** 1727–1739.
58. Holopainen, M. J., Metso, A. J., Mattila, J. P., Jutila, A., and Kinnunen, P. K. (2004) Evidence for the lack of specific interaction between cholesterol and sphingomyelin. *Biophys. J.* **86(3),** 1510–1520.
59. Pandit, S. A., Jakobsson, E., and Scott, H. L. (2004) Simulation of the early stages of nano-domain formation in mixed bilayers of sphingomyelin, cholesterol, and dioleylphosphatidylcholine. *Biophys. J.* **87(5),** 3312–3322.
60. Khelashvili, G. A., Pandit, S. A., and Scott, H. L. (2005) Self-consistent mean field model based on molecular dynamics: Application to lipid-cholesterol bilayers. *J. Chem. Phys.* **123,** 034910.
61. Marrink, S. J., de Vries, A. H., and Mark, A. E. (2004) Coarse-Grained Model for Semuqalitative Lipid Simulations. *J. Phys. Chem. B* **108,** 750–760.
62. Jedlovszky, P., Medvedev, N. N., and Mihaly, M. (2004) Effect of cholesterol on the properties of phospholipid membranes. 3. local lateral structure. *J. Phys. Chem. B* **108(1),** 465–472.
63. Falck, E., Patra, M., Karttunen, M., Hyvönen, MT., and Vattulainen, I. (2004) Lessons of slicing membranes: Interplay of packing, free area, and lateral diffusion in Phospholipid/Cholesterol bilayers. http://www.Biophys.J.**87**, 1076–1091.
64. Gurtovenko, A. A., Patra, M., Karttunen, M., and Vattulainen, I. (2004) Cationic DMPC/DMTAP lipid bilayers: Molecular dynamics study. http://www.Biophys.J. 86, 3461–3472.
65. Shinoda, W. and Okazaki, S. (1998) A voronoi analysis of lipid area fluctuation in a bilayer. *J. Chem. Phys.* **109,** 1517–1521.

20

A Microscopic Model Calculation of the Phase Diagram of Ternary Mixtures of Cholesterol and Saturated and Unsaturated Phospholipids

R. Elliott, I. Szleifer, and M. Schick

Summary

The authors solved a microscopic model that describes mixtures of a saturated phospholipid, an unsaturated phospholipid, and cholesterol. The method employed was the self-consistent field approximation. The model was capable of producing several classes of phase diagram, but only one of them showed a liquid–liquid coexistence region. The phospholipids in the cholesterol-rich liquid are more ordered than those in the cholesterol-poor liquid. Within this model, coexistence of two liquids in the ternary system is intimately tied to such coexistence in the binary cholesterol-saturated phospholipid system.

Key Words: Rafts; ternary mixtures; phase diagrams; phase coexistance; theoretical models; self-consistent field theory.

1. Introduction

The importance of lipid rafts is clearly indicated by the very large number of experimental studies devoted to them. Remarkably, very few theoretical studies have addressed the subject. Underlying almost all of them is the assumption, which the authors share, that such rafts are a consequence of the equilibrium behavior of membranes made up of phospholipids and cholesterol. Of these studies, half are phenomenological in character *(1,2)*, as opposed to more microscopic treatments based on molecular models and statistical mechanics *(3,4)*. The behavior of a membrane of phospholipids and cholesterol is sufficiently complex that it does not, as yet, lend itself to detailed computational simulation over time-scales sufficient to investigate raft formation and structure *(5)*.

The authors have tried to explore a middle course between phenomenology, whose strength *and* weakness is the generality of its description, and detailed, computational modeling whose strength *and* weakness is its specificity *(6)*. A molecular model was used, which concentrates on the hydrophobic interior of a bilayer membrane, which describes phospholipid chains reasonably well, but which ignores almost completely the interaction of headgroups. Therefore, the approach highlights the importance of packing effects. Because the chains are described well, one can determine the degree of chain ordering in any particular phase. The model describes a ternary mixture of saturated and unsaturated phospholipids and cholesterol, and considers only *local* binary interactions between them. Furthermore, the model is solved within self-consistent field theory, which ignores the correlations between entities.

Given these limitations, the conditions can be explored under which two liquid phases can coexist, of which one would represent the "raft" domain, and the other the sea in which the raft floats. In a ternary mixture of cholesterol, saturated, and unsaturated phospholipids, the following are found. If one ignores the interactions between the cholesterol and the phospholipids, so that the cholesterol simply takes up space in the bilayer interior, then one finds only a saturated phospholipid-rich gel phase below the main chain transition of the saturated phospholipid. The gel phase coexists with a *single* liquid phase. There is no liquid–liquid coexistence. If one now turns on an interaction between cholesterol and phospholipids, which causes the former to help order the latter, one finds that the gel phase becomes swollen with cholesterol, a reasonable result, but not one observed experimentally. If one now considers the interaction between cholesterols as well, one finds in addition to the gel phase *two* liquid phases, one which is cholesterol-rich and relatively well ordered, the other cholesterol-poor and poorly ordered. Furthermore, these two liquid phases extend all the way to the binary cholesterol and saturated phospholipid system, as in the experimental work of Vist and Davis *(7)*. The phase diagram calculated by the authors is that of a system of a saturated phospholipid with two tails of sixteen carbons, C16:0, an unsaturated phospholipid with two monounsaturated tails of eighteen carbons, C18:1, and cholesterol. It is shown in **Fig. 1**, and is quite similar to those that have been reported experimentally for the system sphingomyelin, palmitoyloleoylsphatidylcholine, and cholesterol *(8–10)*.

The detailed procedures by which the results have been obtained will not be emphasized. Descriptions of them are readily available *(11–14)*. Rather, in the following, the authors will try to convey the character of what has been done so that the reader can better appreciate what the calculation has produced, and better judge its merits as well as its limitations.

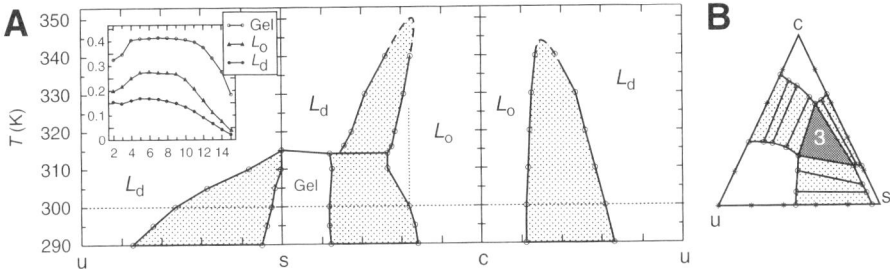

Fig. 1. Binary phase diagrams of the C16:0, C18:1, cholesterol system. The saturated phospholipid-cholesterol mixture has a triple point very near the main chain-transition temperature, so that the gel, L_d, coexistence region is very narrow. Dashed lines are extrapolations. The ternary mixture at $T = 300$ K is shown in **Fig. 1B**. **(Inset)** Order parameters of saturated tails in the three coexisting phases at 300 K.

2. Methods
2.1. The Model and its Self-Consistent Field Solution
2.1.1. Theory of the Disordered Liquid

It is useful to first recall the earlier theory upon which the authors' treatment is a logical extension. Consider the problem of calculating the area per headgroup in the disordered liquid phase of a bilayer, which consists of a single kind of phospholipid. The problem is somewhat analogous to calculating the volume per atom of liquid argon. The latter is a difficult problem owing to the competition between the long-range, attractive, van der Waals interactions, and the short-range, hard-core, repulsive interactions. The former is even more difficult owing to the greater number of internal degrees of freedom of the lipid molecule compared with an argon atom. The approach to this problem taken by Ben-Shaul et al. *(11,14)* was as follows. One knows that the effect of the competing interactions is to produce in the bilayer an interior, which is very much like an incompressible liquid. So why not replace the original system of interacting lipids with another system of *noninteracting* lipids, but one in which the density in the interior of the bilayer is constrained to be constant? This amounts to replacing the interacting, unconstrained system by one which is noninteracting, but constrained. The Hamiltonian of the new noninteracting system is very simple of course:

$$H = \sum_{\gamma=1}^{n} \sum_{k=1}^{N} h_{\gamma,k} \tag{1}$$

where n is the number of lipid tails, N is the number of segments in each chain, and $h_{\gamma,k}$ is the Hamiltonian of a single chain, γ, which now only contains the energy of *gauche* bonds at segment k. The problem looks easy, but there is a difficulty, which is encountered when the Helmholtz free energy is calculated.

$$F(T,n,A,V) = Tr' P'[H + k_B T \ln P'], \qquad (2)$$

where A is the area of the membrane and V its volume. Here,

$$P' = \exp[-\beta H]/Q', \qquad (3)$$

is the probability to observe a configuration of energy H, $\beta = 1/k_B T$, and

$$Q' = Tr' \exp[-\beta H], \qquad (4)$$

is the partition function of the system. The prime indicates that *only chain configurations, that contribute a uniform core density are to be considered in the trace.* This is the difficulty, but it is easily surmounted by going to a different ensemble, one in which the local density is *not* specified, but its conjugate, essentially the local pressure, is. To do this a local volume fraction is defined, $\hat{\Phi}(z)$.

$$\hat{\Phi}(z) = \frac{1}{A}\sum_{\gamma=1}^{n}\hat{\phi}_\gamma(z) \qquad (5)$$

$$\hat{\phi}_\gamma(z) = \sum_{k=1}^{N} v_k \delta(z - z_{\gamma,k}) \qquad (6)$$

$$\frac{A}{V}\int dz\, \hat{\Phi}(z) = \frac{n\sum v_k}{V} = 1 \qquad (7)$$

where z is the coordinate normal to the bilayer surface and the v_k are the volumes of the kth segment. In **Eq. 7** the hydrophobic core of volume V consists only of the sum of its monomeric volumes. This *incompressibility constraint* requires the noninteracting system of lipid chains to occupy a liquid-like slab of fixed density n/V. Under this constraint, the free energy of **Eq. 2** reduces to a natural function of two extensive quantities, $F(T,n,A,V) \to F(T,n,A)$.

Now the probability of a configuration in an external field, $\Pi(z)$, can be written

$$P = \frac{1}{Q}\exp[-\beta H - \frac{A}{v_0}\int dz \Pi(z)\hat{\Phi}(z)] \qquad (8)$$

$$Q = Tr\, \exp[-\beta H - \frac{A}{v_0}\int dz \Pi(z)\hat{\Phi}(z)] \qquad (9)$$

where v_0 is any convenient molecular volume, one which has simply been introduced to make the field $\Pi(z)$ dimensionless. The free energy in this ensemble, which is simply the Legendre transform with respect to A of, $F(T,n,A)$ is

$$G(T,n,\Pi) = F(T,n,A) + \frac{Ak_B T}{v_0} \int \Pi(z)dz \qquad (10)$$

and is given by

$$G = -k_B T \log Q \qquad (11)$$

$$= \mathrm{Tr}\, P[H + \frac{Ak_B T}{v_0} \int dz \Pi(z)\hat{\phi}(z) + k_B T \ln P]. \qquad (12)$$

Because the chains are noninteracting, the probability of a given configuration of all chains is simply the product of the probabilities of each chain to be found in their individual configuration, $P = cP_1^n$, where c is an uninteresting constant, and

$$P_1 = \frac{1}{Q_1} \exp[-\beta \sum_{k=1}^{N} h_k - \frac{1}{v_0} \int dz \Pi(z)\hat{\phi}(z)] \qquad (13)$$

$$Q_1 = \mathrm{Tr}\, \exp[-\beta \sum_{k=1}^{N} h_k - \frac{1}{v_0} \int dz \Pi(z)\hat{\phi}(z)] \qquad (14)$$

and h_k contains the energy of the *gauche* bonds of the single chain. The field $\Pi(z)$ is then determined by requiring that the ensemble-average, local volume fraction, $<\hat{\Phi}(z)>$, be a prescribed constant value at *all z*.

$$<\hat{\Phi}(z)> = \frac{\beta v_0}{A} \frac{\delta G}{\delta \Pi(z)} = 1. \qquad (15)$$

One observes from **Eq. 14** that Q_1 is the partition function of a single chain in the external field $\Pi(z)$. So the central assumption of Ben-Shaul et al. has reduced the many chain problem to that of calculating the one-chain partition function in an external field. The calculation of this partition function is the essential problem in this method, and its difficulty depends on how realistic a description of the chains one takes. Ben Shaul et al. took Flory's rotational isomeric states model *(15)* in which each bond between CH_2 groups can take one of three configurations; *gauche*-plus, *gauche*-minus, or *trans*. For *m* independent bonds, this is only 3^m configurations. However, one also has to specify an origin and *direction* of the chain. This leads to many more configurations. Typically, one enumerates on the order of 10^7 chain configurations. Any configurations that intersect themselves, or that break the planar boundary between the bilayer and its surrounding water, are discarded. The remaining contributions to the partition function are weighted by the field, $\Pi(z)$, and from the partition function one calculates the density. The field $\Pi(z)$ is adjusted until the density is uniform inside the core of the bilayer.

The field $\Pi(z)$ accounts in an average way, for the local intermolecular repulsions that are needed to keep the chain at constant density at each z. Thus, the replacement of the multichain partition function by one of noninteracting, but constrained chains, is in essence a mean-field approximation of the effect of the intermolecular repulsions, one which uses the field, $\Pi(z)$, conjugate to the local density. The van der Waals attractions hold the hydrophobic liquid core together. Because the density is homogeneous, the contribution of the van der Waals interactions to the free energy is simply a constant, which is ignored.

Utilizing the field, $\Pi(z)$, one calculates the single chain partition function, **Eq. 14**, and the Helmholtz free energy per chain, which is the Legendre transform of **Eq. 12**

$$f(T,a) = g - \int dz \Pi(z)\phi(z) \equiv F(T,a)/n \qquad (16)$$

where $g \equiv G/n$ and $a \equiv A/n$ are the Gibbs-like free energy and the area per chain, respectively.

The surface tension of the bilayer is, $\sigma = \partial f/\partial a$, which vanishes at a minimum of $f(T,a)$. One requires the system to maintain zero tension as one expects the real system does so. There *is* a minimum in this model because at large areas per chain, the chains must have many *gauche* bonds in order to fill the space, and these *gauche* bonds cost energy. At small values of a, the chains must be tightly packed, with the consequence that there are few *gauche* bonds, and little entropy. The minimum occurs at the optimum trade-off of these two effects. The area per chain one obtains from this is a bit large compared with experiment, so Ben Shaul et al. also included the effect of the repulsive interaction between water and the hydrophobic chains. They take this contribution to the free energy per chain to be $\sigma_0 a$ with σ_0 the usual oil–water interfacial tension. This term shifts the minimum to smaller a and one finds the minimum to occur at 0.64 nm² for a *two-chain* phospholipid, which is in satisfactory agreement with experiment.

The order parameter profile of the chains, which is essentially the angle between CH_2 planes and the bilayer normal, can be calculated because one knows the equilibrium probability of each configuration of the chains. The order parameter can be measured by nuclear magnetic resonance. There is good agreement between theory and experiment. A deficiency of this theory, which must be addressed in order to apply it to the phenomenon of rafts is that it is not capable of addressing the issue of how local chain order can be affected by the interactions with cholesterol, an interaction which probably is crucial in distinguishing liquid-ordered phases from liquid-disordered phases. If one cannot produce two such liquid phases, one will certainly not be able to describe the existence of rafts as a coexistence phenomena. Another issue related to interactions and local order is the lack of the possibility of describing a gel phase in this theory. To address these shortcomings, the theory was extended as described in the following section *(12)*.

2.1.2. Description of Orientational Order

It is not difficult to understand why the theory presented earlier does not produce a gel phase. In describing packing effects, only a local volume fraction, **Eq. 5** was introduced. As a consequence, all information is lost about the local orientation of bonds between adjacent acyl groups in a chain. In other words, the theory as presented only requires that the density in the interior be uniform. It does not give greater weight to configurations which are more ordered than average and which could therefore more easily fulfill the constraints of packing.

To remedy this defect, the local orientation of the chain is specified by the normal to the plane determined by the kth CH_2 group

$$u_k = \frac{r_{k-1} - r_{k+1}}{|r_{k-1} - r_{k+1}|}, \quad k = 1 \cdots N-1 \tag{17}$$

where the r_k are the position vectors of the kth segment in the chain. In analogy to the number densities, **Eqs. 5** and **6**, the *bond* densities are defined.

$$\hat{\Xi}(z) = \frac{1}{A}\sum_{\gamma=1}^{n}\hat{\xi}_\sigma(z) \tag{18}$$

$$\hat{\xi}(z) = \sum_{k=1}^{N-1} v(k)\delta(z-z_k)g(u \cdot c) \tag{19}$$

which tells us how well these local bonds are oriented with respect to the bilayer normal c. For the function g, the author chose.

$$g(u \cdot c) \equiv (m + 1/2)(u \cdot c)^{2m} \tag{20}$$

For large m, $g \approx m\exp(-m\theta^2)$ where θ is the angle between the two unit vectors. The authors found $m = 18$ to be reasonable. Note that $g(u \cdot c)$ is unity if the bond vector u is aligned with the bilayer, and falls exponentially with the angle between the two vectors. To express the fact that it is energetically favorable for bonds, which are in the same local region to be aligned with one another, and with the bilayer normal, one adds to the system's Hamiltonian a simple interaction

$$V(u,u') = -(J/v_0)g(u \cdot c)g(u' \cdot c) \tag{21}$$

This interaction favors neighboring bond orientations which are aligned with one another and with the bilayer normal, with the strength of the interaction falling exponentially if either bond deviates from the normal. This interaction between bonds is *not* rotationally invariant. One does not expect it to be because the environment is not rotationally invariant as the bilayer normal provides a particular direction in space. The extension of the probability distribution function, **Eqs. 13** and **14**, is immediate *(12)*

$$P_1 = \frac{1}{Q_1}\exp\{-\beta\sum_{k=1}^{N}h_k - \frac{1}{v_0}\int dz[\Pi(z)\hat{\phi}(z) + B(z)\hat{\xi}(z)]\} \tag{22}$$

$$Q_1 = \mathrm{Tr}\exp\{-\beta\sum_{k=1}^{N}h_k - \frac{1}{v_0}\int dz[\Pi(z)\hat{\phi}(z) + B(z)\hat{\xi}(z)]\}. \tag{23}$$

The self consistent equations, which determine the two unknown fields, $\Pi(z)$ and $B(z)$ are now

$$<\hat{\Phi}(z)> = 1, \tag{24}$$

$$<\hat{\Xi}(z)> = -\frac{k_B T B(z)}{J}. \tag{25}$$

3. Results

3.1. One Component and Mixed Phospholipid Systems

The first result that emerges from this extension is that one finds, for a system of phospholipids with two saturated chains, C16:0, a first-order main chain transition from a disordered liquid phase to a more ordered one, which one identifies with the gel phase. Again the distinction between more and less ordered follows from examination of the calculated chain order parameters. The interaction strength, J, is fixed so that the transition temperature occurs at that measured *(16)* for dipalmitoyl phosphatidyl choline (DPPC), 315 K. The area per head group in the fluid phase at 323 K is calculated to be 67 Å^2, compared with the experimental value *(17)* of 64 Å^2, whereas that in the gelphase at 293 K is calculated to be 49.9 Å^2, compared with the experimental value *(17)* of 47.9 Å^2. At the transition, the average number of *gauche* bonds is calculated to be reduced from 4.3 in the liquid to 2.1 in the gel phase.

It is straightforward to repeat the calculation for a system consisting of a monounsaturated phospholipid, C18:1. Nothing in the Hamiltonian, or the interactions, changes at all. In particular, the strength, *J*, of the interaction of **Eq. 21** remains the same. What does change is the ensemble of configurations of the chains, for they all now have a single double-bond in them which causes a kink at its location. The consequence is that there are fewer configurations that can take advantage of the interaction between chains of **Eq. 21**, which lowers the interaction energy between chains that are well aligned with one another and the bilayer normal. As a consequence, one finds for this system that the main chain transition now occurs at a temperature below 0°C in agreement with experiment. This is very nice, because the one parameter that could be played with, the strength of the interaction, had already been set by the transition temperature of the saturated system. So there was nothing to adjust in order to get the main chain temperature of the unsaturated system correct. Nonetheless the theory gets it right.

The phase diagram of a mixture of the saturated and unsaturated chains is easily obtained *(12)*, and is shown in the left-hand panel of **Fig. 1A**. One obtains a liquid phase above the first-order, main chain transition of the saturated phospholipid, and coexistence between liquid and gel phases below it.

3.2. Ternary System Containing Cholesterol

The last component to be introduced in order to make this a ternary system is cholesterol *(6)*. One knows where all the atoms in the cholesterol molecule are, just as one knows where all the atoms are in the phospholipid chains. Thus, one can easily take into account the contribution of cholesterol to the volume of the bilayer interior, which is constrained to be constant. One also has to consider the binary interactions between cholesterol and the phospholipids, and between the cholesterol themselves. The same kind of interaction has been chosen between these pairs as between phospholipid chains, i.e., an interaction that favors *local* alignment of the elements in the binary pair with one another and with the normal to the bilayer normal. One has to identify what one means by the alignment of cholesterol. The alignment of its small acyl tail is defined as for the phospholipid chains. For the rigid part of the cholesterol, a unit vector, u_c, is introduced, which extends from the third to the 17th carbon in the molecule, using the conventional labeling (*see* the **inset** to **Fig. 2**), and a second normal perpendicular to the planes of the rings. All interactions can now be written

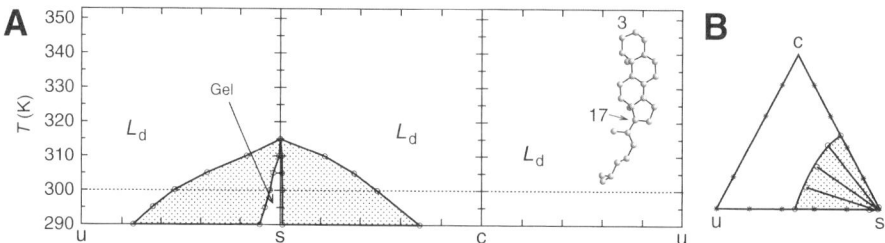

Fig. 2. Calculated phase diagrams of the three binary mixtures of cholesterol (c), saturated (s), and unsaturated (u) phospholipids in temperature-composition space for $J_\parallel (m + 1/2)^2/k_B T^* = 1.44$ and $J_{lc} = J_{cc} = 0$. These binary diagrams form the sides of the Gibbs prism, a cut through which at 300 K produces the Gibbs triangle shown in **Fig. 1B**. Regions of two-phase coexistence are shaded, and some tie lines are shown. (**Inset**) One of the configurations of the model cholesterol.

$$V_{\sigma,\sigma'} = -(J_{\sigma,\sigma'}/v_0) g(u_\sigma \cdot c) g(u_{\sigma'} \cdot c) \tag{26}$$

where σ is an index taking the values, s, u, and c denoting saturated, unsaturated, and cholesterol, respectively. The strengths of the interactions between lipid segments, $J_{s,s}$, $J_{s,u}$, and $J_{u,u}$ are taken to be identical, $J_{l,l}$ because the authors believed that the phospholipid's chains are distinguished by their configurations, not by their interaction strengths. With the addition of cholesterol, there are additional interactions. The strengths of the interactions between cholesterol and both phospholipids are taken to be identical, and for the same reason as aforementioned: $J_{s,c} = J_{u,c} \equiv J_{l,c}$. Finally, there is an interaction between cholesterols, of strength $J_{c,c}$. Thus, there are now two strengths that can be varied, the strength of the cholesterol–phospholipid interaction ($J_{l,c}$), and the strength of the cholesterol–cholesterol interaction ($J_{c,c}$). The effect of these two interactions has been systematically explored with the following results.

First, one asks what is the phase diagram, if the strongest interaction of the three is that between phospholipids themselves, so that the cholesterol interactions are weak, and cholesterol affects the system only through its volume and the constraint that the density in the interior of the bilayer be constant. **Figure 2** shows the result for the limiting case in which $J_{l,c} = J_{cu,c} = 0$. The only aligning interaction is between phospholipids, and its strength is again set so that the main chain transition of the saturated phospholipid is that of DPPC. One observes that there is a gel phase and only one liquid phase. Thus, there can be no liquid–liquid coexistence.

Next, it was reasoned that perhaps the cholesterol–phospholipid interaction should be enhanced. This would have the effect of making the phospholipids more ordered, and might bring about a phase separation between the more ordered phospholipids near the cholesterol and the disordered phospholipids. Therefore, the strength of the cholesterol–phospholipid interaction ($J_{1,c}$) was increased, keeping the cholesterol–cholesterol interaction ($J_{c,c}$) small. However the desired outcome did not come about. Instead it was found that the gel phase became swollen with cholesterol. Furthermore, if the cholesterol–phospholipid interaction became too strong, the main chain temperature increased with cholesterol composition, rather than decreased. All of these are understandable effects of the increased cholesterol–phospholipid interaction, but do not seem to be observed in experiment.

Therefore, the authors turned on the interactions between cholesterol such that the cholesterols would, at reasonable temperatures, phase separate from the phospholipids even in the binary cholesterol, saturated phospholipid system. The phase diagram in **Fig. 1** is obtained this way with $J_{c,c} = 0.73 J_{1,1}$ and $J_{1,c} = 0.78 J_{1,1}$. One observes that there are now two liquidphases; one is cholesterol-rich, the other cholesterol-poor. In the former, the phospholipids are relatively well ordered, whereas in the cholesterol-poor phase, the phospholipids are not so well ordered. The region in which these two liquids coexist would be the regions in which rafts would occur. The order parameters of the three phases which coexist at 300 K are shown in the inset.

The origin of the two liquid phases in this model is clear. They arise from the tendency of cholesterol to phase separate from the phospholipids, creating a cholesterol-rich liquid, the ordered liquid, and a cholesterol-poor one, the disordered liquid. At temperatures above the main chain transition, a coexistence region between the two liquids extends across the ternary diagram from the cholesterol, saturated phospholipid binary axis, to the cholesterol, unsaturated phospholipid binary axis, as shown in **Fig. 3**.

As the temperature is increased from 320 K, the liquid–liquid coexistence region eventually detaches from the binary cholesterol, unsaturated phospholipid axis at a critical point, and recedes toward the cholesterol, saturated phospholipid axis, vanishing there at a critical point. As the temperature is decreased from 320 K, the gel phase appears at 315 K. As the temperature is lowered, the region of liquid-disordered phase on the binary cholesterol, saturated phospholipid axis decreases. When the two coexistence regions, gel, liquid-disordered, and liquid-ordered, liquid-disordered, just touch one another at a critical tie line, a three-phase coexistence region begins and grows with further decrease of temperature. Note that in this model, the gel phase does nothing of significance other than to take up space in the ternary diagram.

4. Discussion

The phase diagram obtained in this calculation for the ternary system of a saturated phospholipid, and unsaturated phospholipid, and cholesterol, is the first

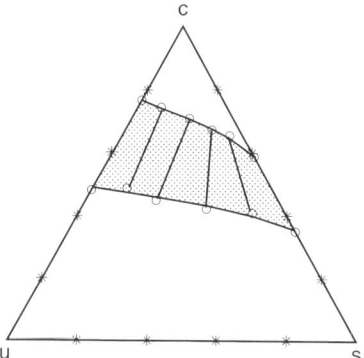

Fig. 3. Phase diagram of the ternary mixture at $T = 320$ K. Interactions are the same as in **Fig. 1**.

obtained from a microscopic model. It has the nice feature of displaying a region of coexistence between two liquid phases, one rich in cholesterol and relatively well-ordered saturated phospholipids, the other rich in relatively poorly ordered unsaturated phospholipids. This region of coexistence would be the locus of "rafts." It resembles the phase diagram observed experimentally for the system sphingomyelin, palmitoyloleoylsphatidylcholine, and cholesterol *(8–10)*.

However, it is not without its faults. In particular, the regions of phase coexistence in the binary system of saturated phospholipid and cholesterol are too wide, and occur at a concentration of cholesterol that is much higher than that observed by Vist and Davis *(7)*. This might be only because of the form of the interaction taken, **Eq. 21**, between cholesterol and phospholipid, which falls exponentially with the local angle between cholesterol and phospholipid. Indeed a recent calculation *(19)*, which is similar in spirit to that described herein but utilizes an interaction in which the ordering effect of cholesterol on the chains remains strong over a much larger range of angles obtains a coexistence region in which the cholesterol concentration is much smaller, and is in reasonable agreement with the experiment. The calculation also utilizes the rotational isomeric description of the chains owing to Flory *(15)*. More accurate configurations generated from molecular dynamics simulations can be used instead *(18)*. Perhaps more important is that mean-field calculations, such as the authors' and others *(19)*, ignore the correlation between phospholipid and cholesterol; i.e., the rough and smooth faces of the cholesterol ring structure presumably interact rather differently with the chains. Similar preferential interaction of the saturated phospholipid with the smooth face of cholesterol was observed in simulations *(5)*. Furthermore, the calculation has largely neglected the effects of headgroups.

The headgroups may play an important role in packing with cholesterol, whose hydrophilic volume is proportionally small.

Irrespective of these shortcomings, the calculation teaches several things. First, it indicates that the main chain transition and the gel phase, are almost certainly irrelevant to the phenomenon of rafts. The presence of a gel phase and its associated three-phase triangle, of course, take up room in the Gibbs triangle, thereby limiting the region in which rafts could exist. But this is a negative role. The irrelevance of the gel phase is also indicated by the observation of liquid–liquid coexistence in some systems well above any main chain transition *(20)*. Second, if one believes that binary interactions are sufficient to describe the system, then the calculation indicates that liquid–liquid coexistence in the ternary system is intimately tied to its existence in at least one of the binary systems. In retrospect, this is not difficult to understand. Suppose that in a ternary system of *s*, *u*, and *c*, the interaction between *s* and *c* were the most repulsive. Then the addition of *u* to the system, with its weaker interactions with the other two components, cannot cause phase separation to occur at a higher temperature. Thus, the separation occurs first in the binary system. There is experimental evidence that phase separation does extend out to the binary systems *(8–10)*, and evidence that it does not *(20)*. Third, in the model, the basic mechanism of the phase separation is simply that cholesterol prefers an environment of itself to that of phospholipids. If this be true, then one expects tie lines between the two liquid phases, which are relatively vertical, as observed in **Fig. 3**. There is some evidence for this *(21)*. On the other hand, there is also evidence against it *(20,22)*. If experiment ultimately concludes that the predictions of the model calculation are not correct, one has to consider what additions are needed to the basic model one has presented. Fortunately, it is likely that the efficacy of possible, additional effects can be judged by utilization of the techniques presented herein.

Acknowledgments

We are grateful to Sarah Keller and Sarah Veatch for numerous profitable discussions. This work was supported by the National Science Foundation under Grant nos. DMR-0140500, 0503752, and CTS-0338377.

References

1. Komura, S., Shirotori, H., Olmsted, P. D., and Andelman, D. (2004) Lateral phase separation in mixtures of lipids and cholesterol. *Europhys. Lett.* **67,** 321–327.
2. Radhakrishnan, A. and McConnell, H. (2005) Condensed complexes in vesicles containing cholesterol and phospholipids. *Proc. Nat. Acad. Sci.* **102,** 12,662–12,666.
3. Ipsen, J. H., Karlstrom, G., Mouritsen, O. G., Wennerstrom, H., and Zuckermann, M. J. (1987) Phase equilibria in the phosphatidylcholine-cholesterol system. *Biochim. Biophys. Acta* **905,** 162–172.

4. Nielsen, M., Miao, L., Ipsen, J. H., Zuckermann, M. J., and Mouritsen, O. G. (1999) Off-lattice model for the phase behavior of lipid-cholesterol bilayers. *Phys. Rev. E* **59**, 5790–5803.
5. Pandit, S., Jakobsson, E., and Scott, H. (2004) Simulation of the early stages of nano-domain formation in mixed bilayers of sphingomyelin, cholesterol and dioleylphosphatidylycholine. *Biophys. J.* **87**, 3312–3322.
6. Elliott, R., Szleifer, I., and Schick, M. (2006) Phase diagram of a ternary mixture of cholesterol and saturated and unsaturated lipids calculated from a microscopic model. *Phys. Rev. Lett.* **96**, 098101.
7. Vist, M. and Davis, J. (1990) Phase equilibria of cholesterol/dipalmitoylphosphatidylcholine mixtures:nuclear magnetic resonance and differential scanning calorimetry. *Biochemistry* **29**, 451–464.
8. de Almeida, R., Fedorov, A., and Prieto, M. (2003) Sphingomyelin/phosphatidylcholine/cholesterol phase diagram: boundaries and composition of lipid rafts. *Biophys. J.* **85**, 2406–2416.
9. Pokorny, A., Yandek, L., Elegbede, A., Hinderliter, A., and Almeida, P. (2006) Temperature and composition dependence of the interaction of d-lysin with ternary mixtures of sphingomyelin/cholesterol/POPC. *Biophys. J.* **91**, 2184–2197.
10. Veatch, S. and Keller, S. (2005) Miscibility phase diagrams of giant vesicles containing sphingomyelin. *Phys. Rev. Lett.* **94**, 148101.
11. Ben-Shaul, A., Szleifer, I., and Gelbart, W. (1985) Chain organization and thermodynamics in micelles and bilayers. I. Theory. *J. Chem. Phys.* **83**, 3597–3611.
12. Elliott, R., Katsov, K., Schick, M., and Szleifer, I. (2005) Phase separation of saturated and mono-unsaturated lipids as determined from a microscopic model. *J. Chem. Phys.* **122**, 144904.
13. Fattal, D. R. and Ben-Shaul, A. (1994) Mean-field calculations of chain packing and conformational statistics in lipid bilayers: Comparison with experiments and molecular dynamics studies. *Biophys. J.* **67**, 983–995.
14. Szleifer, I., Ben-Shaul, A., and Gelbart, W. M. (1986) Chain statistics in micelles and bilayers: effects of surface roughness and internal energy. *J. Chem. Phys.* **85**, 5345–5358.
15. Flory, P. J. (1969) *Statistical Mechanics of Chain Molecules*. Wiley-Interscience, New York.
16. Finegold, L. and Singer, M. (1986) The metastability of saturated phosphatidylcholines depends on the acyl chain length. *Biochim. Biophys. Acta* **855**, 417–420.
17. Nagle, J. F. and Tristam-Nagle, S. (2000) Structure of lipid bilayers. *Biochim. Biophys. Acta* **1469**, 159–195.
18. Khelashvili, G., Pandit, S., and Scott, H. (2005) Self-consistent mean-field model based on molecular dynamics: Application to lipid-cholesterol bilayers. *J. Chem. Phys.* **123**, 034910.
19. Pandit, S., Khelashvili, G., Jakobsson, E., Grama, A., and Scott, H. (2006) Lateral organization in lipid-cholesterol mixed bilayers. *Biophys. J.* **92**, 440–447.
20. Veatch, S., Gawrisch, K., and Keller, S. (2006) Closed-loop miscibility gap and quantitative tie-lines in ternary membranes containing diphytanol PC. *Biophys. J.* **90**, 4428–4436.

21. Swamy, M., Ciani, L., Ge, M, et al. (2006) Coexisting domains in the plasma membranes of live cells characterized by spin label esr spectroscopy. *Biophys. J.* **90,** 4466–4478.
22. Veatch, S. L., Polozov, I., Gawrisch, K., and Keller, S. (2004) Liquid domains in vesicles investigated by nmr and fluorescence microscopy. *Biophys. J.* **86,** 2910–2922.

Index

A

AFM, *see* Atomic force microscopy
Amplitude modulation atomic force microscopy, *see* Atomic force microscopy
Atomic force microscopy (AFM), lipid domains
   amplitude modulation imaging, 278–280, 283
   calibration
      cantilever force constant, 273–275, 282
      piezoelectric motion, 273
   cantilever preparation and sample mounting, 276, 277, 282, 283
   contact mode imaging, 268, 277, 278, 283
   electrostatic shielding, 276
   image analysis, 280, 281
   materials, 269, 270
   overview, 265, 266
   principles, 266–269
   resolution, 266
   supported lipid bilayer preparation
      overview, 270, 271, 281
      vesicle fusion on substrate, 272, 273, 281
      vesicle preparation, 271, 272

B

Bilayer
   domains, *see* Lipid domains
   nuclear magnetic resonance, *see* Nuclear magnetic resonance
   plasmon-waveguide resonance spectroscopy, *see* Plasmon-waveguide resonance spectroscopy
   raft formation, 2
   thickness, *see* X-ray diffraction

Brownian motion, *see* Single-molecule tracking

C

CHAPS, liquid-disordered membrane solubilization, 9
CHARMM, molecular dynamics simulations, 287
Cholesterol, *see* Computer simulations; Fluorescence recovery after photobleaching
Computer simulations, lipid domains
   coarse-grain models
      analysis, 298, 300
      pseudoatom models, 298
      self-consistent mean field theory models, 297, 298, 300
   materials, 287, 288
   molecular dynamics
      analysis, 291–293, 295, 296, 299, 300
      assumptions, 288
      force field parameters, 288, 289, 299
      GROMACS setup, 289, 290, 299
      running, 290, 291, 299
   overview, 285–287
   phase diagram analysis of ternary mixtures of cholesterol, saturated, and unsaturated phospholipids
      disordered liquid theory, 307–311
      one component and mixed phospholipid system analysis, 312, 313
      orientational order, 311, 312
      overview, 305, 306
      ternary system analysis, 313–317
Confocal microscopy, *see* Giant unilamellar vesicles

Contact mode atomic force microscopy, *see* Atomic force microscopy

D

Dehydroergosterol, *see* Multiphoton laser-scanning microscopy
Detergent-resistant membrane (DRM)
  features, 2
  glycophosphatidylinositol-anchored protein-associated membrane extraction
    denaturing gel electrophoresis, 15, 18
    gradient ultracentrifugation and harvesting, 14, 16
    materials, 10–12, 17, 18
    overview, 10, 12, 13, 18
    transfection and sample preparation, 13, 14
    Western blot analysis, 15–18
  liquid-ordered membranes, 9, 10
Diffusion coefficient
  fluorescence recovery after photobleaching calculation, 190, 191
  lipid raft lateral diffusion coefficient determination from pulsed field gradient NMR
    data acquisition, 130–133
    data analysis
      component resolved spectroscopy, 133, 135, 139, 140
      diffusion component number, 135
      peak intensity, 133
    domain formation, 128
    interpretation
      intermediate exchange, 138
      large domains and slow exchange, 137, 138
      no domains, 136, 137
      overview, 135, 136
      small domains and fast exchange, 137
    materials, 128, 129
    oriented bilayer preparation, 129, 130, 138, 139
    overview, 127, 128
    single-molecule tracking
      calculation, 204, 218, 219
      histograms, 204, 205, 219
Diphenylhexatriene fluorescence quenching, *see* Tempo
Discrimination by oxygen transport (DOT), membrane domain characterization
  electron paramagnetic resonance
    conventional spectra acquisition, 151
    saturation-recovery electron paramagnetic resonance, 151, 152, 155, 156
  materials, 149
  oxygen equilibration, 150, 154, 155
  oxygen transport parameter
    calculation, 152, 156
    comparison in different membrane domains, 152–154, 156, 157
  principles, 146, 148, 149
  sample preparation, 149, 150
  spin labels
    labeling, 149
    structures, 147
Domains, *see* Lipid domains
DOT, *see* Discrimination by oxygen transport
DRM, *see* Detergent-resistant membrane

E

Electron microscopy, *see* Transmission electron microscopy
Electron paramagnetic resonance, *see* Discrimination by oxygen transport

F

FCS, *see* Fluorescence correlation spectroscopy

Index

Film balance, lipid lateral interaction characterization
  materials, 42, 52
  monolayers
    isotherm
      acquisition, 46, 48, 53
      average area molecular versus composition analysis, 48–50, 53
      interfacial potential, 51, 52, 54
      lateral compressibility analysis, 50, 51
      preparation containing lipid rafts, 48, 53
    overview, 41, 42
    principles, 43–45
    setup conditions, 45, 46, 53
Fluorescence correlation spectroscopy (FCS), giant unilamellar vesicles in local lipid structures
  calibration, 78–81
  cell culture and labeling, 75, 79
  confocal microscopy, 77–79, 81–83
  data analysis, 79, 81, 82
  focus positioning, 78, 82
  image acquisition, 78, 83
  materials, 75, 76, 79
  overview, 73, 74
  vesicle electroformation, 76, 77, 79, 80
Fluorescence microscopy, see Fluorescence recovery after photobleaching; Giant unilamellar vesicles; Single-molecule tracking
Fluorescence quenching, see Tempo
Fluorescence recovery after photobleaching (FRAP), lipid raft analysis
  cholesterol manipulation in cells
    depletion, 186
    loading, 186
    mock-treated cells, 186
    repletion, 186, 187
  data analysis
    diffusion coefficient calculation, 190, 191
    fraction of molecules recovered, 190, 192
    normalization, 189, 190
    quantitative fluorescence intensities, 188, 189
  data collection
    bleach region selection, 187, 192
    bleaching conditions, 187, 188
    imaging, 188, 192
    setting optimization, 187
  materials, 183, 184, 191
  membrane sheets, 251
  overview, 181–183
  probe labeling
    CTXB, 185, 191, 192
    DilC$_{16}$, 185, 192
    DilC$_{18}$, 185, 192
  transfection, 184, 185, 191
Fluorescence resonance energy transfer, see Tempo
FRAP, see Fluorescence recovery after photobleaching

G

GAUSSIAN, force field parameter calculation, 288, 289
Giant unilamellar vesicles (GUVs), fluorescence microscopy
  coexisting liquid phase analysis
    fluorescence microscopy, 64, 65, 68–70
    materials, 61
    overview, 59–61
    vesicle electroformation, 61–64, 66, 67
  fluorescence correlation spectroscopy of local lipid structures
    calibration, 78–81
    cell culture and labeling, 75, 79
    confocal microscopy, 77–79, 81–83
    data analysis, 79, 81, 82

focus positioning, 78, 82
image acquisition, 78, 83
materials, 75, 76, 79
overview, 73, 74
vesicle electroformation, 76, 77, 79, 80
Glycophosphatidylinositol-anchored protein
detergent-resistant membrane extraction
denaturing gel electrophoresis, 15, 18
gradient ultracentrifugation and harvesting, 14, 16
materials, 10–12, 17, 18
overview, 10, 12, 13, 18
transfection and sample preparation, 13, 14
Western blot analysis, 15–18
plasmon-waveguide resonance spectroscopy of segregation into microdomains, *see* Plasmon-waveguide resonance spectroscopy
GROMACS, *see* Molecular dynamics
GUVs, *see* Giant unilamellar vesicles

H

High-performance liquid chromatography (HPLC), dehydroergosterol, 90
HPLC, *see* High-performance liquid chromatography

I

Immunoelectron microscopy, *see* Transmission electron microscopy
Immunoisolation, lipid rafts, 23, 25–28

L

Large unilamellar vesicles, *see* Multiphoton laser-scanning microscopy
Lipid bilayer, *see* Bilayer
Lipid domains
atomic force microscopy, *see* Atomic force microscopy
computer simulations, *see* Computer simulations
electron microscopy, *see* Transmission electron microscopy
electron paramagnetic resonance, *see* Discrimination by oxygen transport
nuclear magnetic resonance, *see* Nuclear magnetic resonance
plasmon-waveguide resonance spectroscopy of segregation into microdomains, *see* Plasmon-waveguide resonance spectroscopy
single-molecule tracking, *see* Single-molecule tracking
small-angle neutron scattering, *see* Small-angle neutron scattering
types, 195, 196
Lipid raft
atomic force microscopy, *see* Atomic force microscopy
computer simulations, *see* Computer simulations
definition, 21, 22
detergent-free isolation, *see* Immunoisolation; Optiprep; Sodium carbonate
detergent-resistant membrane extraction, *see* Detergent-resistant membrane
electron microscopy, *see* Transmission electron microscopy
electron paramagnetic resonance, *see* Discrimination by oxygen transport
film balance, *see* Film balance
fluorescence recovery after photobleaching, *see* Fluorescence recovery after photobleaching
nuclear magnetic resonance, *see* Nuclear magnetic resonance

Index

plasmon-waveguide resonance spectroscopy, *see* Plasmon-waveguide resonance spectroscopy
single-molecule tracking, *see* Single-molecule tracking
small-angle neutron scattering, *see* Small-angle neutron scattering
tempo fluorescence quenching detection, *see* Tempo
thickness, *see* X-ray diffraction
Low-angle X-ray diffraction, *see* X-ray diffraction

M

Magic angle spinning, *see* Nuclear magnetic resonance
Mass spectrometry, dehydroergosterol, 90
MD, *see* Molecular dynamics
Molecular dynamics (MD), lipid domain simulations
  analysis, 291–293, 295, 296, 299, 300
  assumptions, 288
  force field parameters, 288, 289
  GROMACS setup, 289, 290, 299
  materials, 287, 288
  overview, 285–287
  running, 290, 291, 299
MPLSM, *see* Multiphoton laser-scanning microscopy
Multiphoton laser-scanning microscopy (MPLSM), dehydroergosterol distribution on living cell membranes
  dehydroergosterol
    high-performance liquid chromatography, 90
    mass spectrometry, 90
    spectroscopic analysis, 89, 101
    synthesis, 89
  image acquisition, 91–93, 101
  image analysis
    spatial pattern analysis, 97, 99–101
    statistical analysis, 93–95, 97, 101
  large unilamellar vesicle preparation, 90
  materials, 87, 88
  overview, 85, 96
  probe labeling and cell culture, 90, 91

N

NMR, *see* Nuclear magnetic resonance
Nuclear magnetic resonance (NMR), lipid domain analysis
  deuterium solid-state nuclear magnetic resonance
    data acquisition, 111, 112
    spectra interpretation, 112–115, 123
  lipid raft lateral diffusion coefficient determination from pulsed field gradient NMR
    data acquisition, 130–133
    data analysis
      component resolved spectroscopy, 133, 135, 139, 140
      diffusion component number, 135
      peak intensity, 133
    domain formation, 128
    interpretation
      intermediate exchange, 138
      large domains and slow exchange, 137, 138
      no domains, 136, 137
      overview, 135, 136
      small domains and fast exchange, 137
    materials, 128, 129
    oriented bilayer preparation, 129, 130, 138, 139
    overview, 127, 128
  materials, 108, 109, 122, 123
  overview, 107, 108
  proton solid-state nuclear magnetic resonance
    data acquisition, 115, 116, 123, 124

pulsed field gradient magic angle spinning studies
  data acquisition, 119, 124, 125
  domain size determination, 120–122
  spectra interpretation, 116–119
  sample preparation
    deuterated lipid samples, 110, 123
    proton magic angle spinning samples, 111, 123

O

Optiprep, lipid raft isolation, 23–27
Oxygen transport, *see* Discrimination by oxygen transport

P

PLAP, *see* Glycophosphatidylinositol-anchored protein
Plasmon-waveguide resonance (PWR) spectroscopy, lateral segregation in solid-supported proteolipid bilayers
  imperfections in bilayer and microdomains, 167–169
  instrumentation, 162, 163
  materials, 162, 177, 178
  overview, 161, 162
  segregation of lipids and proteins into microdomains
    examples, 170–175
    protein segregation, 175–177
  single planar lipid membrane formation on resonator surface, 165, 167
  spectra
    properties, 164, 165, 178
    simulation, 169, 170
Protein trafficking, *see* Glycophosphatidylinositol-anchored protein; Plasmon-waveguide resonance spectroscopy
Pulsed field gradient NMR, *see* Nuclear magnetic resonance

PWR spectroscopy, *see* Plasmon-waveguide resonance spectroscopy

R

Radial distribution function
  molecular dynamics simulation, 295, 296
  single-molecule tracking, 205, 207

S

SANS, *see* Small-angle neutron scattering
Saturation-recovery electron paramagnetic resonance, *see* Discrimination by oxygen transport
Scattering length density, *see* Small-angle neutron scattering
SCMFT, *see* Self-consistent mean field theory
Self-consistent mean field theory (SCMFT)
  coarse-grain models of lipid domains, 297, 298, 300
  phase diagram analysis of ternary mixtures of cholesterol, saturated, and unsaturated phospholipids
    disordered liquid theory, 307–311
    one component and mixed phospholipid system analysis, 312, 313
    orientational order, 311, 312
    overview, 305, 306
    ternary system analysis, 313–317
Single-molecule tracking (SMT)
  cell culture and sample preparation, 198, 199, 214
  data analysis
    Brownian versus anomalous diffusion, 207–210
    confinement region identification, 210, 211
    cumulative radial distribution function, 205, 207
    diffusion coefficients

Index

calculation, 204, 218, 219
histograms, 204, 205, 219
radial displacements
  calculation, 202, 203, 218
  probability distribution, 200, 201
  random walk simulation, 201, 211
  relative diffusion between two molecules, 211–213
  trajectory recording, 201, 202, 211
imaging setup, 199, 200, 216
label selection, 214–216
lipid domain types, 195, 196
materials, 197, 198, 214
non-Brownian behavior, 197, 207–210
principles, 196, 197
Small-angle neutron scattering (SANS)
  contrast matching conditions, 235, 236, 239, 241, 245
  data acquisition, 238, 239
  lateral segregation detection, 241–243
  lipid domain detection, 233, 234
  lipid preparation
    film deposition, 237
    lipid extrusion, 237, 238, 244
    rapid solvent exchange, 236, 244
  materials, 234, 235
  model-dependent analysis, 243
  scattering length density, 234–237
Small-angle X-ray scattering, form factors and molecular dynamics simulation, 293
SMT, *see* Single-molecule tracking
Sodium carbonate, lipid raft isolation, 23, 24, 26, 27
Solid-state nuclear magnetic resonance, *see* Nuclear magnetic resonance

T

TEM, *see* Transmission electron microscopy

Tempo, fluorescence quenching of lipid rafts
  controls, 35, 36
  fluorescence measurement, 35, 38
  lipid stock solution preparation, 31, 33, 34, 38
  materials, 31, 38
  melting curve data analysis, 36, 37
  principles, 29–33
  sample preparation, 34, 38
Thickness
  molecular dynamics simulations, 291, 292
  X-ray diffraction studies, *see* X-ray diffraction
Thin-layer chromatography (TLC), membranes, 225
TLC, *see* Thin-layer chromatography
Transferrin receptor, *see* Glycophosphatidylinositol-anchored protein
Transmission electron microscopy (TEM), lipid domains
  clustering analysis, 258–260
  fluorescence recovery after photobleaching of samples, 251
  gold probes, 255, 256
  immunoprecipitation, 260, 261
  materials, 248
  membrane sheet viewing, 251–254, 257–260
  overview, 247, 248
  specimen preparation, 249, 250, 254–256
Triton X-100, liquid-disordered membrane solubilization, 9

V

Voronoi tessellation, molecular dynamics simulations, 291, 300

W

Western blot, glycophosphatidylinositol-anchored protein analysis, 15–18

X

X-ray diffraction, bilayer thickness measurement
  data analysis
    phase angle determination, 226, 227
    electron density profiles, 227–229
    thickness determination, 228–230
  data collection, 225, 226
  detergent extraction of membranes, 224, 225
  materials, 224
  multilamellar vesicle preparation containing rafts, 224
  overview, 223, 224
  specimen preparation, 225
  thin-layer chromatography of membranes, 225